T0233793

Economic Complexity and Evolution

Series Editors

Uwe Cantner
Jena, Germany

Kurt Dopfer
St. Gallen, Switzerland

John Foster
Brisbane, Australia

Andreas Pyka
Stuttgart, Germany

Paolo Saviotti
Grenoble, France

More information about this series at
http://www.springer.com/series/11583

Muhamed Kudic

Innovation Networks in the German Laser Industry

Evolutionary Change, Strategic Positioning, and Firm Innovativeness

 Springer

Muhamed Kudic
Stifterverband
Essen, Germany
and
Halle Institute for Economic Research
Halle (Saale), Germany

ISSN 2199-3173   ISSN 2199-3181 (electronic)
ISBN 978-3-319-35573-3   ISBN 978-3-319-07935-6 (eBook)
DOI 10.1007/978-3-319-07935-6
Springer Cham Heidelberg New York Dordrecht London

To
My parents,
Katharina, Helena and Luisa

Preface and Acknowledgments

The initial idea for this book was motivated by the observation that cooperative activities can be found in all fields of social, political, and economic life. The second, and possibly even more important, observation was that cooperative ventures often outperform solitary actors. R&D cooperation activities among science-driven firms and public research facilities in the field of laser research are certainly no exception. This in turn made me want to understand how more complex cooperation structures, like innovation networks, evolve over time and how these structural change processes impact the innovative performance of the actors involved.

This book is based on my dissertation, submitted and accepted by the Faculty of Business, Economics and Social Sciences at the University of Hohenheim, Germany, and successfully completed in 2012 (cf. Kudic 2012). This revised version differs from the original dissertation in several respects. The original manuscript was divided into six chapters. In this monograph, several chapters were reorganized and subdivided. All of the chapters were carefully reviewed and, if necessary, amended and updated. Some lines of argument, important in the context of the dissertation, were shortened; other passages were extended and developed further to address the needs of a wider audience. Literature was updated to account for the latest developments in interdisciplinary network research. Similarly, we updated the bibliometric analysis in Chap. 1. Finally, the last chapter was revised so that it now points to a number of additional, highly promising research areas in the interdisciplinary field of network research. The result is an entirely new structured monograph encompassing 14 revised and updated chapters.

With hindsight, it becomes obvious to me that the most important lesson that I have learned during my doctoral studies was not academic or scientific in nature. I have faced some unexpected challenges that I would surely not have been able to overcome on my own. However, I was lucky to meet various people who have spent their time and devoted their energies to helping me remove all of the obstacles. I would like to take this opportunity to sincerely thank all those who accompanied me on my journey and encouraged me to stay on track.

First of all, I would like to express my heartfelt gratitude to my parents, Alija and Fazila Kudic, and to my family, Katharina, Helena, and Luisa, for their caring support. A look back at the start of my doctoral studies prompts me to thank my former colleague and friend Benedikt Albrecht for always having a good sense of humor and staying optimistic. My academic development was strongly supported by the Halle Institute of Economic Research (IWH) and the colleagues and friends I got to know there. In particular, I'm deeply indebted to Professor Jutta Guenther, who opened many doors for me and paved the way for my academic work. In addition, I would like to thank my office mates Katja Guhr and Andrea Gauselmann for the very pleasant atmosphere in our office. Moreover, I would like to thank the following colleagues for their academic guidance, critical discussions, and many good conversations during the past few years: Dr. Wilfried Ehrenfeld, Dr. Toralf Pusch, Dr. Marc Banaszak, Matthias Mueller, Tobias Buchmann, Dr. Bjoern Jindra, and Philip Marek.

The kind support of the LASSSIE project consortium, especially from Professor Guido Buenstorf, Dr. Matthias Geissler, and Professor Michael Fritsch, is gratefully acknowledged. This dissertation project would not have been possible without access to raw data sources and valuable information on the German laser industry. Moreover, I pay due tribute to the CORDIS support team, especially Evi Guinou, for their friendly support and for providing an extraction of the CORDIS project database. My thanks also go to Markus Bachmeyer from the Creditreform company for providing me with archival firm-level data. Likewise, I would like to thank Norbert Schoene from the AT publishing company, Gerhard Hein from the VDMA association, and Ms. Schamberger from the B-Quadrat publishing company for providing me with historical laser technology journals, industry brochures, and business directories on the German laser industry. I would also like to thank Michael Barkholz for his methodological support and the following interns and students for their excellent research assistance: Sarah Langlotz, Nicolas Listel, Ann Weiland, Irene Bullmer, Jaqueline Kattner, Natasa Randjelovic, Regina Hilz, and many others. My special thanks go to Lorri King, Natalie Oliver, and Aida Inwood for proofreading and all last-minute corrections.

I am most grateful to Professor Andreas Pyka for his careful guidance and diligent Ph.D. supervision. He inspired me to look beyond the boundaries of mainstream economics and made me aware of the pathbreaking contributions Joseph Alois Schumpeter made to economics. He offered me excellent support and the freedom to do the research that I am really interested in. In the same breath, I would sincerely like to thank my co-supervisor Professor Guido Buenstorf not only for his continuous support but also for his many helpful comments and suggestions.

This book would certainly not have been possible without the kind support of Professor Uwe Cantner and Professor Andreas Pyka. I would also like to thank Mrs. Wentzel-Vandai for her professional guidance throughout the last few months. Last but not least, I would like to gratefully acknowledge the kind support of the Stifterverband. My thanks go in particular to Dr. Gero Stenke and Dr. Andreas Kladroba for providing me with time and resources during the completion phase of the manuscript.

Essen   Muhamed Kudic
June 2014

Contents

List of Figures

List of Tables

Part I
Introduction and Theoretical Background

Chapter 1
Introduction

At its heart, economic theory is about individuals and their interactions on markets or other social systems.

(James J. Heckman 2000)

Abstract Without a doubt, the twentieth century saw some of the most notable innovations in world history and the laser can certainly be counted among them. The aim of this study is to contribute to an in-depth understanding of collective innovation processes by analyzing R&D cooperation and innovation networks in the German laser industry. Following the neo-Schumpeterian tradition, it employs interdisciplinary analytical concepts and draws upon a unique longitudinal dataset from the laser industry that covers more than two decades of observations. The first chapter provides a general introduction to the subject and is structured as follows: Sect. 1.1 starts with a brief introduction of the laser and its roots in Germany. In Sect. 1.2 we raise awareness of the importance of R&D cooperation and innovation networks in science-driven industries. In Sect. 1.3 the overall research questions underlying this study are presented. And finally, the research design and the plan of the book are outlined in Sect. 1.4. In short, our aim with this study is to contribute to the existing body of literature by exploring how and why firm-specific R&D cooperation activities and network positions, large-scale network patterns, and evolutionary network change processes affect the innovative performance of laser source manufacturers in Germany.

1.1 The Laser and its Beginnings

The twentieth century saw some of the most notable innovations in world history. Many of these innovations led to the emergence of entirely new technological fields which have affected our lives and habits in a remarkable way. For instance, the development of novel means of transportation has enabled the world to grow closer together and has paved the way for trade between nations. The development of the transistor has revolutionized the field of electronic engineering and enabled pocket

© Springer International Publishing Switzerland 2015
M. Kudic, *Innovation Networks in the German Laser Industry*,
Economic Complexity and Evolution, DOI 10.1007/978-3-319-07935-6_1

calculators, personal computers and countless other electronic devices to be developed. New information and communication technologies, such as the Internet and other mobile communication devices, have changed the way people interact in their private and professional lives. This small selection of examples illustrates how tremendously new ideas can influence the social and economic life of individuals in modern societies.

The invention of the laser in the late 1950s can also be included in the list above. The acronym *laser* was originally coined by Gordon (1959) and stands for *"Light Amplification by Stimulated Emission of Radiation"*. At the onset of laser research several competing research groups were working under extreme pressure to secure their supremacy in this vibrant research field. Only one year after Gould's seminal article was presented at the *"Ann Arbor Conference on Optical Pumping"*, Maiman (1960) commenced operation of the first stable laser device.

Almost instantly the commercial sector took notice of the new technology and numerous laser source manufacturers (LSMs) entered the scene, not only in the United States but also in Germany. In the early 1960s, the Siemens Group, whose headquarters were located in Munich at that time, started to play a dominant role in the development and manufacturing of lasers in Germany. Shortly afterwards, an entire industry started to emerge that was characterized by its high number of micro and small businesses (Buenstorf 2007). Expertise in electrical engineering, physical and technical skills as well as access to cutting-edge technologies and new sources of scientific knowledge are essential for LSMs to keep pace with competitors. As a consequence, the demand began to increase for both applied and basic research into novel laser operating principles, gain media and laser components. Numerous public funding initiatives were launched to promote research in this field (Fabian 2011). New laser-related research facilities were founded and entered the German research landscape. Physics departments at universities and other publicly funded research organizations (PROs) started to intensify their efforts with regard to laser research. When Germany was reunified in 1989, the leading laser research facilities in the former German Democratic Republic (GDR) were integrated into the German laser innovation system. All in all, these efforts led to substantial refinements in the initial laser devices and were accompanied by groundbreaking technological advances in modern laser research carried out over the past half-century.

Today, laser applications can be found in nearly every walk of life. Their output power range from 1 to 5 milliwatt (10^{-3} W) for DVD-ROM drives and laser pointers, to 1–5 kW lasers (10^3 W) commonly used for industrial laser cutting and petawatt lasers (10^{15} W) used for experiments in plasma and atomic physics. The economic potential of laser technology has increased significantly over the past decades. In 2006 the revenue of German producers of laser sources and optical components amounted to approximately EUR 8 billion and about 45,000 people were employed in the industry (Giesekus 2007, p. 11).

1.2 Why Study Innovation Networks?

The previous reflections illustrate the enormous economic potential of new and innovative ideas. One of the first scholars to recognize the importance of innovations for economic welfare was Schumpeter (1912, 1939, 1942). He emphasized the role of entrepreneurs and their innovative ideas as the driving forces behind economic change processes in capitalist societies. Nowadays, it is widely accepted that technological progress is fundamental to economic growth and the prosperity of nations (Graf 2006). In this context at least one central question arises that also constitutes the initial starting point for this book: *what are the factors that affect a firm's ability to generate novelty and innovate over time?*

The search for an answer to this question is anything but new. Over the past decades scholars of economics and related disciplines have addressed this question (cf. Sect. 2.2). Previous research on the very nature of innovation processes (cf. Sect. 2.3) teaches us that the innovation process itself is neither linear in nature nor is it limited to the individual efforts of single economic entities. Instead, it is characterized by small incremental steps and accompanied by multiple feedback loops. The generation of novelty is a highly uncertain and, in most cases, collective process which is characterized by multiple interactions of independent but hetero-geneous economic actors with different capabilities, goals and strategies. Neo-Schumpeterian scholars (Freeman 1988; Lundvall 1988, 1992; Nelson 1992) explicitly addressed the collective nature of innovation processes by introducing the concept of "national innovation systems". Since then, several refinements to the originally proposed concept have been discussed in the literature (cf. Sect. 2.3). The common ground shared by all systemic concepts is that: **(I)** they involve creation, diffusion and use of knowledge, **(II)** feedback mechanisms are inherently built in, **(III)** they can be fully described by a set of components and relationships among these components, and **(IV)** the configuration of components, attributes, and rela-tionships is constantly changing (Carlsson et al. 2002).

The overlapping of systemic concepts and network concepts is obvious. How-ever, the systemic approach can be seen as a broader and more general approach that inherently entails innovation networks. It has been argued that innovation is the outcome of the interaction between a wide range of heterogeneous economic actors (Pyka 2002, 2007). These actors are, in many cases, connected through formal agreements[1] such as cooperation in research and development (R&D) (Brenner et al. 2011, p. 1). Innovation networks allow organizations to exchange existing information, knowledge and expertise (Cantner and Graf 2011, p. 373). At the same time, innovation networks provide the basis to commonly generate new knowledge which can be embodied in new products, services or processes (ibid).

The aim of this book is to analyze innovation networks in the German laser industry from various angles. More precisely, the investigations below seek to

---

[1] It is important to note that informal cooperation is not a subject of this investigation. Others have addressed this mode of cooperation in detail (Pyka 1997).

contribute to the existing body of literature about innovation networks by exploring how and why firm-specific cooperation activities, structural network patterns, strategic network positioning, and network evolution affect the innovative performance of firms at the micro-level.

1.3   The Current State of Scientific Research and Research Questions

The initial starting point for every research project is to conduct a comprehensive literature review. We carried out a bibliometric analysis[2] to gain an overall picture of previous theoretical and empirical contributions in the field of alliances and networks. In an initial step, we systematically screened various databases in order to identify all of the relevant articles on cooperation, alliances and networks. We identified a total of 3,694 publications between 1937 and 2014 from 242 academic journals. In a second step, we excluded all publications in which alliances and networks were used in another context or only mentioned in passing. We ended up with a collection of 2,103 scientific publications for the period between 1980 and 2013.[3] In a third step, we explored the bibliometric data from various angles. The results of this analysis revealed some interesting insights. Figure 1.1 provides a general overview of alliance and network research over the past three decades.

The solid black line illustrates the full set of empirical and theoretical publications that focus on interfirm or interorganizational alliances, networks and other collaborative forms. Hence, this category also includes publications that deal with a wider range of hybrid organizational structures, such as joint ventures, licensing or franchising agreements. The dotted black line represents publications that concentrate mainly on interfirm and interorganizational networks in the narrow sense. This category also includes a small number of publications on complex cooperation structures like, for instance, core-periphery and small-world patterns at the overall network-level. The solid gray line represents all of the publications that focus primarily on dyadic strategic alliances or bilateral partnerships. Finally, the dotted gray line represents publications with a clear emphasis on firm-specific networks like, for instance, ego networks, alliance constellations, multi-partner alliances and alliance portfolios.

In conclusion, the early period between 1980 and 1990 is characterized by a very small number of relevant publications. In the mid-1990s alliance and network

---

[2] This exploration does not claim to be complete or exhaustive. Instead it aims to uncover general trends in the literature. Appendix 1 provides an overview of bibliometric data sources, a full list of the evaluated academic journals and a brief description of the applied conventions and search methods.

[3] We restricted the time period for two reasons: (I) research on alliances and networks before 1980 is rare, and (II) due to the time of evaluation, data for 2014 was incomplete.

Total number of
publications

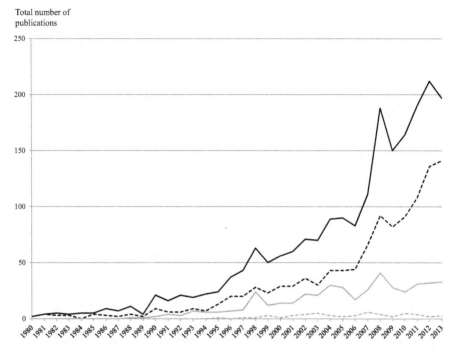

Fig. 1.1 Alliance and network research, 1980–2013 (Source: Author's own illustration)

research starts gaining momentum. This trend is set to continue over the next decades. Figure 1.1 clearly shows that the total number of publications has increased significantly in all areas of alliance and network research. Over the last few years we observe a strong increase, especially in network-related publications.

Our next analysis (cf. Fig. 1.2) explores alliance and network research broken down by scientific field. Our initial bibliometric exploration is based on three periods: 1980–1993; 1994–2003; 2004–2013. Figure 1.2 illustrates our findings for each of the three observation windows. In addition, particular attention was paid to the exploration of scientific publications over the entire observation period between 1980 and 2013 (results are reported using a log-scale and in percentage terms).

In the early phases of alliance and network research (cf. Fig. 1.2, dotted black line, solid gray line), we only found a relatively small number of papers in mainstream economics, economic geography, international business, marketing and entrepreneurship literature. Not surprisingly, we found a relatively high proportion of cooperation and network-related articles in typical sociological journals. The majority of publications fall into three groups: management science, organization science and innovation economics.

The dotted gray line (Fig. 1.2) represents the total number of scientific publications in the most recent period between 2004 and 2013. The findings confirm most

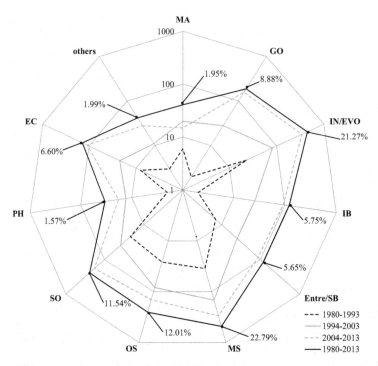

Fig. 1.2 Interdisciplinarity of alliance and network research (Source: Author's own illustration)

of the patterns identified before. However, there are at least two notable exceptions. Firstly, we observe a decreasing number of alliance and network-related publications in the field of marketing research. Secondly, there is a growing interest in alliances and networks in the field of mainstream economics and economic geography.

Finally, a closer look at percentage terms for the entire observation period between 1980 and 2013 reveals some interesting insights (cf. Fig. 1.2, solid black line). Alliance and network-related publications in the field of management science make up the largest percentage in our sample at about 23 %. The proportion of publications in the field of innovation economics, organization science, and sociology was 21.3 %, 12.0 %, and 11.5 %, respectively. About 8.8 % of all alliance and network-related papers were published in typical geographical journals and only 6.6 % of all papers appeared in mainstream economic journals.

Our final analysis explores how many of these publications address dynamic or evolutionary issues. Figure 1.3 illustrates the results of this exploration. As before, the solid black line represents the full set of papers on alliances, networks and other collaborative forms. The dotted black line illustrates the proportion of publications that focus explicitly on dynamic or evolutionary issues.

What do these initial investigations tell us? Firstly, alliance and network research is a vibrant and still growing field of research. Nonetheless, papers with

Total number of
publications

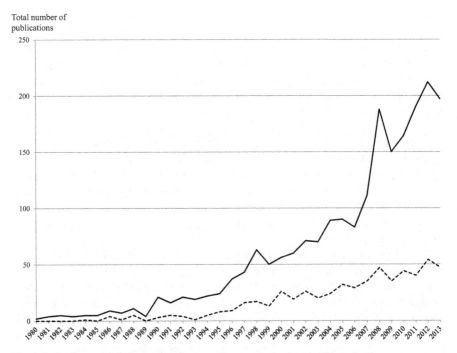

Fig. 1.3 Static versus dynamic contributions (Source: Author's own illustration)

a clear emphasis on firm-specific ego networks and on large-scale network properties are still rare. Secondly, alliance and network research is a highly interdisciplinary area. To illustrate this point, a notable number of relevant publications can
be found in more than ten scientific disciplines and about 1.6 % of all relevant
publications between 1980 and 2013 were published in typical physical journals.
Finally, we found a large amount of papers that focus on interfirm or interorganizational networks. However, publications that explicitly address evolutionary or
dynamic issues are clearly underrepresented in alliance and network research.
Despite these interesting findings, a closer look at the literature is needed to identify
research areas in the field that are still widely unexplored.

We will start by addressing the most general issues. By now, it is well recognized that a firm's position in the network affects its innovative performance in
various ways. Previous studies have explored the important role that structural
network characteristics play in a firm's innovation generating process (Shan
et al. 1994; Podolny and Stuart 1995). These early studies did not directly examine
the role of strategic positions in the network structure as predictors of firm-level
innovation output. Over the past few years, scholars have started to analyze how a
firm's innovative performance is impacted by the various types of network positions in interfirm or interorganizational network structures (Powell et al. 1996;
Ahuja 2000; Stuart 2000; Baum et al. 2000; Gilsing et al. 2008). However, it is

important to note that the type of network positioning that matters for a firm in its efforts to innovate can differ significantly from industry to industry. Accordingly, we reviewed the alliance and network literature that specifically focused on optical or laser-related technologies. We found very few publications that have explicitly analyzed the relatedness between network positioning and innovative performance in the optical industry. For instance, Ouiment and his colleagues (2007) have explored the relationship between a firm's network position and its innovativeness in small Canadian optics and photonics clusters. Lerch (2009) has investigated network dynamics in the optical cluster in the Berlin-Brandenburg region in Germany. Similarly, Sydow et al. (2010) have studied path dependencies in a network context in the Berlin-Brandenburg optics cluster. Joshi and Nerkar (2011) have analyzed the performance consequences of participating in patent pools, a unique form of R&D consortia, in the global optical disc industry. They found that patent pool participation decreases both the quantity and quality of patents of participating firms.

Even less research has been conducted on interorganizational networks in the laser industry. In a very early piece of work, Noyons et al. (1994) explored the science and technology interface by addressing inventor-author relations in laser medicine research. Shimizu and Hirao (2009) have analyzed interorganizational networks in the semiconductor laser industry in North America, Europe and Asia between 1975 and 1994. The two latter studies build upon patent data and bibliometric data, respectively. The results of both analyses are exploratory in nature. In summary, to the best of our knowledge, there is currently no longitudinal empirical study that has analyzed the collective nature of innovation processes in the German laser industry over a time span of two decades.

Another critical issue is that the majority of the network studies outlined above are static. The few longitudinal studies that are concerned with performance outcomes in evolving networks have quite contradictory findings. For instance, it is still unclear the extent to which network-hub positions or broker-positions are most beneficial in terms of performance outcomes (Rowley et al. 2000; Gargiulo and Benassi 2000; Burt 2005). Researchers from various disciplines have called for more dynamic-oriented alliance and network research (Parkhe et al. 2006; Cantner and Graf 2011; Ahuja et al. 2012). In general, networks are subject to change due to multiple network change processes at the micro-level (cf. Chap. 9). Tie formations or tie terminations as well as node entries or node exits affect the structural configuration of overall networks over time. These processes of "creative destruction" are clearly Schumpeterian in nature and provide the basis for explaining the evolution of networks (Boschma and Frenken 2010, p. 129). The dynamic nature of networks implies that a firm's structural positioning within such a network is by no means static. In other words, neither single cooperation events nor static network positions should be considered at given points in time but rather cooperation sequences or positioning paths should be taken into consideration in future research. This recognition brings us to one of the most crucial points in this section. A comprehensive analysis of how and why firm-specific cooperation activities, structural network patterns and network positions are related to innovative out-comes at the micro-level requires a dynamic research setting.

Keeping in mind the considerations above, we turn our attention now to more specific issues in order to stipulate our research questions. An in-depth evaluation of the literature[4] reveals some interesting findings and allows us to extract four widely unexplored research areas: **(I)** causes and consequences of evolutionary network change, **(II)** cooperation events, ego networks and firm innovativeness, **(III)** large-scale network properties and micro-level innovation outcomes, **(IV)** network prox-imity, geographical proximity and firm innovativeness.

Research area **(I)** addresses the dynamic nature of networks. The evaluation of the literature shows[5] that we still have a rather incomplete understanding of the drivers and mechanisms that cause evolutionary change in complex interorgani-zational networks. For instance, considerably little research has been conducted on network formation processes affected by both endogenous and exogenous factors. In addition, there is a strong bias in the literature towards the presence rather than the absence of relationships (Kenis and Oerlmans 2008, p. 299). To enhance our understanding of how and why networks change over time, we propose a concep-tual network evolution framework and empirically analyze a still widely neglected facet of network dynamics, i.e. the propensity and timing of network entry pro-cesses. More precisely, we seek to answer the following two research questions:

How can we explain the network evolution process and its structural implications in a theoretical way? What are the endogenous or exogenous determinants affect-ing a firm's propensity and timing to cooperate for the first time and enter the industry's innovation network?

Research area **(II)** focuses on the innovative performance of firms and seeks to disentangle the relationship between cooperation events, ego network characteris-tics and firm innovativeness.[6] An essential question that arises in this context is whether the innovativeness of firms in the German laser industry is directly affected by individual R&D cooperation events or more indirectly by structure and struc-tural change in firm-specific ego network characteristics over time. In other words, through which transmission channels do cooperation events affect a firm's subse-quent innovative performance? This dual character of individual R&D cooperation events has been widely neglected in previous research on ego networks and constitutes the core of this investigation. Consequently, we address the following research questions:

---

[4] Each chapter in Part IV provides a comprehensive literature review for each of the four research areas.

[5] For a literature review on the dynamics of alliances and networks, see Sects. 9.1 and 9.2.

[6] A literature review on R&D alliances, networks and innovation output is provided in Sect. 10.2.1. Previous research on the relationship between ego network structure and innovation output is discussed in Sect. 10.2.2.

Can we identify a significant relationship between individual cooperation events (i.e. "direct effects") or ego network characteristics (i.e. "indirect effects") and firm innovativeness over time? How do individual cooperation events affect the structural configuration of the focal actor's ego network and which structural features affect its subsequent innovation output?

Research area (III) turns our attention to the overall network level. Contemporary research on large-scale network properties implies that the network topology itself is likely to affect the exchange of knowledge in innovation networks.[7] It is, however, important to note that the relationship between large-scale network properties at the macro-level and innovation outcomes at the micro-level have been widely neglected in the field of interorganizational alliance and network research. In this study we focus on small-world properties of large-scale industry networks. We propose a theoretical framework that draws upon a reconceptualization (Zahra and George 2002) of the absorptive capacity (Cohen and Levinthal 1990) in order to provide the missing link between overall network characteristics and a firm's innovative performance. More precisely, we raise the following research question:

Can we identify a significant relationship between distinct large-scale network characteristics (i.e. a "high degree of clustering" *or* "short average paths") or small-world properties (i.e. a "high degree of clustering" *and* "short average paths") and firm innovativeness over time?

Research area (IV) addresses the fact that firms are concurrently exposed to various proximity dimensions. Boschma (2005, pp. 63–71) and Boschma and Frenken (2010, pp. 122–124) have proposed a theoretical concept that allows for an unambiguous definition and a clear-cut distinction of five proximity dimensions. In this study we seek to disentangle the relationship between network positioning, geographical co-location and firm innovativeness. The literature review reveals[8] that integrative research addressing both distinct and combined proximity effects remains rare. This is in line with the observation made by Whittington et al. (2009). Thus, we address the following research question:

Are firm-level innovation outcomes positively or negatively related to network positioning effects, geographical co-location effects or combined proximity effects; and if the latter case is true, are the combined effects substitutional or complementary in nature?

---

[7] For an overview of previous research on small world characteristics in an interorganizational context, see Sect. 11.1. Previous research on the graph theoretical foundations of the "small-world" phenomenon is discussed in Sect. 11.2.1.

[8] For a literature review and discussion of contemporary research in the field, see Sect. 12.2.

1.4 Research Design and Plan of the Book

The research design guides and structures the entire research process. Designing the research project requires some fundamental decisions at quite an early stage. The initial question that needs to be addressed is whether to apply a theory-building or a theory-testing strategy (De Vaus 2001, p. 5).

This book is governed by a deductive theory-testing approach. Deductive reasoning starts from a general theoretical framework and the theoretical considerations within this broader framework stipulate which observations are to be made (De Vaus 2001, p. 6). The underlying research logic implies a move from the general to the specific (ibid). The research process involves the deduction of testable hypotheses, data collection and hypotheses testing. The neo-Schumpeterian approach provides the general theoretical framework for all empirical parts in this book. This approach explicitly addresses the importance of knowledge, learning and innovation processes in complex socio-economic systems for the economic performance of economic agents at the micro and macro-level. Even though Schumpeter did not address cooperation or networks explicitly, his writings help to improve the understanding of how interorganizational connections among firms lead to new combinations and innovative endeavors (Dodgson 2011, p. 1142). We concretize our hypotheses by drawing upon theoretical concepts and arguments from economics and related disciplines.

At this point, it is important to note that the applied methods and data are irrelevant to the logic of the research design (De Vaus 2001, p. 8). The very nature of the phenomenon in question guides the selection of data and methods. Collective innovation processes are, as the name already suggests, not a static but rather a dynamic phenomenon. Consequently, this led to the decision in favor of a quantitative approach and a longitudinal data design. For the purpose of this study, multiple streams of archival raw data were exploited to create a comprehensive picture of cooperation and innovation activities for the entire population of German laser source manufacturers between 1990 and 2010. In principle, two strategies can be applied to conduct an empirical research project. A "descriptive approach" allows specific facts and patterns to be identified and explored, whereas an "explanatory approach" answers the question of how and why specific observations came to be the way they are (De Vaus 2001, pp. 1–3). We started with a descriptive analysis to gain a fundamental understanding of industry, firm, cooperation and innovation patterns in the German laser industry. De Vaus (2001, p. 2) notes that good descriptive analysis usually provokes questions for explanatory research. This was certainly also the case here, since many of the descriptive findings triggered several subsequent in-depth analyzes. This brings us to the main body of the book. Each of the four empirical sections explicitly addresses one of the four initially raised questions and draws upon a separate conceptual framework that schematizes the transition from the conceptual level to the empirical level. Finally, the ultimate goal of explanatory methods is to test whether a prediction is correct or incorrect and to support or reject the theoretical argument that underlies this particular hypothesis (De Vaus 2001, p. 7). We applied event history techniques and panel data count models to accomplish this task. Last but not least, the extraction of descriptive and exploratory results is necessary but not sufficient. Results have to be critically discussed and interpreted against the backdrop of a broader theoretical context. Accordingly, this book is divided into five parts and fourteen chapters whose contents are described in more detail below.

Chapter 2 provides the theoretical foundation for this book. We start from a classical-neoclassical perspective and discuss the role of knowledge and innovation in traditional economic approaches. Then we turn our attention to evolutionary approaches in economics and related disciplines. The neo-Schumpeterian approach in evolutionary economics constitutes the core of the theory chapter. Then we continue by introducing theoretical concepts at the firm level, i.e. the "structure conduct performance" (SCP) paradigm, the "resource-based view" (RBV), and the knowledge-based view (KBV), that seek to explain the sources of a firm's competitive advantage. Next, we draw upon interdisciplinary alliance and network research. We conclude the theoretical discussion by exposing how this research project relates to the previously outlined theoretical concepts.

Chapter 3 gives a short introduction of basic laser operating principles and outlines the most notable technological developments over the past 50 years. Subsequently, we focus our attention on the German laser industry and illustrate the configuration of the industry value chain.

Chapter 4 starts with a general discussion of methodological issues and provides a detailed description of the data sources used to construct a unique longitudinal laser industry database. This lays the ground for the analytical parts of this study. The laser industry database covers a time period between 1990 and 2010.

Chapter 5 presents some general graph theoretical concepts and introduces indicators and measures needed for the quantitative description of the industry and the industry's innovation network. Focus is on quantitative network analysis methods and geographical indicators.

Chapter 6 is divided into two sections. First two longitudinal datasets are presented. Then there is an overview and general discussion on estimation methods. In this context, event history analysis methods as well as techniques for analyzing longitudinal count data are addressed.

Chapter 7 reports descriptive findings at the industry level. We analyze geographical concentration patterns for three types of organizations, i.e. laser source manufacturers (LSMs), laser system providers (LSPs) and laser-related public research organizations (PROs). The subsequent explorations concentrate on LSMs which are considered to be at the core of the industry value chain.

Chapter 8 focuses on R&D cooperation and innovation networks. We start with summary statistics on publicly-funded R&D cooperation projects in the German laser industry. The next descriptive analysis explores the organization's involvement in these projects from various angles. Our data allows us to construct innovation networks on an annual basis for the entire population of LSMs and PROs in the German laser industry. This provides the opportunity to analyze basic node-related and tie-related network measures and reveals characteristic network change patterns over time. We supplement this initial longitudinal network exploration by conducting an in-depth analysis of the overall network topology. In the last descriptive analysis, we check for the existence of scale-free patterns, test for small-world properties and analyze the emergence of a core-periphery structure over time.

Chapter 9 focuses on the evolution of innovation networks. The aim of this analysis is to investigate the determinants of evolutionary change processes in innovation networks. We address one particular facet of the network evolution process in the empirical part of this chapter. More precisely, we conduct an event history analysis in order to disentangle the extent to which exogenous or endogenous determinants affect a firm's propensity and timing to cooperate and enter the industry's innovation network.

Chapter 10 points to the importance of firm-specific cooperation strategies. The goal of this investigation is to shed light on the relationship between individual cooperation events, firm-specific ego network characteristics and firm-level innovation outcomes. In short, by using a panel data count model, we explore how a firm's innovativeness is related to its cooperation events on the one hand, and the structural configuration and dynamics of its ego network on the other.

Chapter 11 raises awareness for large-scale network properties. Consequently, we switch the analytical level and turn our attention to systemic level properties. The aim of the third analysis is to understand how the structural network configuration at the macro-level is related to firm-level innovation outcomes at the micro-level. We use longitudinal network data and quantitative network analysis methods to quantify large-scale network properties and put the "small-world" hypothesis to the test in terms of which networks with a high degree of clustering and high reachability provide a superior environment for firm innovativeness.

Chapter 12 draws upon the proximity concept and points to the fact that firms are concurrently exposed to multiple proximity dimensions. We apply panel data methods to find out the extent to which distinct and/or combined effects between network proximity and geographical co-location are positively related to subsequent firm-level innovation outcomes.

Chapter 13 marks the completion of the research project. We summarize the findings and raise awareness of the limitations of our results.

Chapter 14 concludes with some final considerations and critical remarks. It includes suggestions for further research.

References

Ahuja G (2000) Collaboration networks, structural hole, and innovation: a longitudinal study. Adm Sci Q 45(3):425–455

Ahuja G, Soda G, Zaheer A (2012) The genesis and dynamics of organizational networks. Organ Sci 23(2):434–448

Baum JA, Calabrese T, Silverman BS (2000) Don't go it alone: alliance network composition and startup's performance in Canadian biotechnology. Strateg Manag J 21(3):267–294

Boschma R (2005) Proximity and innovation: a critical assessment. Reg Stud 39(1):61–74
Boschma R, Frenken K (2010) The spatial evolution of innovation networks: a proximity perspective. In: Boschma R, Martin R (eds) The handbook of evolutionary economic geography. Edward Elgar, Cheltenham, pp 120–135
Brenner T, Cantner U, Graf H (2011) Innovation networks: measurement, performance and regional dimensions. Ind Innov 18(1):1–5
Buenstorf G (2007) Evolution on the shoulders of giants: entrepreneurship and firm survival in the German laser industry. Rev Ind Organ 30(3):179–202
Burt RS (2005) Brokerage & closure – an introduction to social capital. Oxford University Press, New York
Cantner U, Graf H (2011) Innovation networks: formation, performance and dynamics. In: Antonelli C (ed) Handbook on the economic complexity of technological change. Edward Elgar, Cheltenham, pp 366–394
Carlsson B, Jacobsson S, Holmen M, Rickne A (2002) Innovation systems: analytical and methodological issues. Res Policy 31(2):233–245
Cohen WM, Levinthal DA (1990) Absorptive capacity: a new perspective on learning and innovation. Adm Sci Q 35(3):128–152
De Vaus D (2001) Research design in social science research. Sage, London
Dodgson M (2011) Exploring new combinations in innovation and entrepreneurship: social networks, Schumpeter, and the case of Josiah Wedgwood (1730–1795). Ind Corp Chang 20 (4):1119–1151
Fabian C (2011) Technologieentwicklung im Spannungsfeld von Industrie, Wissenschaft und Staat: Zu den Anfängen des Innovationssystems der Materialbearbeitungslaser in der Bundesrepublik Deutschland 1960 bis 1997. Dissertation, TU Bergakademie Freiberg
Freeman C (1988) Japan: a new national system of innovation. In: Dosi G, Nelson RR, Silverberg G, Soete L (eds) Technical change and economic theory. Pinter, London, pp 330–348
Gargiulo M, Benassi M (2000) Trapped in your own net? Network cohesion, structural holes, and the adaptation of social capital. Organ Sci 11(2):183–196
Giesekus J (2007) Die Industrie für Strahlquellen und optische Komponenten – Eine aktuelle Marktübersicht von SPECTARIS. Laser Technik J 4(5):11–13
Gilsing V, Nooteboom B, Vanhaverbeke W, Duysters G, van den Oord A (2008) Network embeddedness and the exploration of novel technologies: technological distance, betweenness centrality and density. Res Policy 37(10):1717–1731
Gould GR (1959) The laser: light amplification by stimulated emission of radiation. In: Ann Arbor conference on optical pumping, conference proceeding, 15–18 June 1959, pp 128–130
Graf H (2006) Networks in the innovation process. Edward Elgar, Cheltenham
Joshi AM, Nerkar A (2011) When do strategic alliances inhibit innovation by firms? Evidence from patent pools in the global optical disc industry. Strateg Manag J 32(11):1139–1160
Kenis P, Oerlmans L (2008) The social network perspective – understanding the structure of cooperation. In: Cropper S, Ebers M, Huxham C, Ring PS (eds) The Oxford handbook of inter-organizational relations. Oxford University Press, New York, pp 289–312
Lerch F (2009) Netzwerkdynamiken im Cluster: Optische Technologien in der Region Berlin-Brandenburg. Dissertation, Freien Universität Berlin, Berlin
Lundvall B-A (1988) Innovation as an interactive process: from user-producer interaction to the national system of innovation. In: Dosi G, Freeman C, Nelson RR, Silverberg G, Soete L (eds) Technical change and economic theory. Pinter, London, pp 349–369
Lundvall B-A (1992) National systems of innovation – towards a theory of innovation and interactive learning. Pinter, London
Maiman TH (1960) Stimulated optical radiation in ruby. Nature 187(4736):493–494
Nelson RR (1992) National innovation systems: a retrospective on a study. Ind Corp Chang 1 (2):347–374

Noyons E, Raan VA, Grupp H, Schoch U (1994) Exploring the science and technology interface: inventor-author relations in laser medicine research. Res Policy 23(4):443–457

Ouimet M, Landry R, Amara N (2007) Network position and efforts to innovate in small Canadian optics and photonics clusters. Int J Entrep Innov Manag 7:251–271

Parkhe A, Wasserman S, Ralston DA (2006) New frontiers in network theory development. Acad Manag Rev 31(3):560–568

Podolny JM, Stuart TE (1995) A role-based ecology of technological change. Am J Sociol 100 (5):1224–1260

Powell WW, Koput KW, Smith-Doerr L (1996) Interorganizational collaboration and the locus of innovation – networks of learning in biotechnology. Adm Sci Q 41(1):116–145

Pyka A (1997) Informal networking. Technovation 17(4):207–220

Pyka A (2002) Innovation networks in economics: from the incentive-based to the knowledge based approaches. Eur J Innov Manag 5(3):152–163

Pyka A (2007) Innovation networks. In: Hanusch H, Pyka A (eds) Elgar companion to neo-Schumpeterian economics. Edward Elgar, Cheltenham, pp 360–377

Rowley TJ, Behrens D, Krackhardt D (2000) Redundant governance structures: an analysis of structural and relational embeddedness in the steel and semiconductor industries. Strateg Manag J 21(3):369–386

Schumpeter JA (1912) Theorie der wirtschaftlichen Entwicklung (The theory of economic developmnt 1934). Duncker & Humblot, Berlin

Schumpeter JA (1939) Business cycles – a theoretical, historical and statistical analysis of the capitalism process. McGraw-Hill, New York

Schumpeter JA (1942) Kapitalismus, Sozialismus und Demokratie (Capitalism, socialism and democracy, 1950). Harper & Bros, New York

Shan W, Walker G, Kogut B (1994) Interfirm cooperation and startup innovation in the biotechnology industry. Strateg Manag J 15(5):387–394

Shimizu H, Hirao T (2009) Inter-organizational collaborative research networks in semiconductor laser 1975–1994. Soc Sci J 46(2):233–251

Stuart TE (2000) Interorganizational alliances and the performance of firms: a study of growth and innovational rates in a high-technology industry. Strateg Manag J 21(8):791–811

Sydow J, Lerch F, Staber U (2010) Planning for path dependence? The case of a network in the Berlin-Brandenburg optics cluster. Econ Geogr 86(2):173–195

Whittington KB, Owen-Smith J, Powell WW (2009) Networks, propinquity, and innovation in knowledge-intensive industries. Adm Sci Q 54(1):90–122

Zahra SA, George G (2002) Absorptive capacity: a review, reconceptualization, and extension. Acad Manag Rev 27(2):185–203

Chapter 2
Theoretical Background

A theory of essentially complex phenomena must refer to a large number of particular facts; and to derive a prediction from it, or to test it, we have to ascertain all these particular facts.

(Friedrich August von Hayek 1974)

Abstract The aim of this chapter is threefold: firstly, we seek to integrate this study into a broader context of historical and contemporary economic reasoning. Secondly, we introduce the theoretical pillars which this research project is built upon. Due to the interdisciplinary nature of alliance and network research, we refer to concepts from economics and related scientific disciplines. Finally, we discuss the initially raised research questions against the backdrop of our preceding theoretical considerations. Chapter 2 is structured as follows. In Sect. 2.1 we start with a brief introduction of classical and neoclassical economics and outline the contributions but also the limitations of these schools of thought. In Sect. 2.2 we turn our attention to evolutionary approaches in economics and related disciplines. In Sect. 2.3 we introduce the neo-Schumpeterian approach to evolutionary economics. Here, we start by clarifying some basic terminology and concepts. Next we discuss the theoretical cornerstones of the approach. Finally, we address three selected concepts: proximity, innovation systems and innovation networks. In Sect. 2.4 we draw upon the knowledge-based theory of the firm to establish the theoretical linkage between knowledge, competitive advantage and firm performance. In Sect. 2.5 we look at key concepts from interdisciplinary alliance and network research. Finally, in Sect. 2.6 we uncover links to previous research and discuss our own contribution in light of the previously outlined theoretical concepts.

2.1 The Classical-Neoclassical Paradigm in Economics

We'll start by taking a brief look at the classical-neoclassical paradigm in economics. This is important for two reasons. On the one hand, it allows us to acknowledge the long journey in economic history towards a better understanding of the

© Springer International Publishing Switzerland 2015
M. Kudic, *Innovation Networks in the German Laser Industry*,
Economic Complexity and Evolution, DOI 10.1007/978-3-319-07935-6_2

relationship between knowledge, innovation and economic prosperity. On the other hand, it enables us to identify the weaknesses and limitations of classical and neoclassical approaches that paved the way for the appearance of alternative schools of thought in economics.

The dominant paradigm in economics at the end of the nineteenth century was the classical school of thought, strongly influenced by the work by Smith (1776), Ricardo (1817), Malthus (1798, 1820), Mill (1848, 1859) and Marx (1857, 1867). According to Brewer (2010, p. 4), the classical theory of economic growth has the following characteristic features: capital accumulation as the primary source of economic prosperity, endogeneity of population growth, and innovation processes and technological change were largely ignored. Even though knowledge and innovation play, at best, a secondary role in classical mainstream models, several classical authors did address these issues, at least implicitly, in their writings. Ricardo (1817) and Malthus (1820) mentioned technological change without giving it a central role (Brewer 2010, p. 183).

Smith (1776) focused mainly on the efficient organization of production processes and economic growth. "*An Inquiry into the Nature and Causes of the Wealth of Nations*", published in 1776, is still considered by many as one of the most influential contributions to the field of economics. According to Adam Smith's theory, the specialization and division of labor are the main sources of growth in productivity (Antonelli 2009, p. 615). He explicitly addresses the role of inventions in his writings. However, it is important to note that in Smith's theory the invention is the result of a division of labor and not the other way around (Swann 2009, p. 8). In addition, Adam Smith provides us with further important insights by pointing to the increasingly specialized nature of knowledge in industrial production processes (Pavitt 1998, p. 433). Pavitt (1998, pp. 435–436) argues that two types of specialization patterns occurred in parallel which have significantly increased the efficiency of discovery, invention and innovation: (**a**) entirely new scientific disciplines with specialized bodies of knowledge have emerged, (**b**) the rate of technical change has been augmented by the functional division of labor and corporate R&D laboratories have been established within large businesses. Pavitt (1998, p. 447) comes to the conclusion that, as forecasted by Adam Smith, the specialization patterns of knowledge production outlined above are crucial for an in-depth understanding of the innovating firm.

Mill (1848, 1859), another influential classical economist, contributed to the economics of innovation by highlighting the central role invention plays in wealth creation. At the same time he addressed the paradoxical notion that invention did not obviously lead to an improvement in the wealth of the population at large (Swann 2009, p. 8).

Finally, Marx[1] (1857, 1867) significantly improved our understanding of the nature and dynamics of capitalist economic systems and was among the first to

---

[1] Even if one is critical of Marx's doctrine, one should define science by its objects of analysis and not by its methods or assumptions (Hodgson 2006, p. 1). In a comprehensive bibliometric analysis,

correctly predict the increasing globalization of markets and the appearance of giant firms (Hodgson 2006, p. 1). According to Marx, the "bourgeoisie", what we today would call a "company", can only exist by continuously improving the instruments of production (Swann 2009, p. 9). This recognition highlights the role of innovation in ensuring a company's competitiveness (ibid). In a similar vein, Antonelli (2009, pp. 616–618) notes that Marx contributed to the economics of innovation in at least two additional ways. On the one hand, he provides a very first analysis of endogenous technological change processes that are caused by the changing cost of labor due to an intentional process of augmented labor substitution. On the other hand, Marx acknowledges the role of knowledge as an endogenous productive force and emphasizes the importance of learning processes in changing economic systems. The ideas of Marx and Schumpeter are closely related, especially when it comes to the instabilities of capitalist systems. Rosenberg (2011, p. 1216) argues that "[...] they held in common a vision of capitalism as a social system that possessed its own internal logic, and that underwent, over time, a process of self-transformation". The classical theory lost its dominance by the end of the nineteenth century due to the emergence of the neoclassical paradigm.

Neoclassical economists Jevons (1871), Marshall (1890), Menger (1871) and Walras (1874) entered the scene and paved the way for the micro foundation of economics. In general, neoclassical economics is characterized by a high degree of formalization and puts a strong emphasis on optimization calculus. The primary task of neoclassical concepts is to establish an equilibrium state (Rutherford 2007, p. 149). The neoclassical paradigm started to flourish with the onset of the so-called "marginal revolution" in economics. Most concepts that constitute the core of this approach, marginal cost, marginal revenues, marginal utility, etc., were mainly developed by Jevons, Walras and Menger (Rutherford 2007, p. 133). Carl Menger criticized the German historical school of economics in the so-called "*Methodenstreit*"[2] and is regarded as one of the key proponents of the Austrian school of economics.[3]

In addition to these contributions, the writings of Marshall (1890) provide us, in particular, with important insights into the interplay between knowledge and organizations to perpetuate the growth and development of economies (Metcalfe 2010, p. 4). According to Marshall, knowledge is the most powerful source of industrial production (ibid). Knowledge is considered to be a key component of capital and is, in itself, a production factor (Antonelli 2009, p. 629). The generation of technological knowledge is a collective process where a variety of agents, co-localized within the industrial districts, contribute complementary bits of knowledge (ibid). In addition, Marshall acknowledged the importance of knowledge and organization for non-random economic change and accounts for

---

Hodgson (2006, p. 4) demonstrates that Karl Marx was one of the most frequently cited authors in leading mainstream economic journals between 1890 and 1990.

[2] For more details on the discourse between the two schools of thought, see Von Mises (1969).

[3] For an introduction to the Austrian school of economics, see Rutherford (2007, p. 9).

self-transformation processes[4] of capitalist economic systems (Metcalfe 2010, p. 5). Marshall distinguishes between inventions and innovations and emphasizes the prominent role of entrepreneurs in his writings (Metcalfe 2010, p. 6). Moreover, he recognized the importance of interfirm cooperation activities as a way for firms, whose activities are complementary, to jointly coordinate production processes (Corolleur and Courlet 2003, p. 300).

Another major concern, not only of classical but also of neoclassical economists, was economic growth. At the inception of economic growth theory, simple stage models of economic development dominated the research landscape (Rutherford 2007, p. 59). Economic growth models in the neoclassical tradition prospered in the mid-twentieth century. Mainstream economics did not account for technological change processes at all or treated technological change as an exogenous factor. For instance, the "Harrod-Domar" model (Harrod 1948; Domar 1948) explains economic growth on the basis of a maximum long-term growth rate and a warranted growth rate (Rutherford 2007, p. 60).[5] The prevailing economic doctrine at the time was that capital accumulation was much more important than technological progress in explaining economic growth (Swann 2009, p. 14).

The most notable step towards an endogenous growth theory is the "Solow-Swan" model of economic growth (Solow 1956, 1957; Swan 1956). In spite of Solow's merits, one has to note that his writings on growth are based on a simplified neoclassical framework (Nelson 2007, p. 841). Since then, considerable efforts have been made to incorporate technological change in macroeconomic growth models and to develop what we call today endogenous growth theory.[6]

What does this short overview and discussion tell us? First of all, it is essential to note that several protagonists of the classical-neoclassical school of thought provided us with important ideas that are still highly topical. The writings are much richer than mainstream economic models at first suggest. The insights outlined above provide a basis to better understand how knowledge, innovation and technological progress affect economic growth and prosperity. The crucial point, however, is that these leading ideas have been largely overlooked or ignored by mainstream economists.

In traditional mainstream economics, firms are assumed to be perfectly homogenous economic entities. Firms are usually treated as a black box that can be fully described by a simple production function with two input factors: capital and labor.

---

[4] Both variety and selection processes are essential elements in Marshall's notion of competition in a sense that heterogeneous firms (in terms of size, location and efficiency) face each other in the product marketplace and are sorted out by the competitive process (Antonelli 2009, p. 622).

[5] For a comparison of the "Harrod-Domar" model and the neoclassical growth model, see Sato (1964). The first rate is also called the "natural rate of growth" (this rate equals the sum of population growth and technical progress) and the second rate reflects the capital accumulation process (this rate is equal to the ratio of the proportion of income saved to the capital income ratio) (Rutherford 2007, p. 60).

[6] For an overview and in-depth discussion, see Nelson and Winter (1974) and Silverberg and Verspagen (2005).

Goods or services produced by these firms are considered to have the same quality. Consumers are usually described by a utility function. They behave under the conditions of the well-known "homo oeconomicus" postulate.[7] This implies that individuals in markets always act as fully rational utility maximizers. Market actors, such as buyers and sellers, face no trade restrictions in a sense that transactions do not produce additional costs and time plays no significant role. Market actors interact as part of completely transparent markets and share the same information. In general, information is freely available and has the character of a public good. Static as well as comparative-static models in classical and neo-classical economics are able to show, under the restrictive assumption of fully rationally behaving individuals and perfect market conditions, how markets tend to reach an equilibrium, at least in the long run, where supply and demand are balanced.

These strongly simplified and idealized conditions have some important implications. In a world characterized by homogenous firms, freely available information and perfect market conditions, there are no incentives for the generation of novelty. Firms have identical resource endowments and they produce goods and services of homogenous quality. Under these assumtions, information asymmetries do not exist by definition. Generating novelty is costless and is subject to no other restrictions (Boulding 1966, p. 3). As a consequence, there is no need for firms to access or generate new stocks of knowledge. The creativity of inventors and the ingenuity of entrepreneurs play no role in classical-neoclassical theory. Similarly, the ability of firms to generate innovative goods and services as well as the impact of these innovations on technological change is not inherent in traditional economic models. Even though neoclassical growth models incorporate technological change, they fail to explain the factors that cause innovation and technological progress.

Recent developments in mainstream economics might invalidate some of the arguments raised above. Nonetheless, it must be noted that the research program in mainstream economics only addresses, in passing, key topics such as knowledge creation and distribution, invention and innovation, cooperation and networks, and the evolutionary nature of complex economic systems.

2.2 Evolutionary Thinking in Economics and Related Fields

In response to the limitations of the classical-neoclassical paradigm, several alternative economic schools of thought emerged.[8] The onset of evolutionary thinking in economics can be traced back to the mid-nineteenth and early twentieth

---

[7] For a clarification of the concept and critical reflections from an Austrian view, see Alter (1982).

[8] For an overview of evolutionary approaches in economics, see Dopfer (2005), Hanusch and Pyka (2007c) and Witt (2008a, b).

centuries. Based on the ideas of Veblen (1898) and Schumpeter (1912), two independent strands of evolutionary thinking emerged in economics.[9] These two strands moved closer together through the synthesis of Schumpeterian ideas and natural selection analogies in Nelson and Winters' (1982) evolutionary theory of economic change. Today, evolutionary economics can be divided into four distinct research strands (cf. Witt 2008b): Universal Darwinism, Naturalistic, Schumpeterian and neo-Schumpeterian approaches. The common denominator of these approaches is the interest in dynamic change processes in economies occurring at multiple levels of analysis against the backdrop of past events. For the purpose of this study, we mainly focus on the Schumpeterian and neo-Schumpeterian schools of thought. Nonetheless, we should not miss the opportunity to start the discussion by taking a brief look at evolutionary ideas in related disciplines.

Not only economists but also scholars from related disciplines explicitly address evolutionary ideas in their writings. Evolutionary principles have a long tradition in sociology and organization science. These ideas attracted attention and started to thrive in the second half of the twentieth century. A central debate arose among organizational scholars on the very nature of evolutionary change processes. Two alternative perspectives, the adaptation perspective and the selection perspective, dominated the discourse at that time (Carroll 1984, p. 73).[10]

The first perspective is based on the notion that organizations are highly adaptive and that structural changes occur in response to internal and external triggers (Carroll 1984, p. 73). Organizations adapt to their environments by changing routines and standard operating procedures (Bruderer and Singh 1996, p. 1322). The literature on organizational learning[11] is clearly based on the adaptation perspective (Aldrich and Ruef 2006, p. 47). The adaptive learning approach considers organizations to be goal-oriented entities that learn though experience by undergoing repeated or non-repeated trial-and-error processes (ibid). Some path-breaking learning models were developed based on the seminal contribution of Cyert and March (1963). For instance, Agyris and Schön (1978) proposed so-called single-loop learning and double-loop learning models and the concept of deuteron learning.[12] Subsequent contributions have explicitly addressed the constraints of an organization's learning capabilities (Aldrich and Ruef 2006, p. 47). The most notable contributions are those of Levitt and March (1988) on the nature of organizational learning processes, Levinthal (1991) on the interrelatedness of

---

[9] For an overview and discussion, see Witt (2003).

[10] It is important to note that these perspectives are not mutually exclusive. According to Bruderer and Singh (1996, p. 1322), substantial research efforts have been undertaken to reconcile these two perspectives since the early 1990s.

[11] According to Aldrich and Ruef (2006, p. 47) at least two strands of research can be identified within the organizational learning literature: the "knowledge development approach", pioneered by Weick (1979) and the "adaptive learning approach" pioneered by Cyert and March (1963). In this brief review we focus on the latter strand of literature.

[12] For a review and discussion of these concepts, see Dodgson (1993) and Bierly et al. (2000).

adaptation and selection processes, and March (1991) on exploration and exploitation in organizational learning processes. Recently, this strand of literature has strongly influenced a series of learning-oriented alliance and network studies by, for example, Rothaermel and Deeds (2004), Lavie and Rosenkopf (2006), Yamakawa et al. (2011), Bessant et al. (2012).

The second perspective emerged in the late 1970s. In their seminal article on the population ecology of organizations, Hannan and Freeman (1977) argue that change is not primarily driven by adaptation. Organizational structures are considered to be structurally inert (Hannan and Freeman 1984). According to this view, the dominant mechanism of change is not adaptation but rather selection, governed by competition, environmental opportunities and constraints (Carroll 1984, p. 73).[13] The organizational ecology approach explains the outcomes of organizations, such as their survival in terms of the demographic composition of population and the resource environment they are located in (Aldrich and Ruef 2006, p. 35). Organizations are considered to be in a state of competitive interdependence as they compete for scant resources in the same environment (ibid). A broad range of ecological models has been introduced since the beginning of organizational ecology research.[14] These models enable us to explain how both competitive relationships between organizations and cooperative interdependencies jointly affect organizational performance in terms of survival (Aldrich and Ruef 2006, p. 35). As a consequence, a rich body of alliance and network literature began emerging from this area of research. To exemplify this, several studies addressed the relationship between network embeddedness and firm survival (Baum et al. 2000; Staber 1998; Brüderl and Preisendörfer 1998). Others have applied organizational level concepts and adapted them to an interorganizational context. For instance, the structural inertia concept, originally proposed by Hannan and Freeman (1984), was applied by Kim at al. (2006) to explain a network's resistance to change. Finally, considerable efforts were made to address evolutionary change of interorganizational relations and complex networks (Amburgey et al. 2008; Lomi et al. 2008).

Similarly, the importance of evolutionary thinking in economics increased at the beginning of the twentieth century. Not really convinced by the mainstream classical-neoclassical paradigm, it was at this time that Austrian economist, Joseph A. Schumpeter (1908, 1912, 1939, 1942), emerged on the scene. In his early years he had already made several seminal contributions. He was the first to introduce the

---

[13] Organizational sociologists have adapted general evolutionary principles of variation, selection and retention proposed by Campbell (1969) in an organizational context. A strict Darwinian interpretation of organizational evolution implies that traits are inherited through intergenerational processes, whereas the Lamarckian concept of organizational evolution regards traits as being acquired within a generation through learning and imitation (VanDeVen and Poole 1995, p. 519).

[14] An excellent overview of organizational ecology research is provided by Amburgey and Rao (1996) and for some promising future research perspectives, see Van Witteloostuijn (2000). For an overview of population ecology model applications, see Betton and Dess (1985).

concept of "methodological individualism" (Schumpeter 1908, pp. 88–98).[15] With his seminal work *"Theorie der wirtschaftlichen Entwicklung"*, published in 1912, he questioned mainstream economics and, at the same time, provided the theoretical foundation for a new economic paradigm.[16] He emphasized the role of entrepreneurs and their novel ideas as the driving forces behind evolutionary change processes in capitalist societies. Perhaps the most striking difference between evolutionary thinking and classical-neoclassical approaches is the dynamic view of economic systems. Schumpeter was convinced that economic change is generated by the economic system and that there must be some kind of inherent force that destroys every state of equilibrium (Hanusch and Pyka 2007b, p. 21). He pointed to the fact that "new combinations", or what we today would call innovations, are the major source of endogenous economic change (Hanusch and Pyka 2007b, p. 22). Schumpeter called the activity of combing existing resources in order to generate new combinations the "entrepreneurial functions" (Fagerberg 2005, p. 6). Both Marshall and Schumpeter made a distinction between inventions and innovations in their writings (Metcalfe 2010, p. 6); however, Marshall had another notion of innovation. The thing that distinguishes him most sharply from Schumpeter is that he considered innovation to be an intrinsic feature of economic leadership and part of the normal, routine discovery process (Metcalfe 2010, p. 12). In contrast, Schumpeter emphasized the radical nature of innovations and explicitly addressed the disruptive but also creative nature of novelties. Schumpeter was particularly influenced in many ways by the ideas of Marx. According to Tushman and Nelson (1990, p. 1), "Both Schumpeter and Marx saw technology and technical change as central factors underlying organization and political dynamics and as a critical determinant of group power and individual outcomes." In his later writings, Schumpeter (1942) used the metaphor "creative destruction", originally coined by Karl Marx,[17] to typify the process of how existing economic structures get replaced by endogenous innovative forces. Schumpeter's contributions are usually assigned to either his early phase, i.e. "Schumpeter Mark I" or his later phase, i.e. "Schumpeter Mark II". In his early phase, he predominantly focused on the role of individual entrepreneurs and their abilities to generate novelty. In the second phase he focused mainly on the role of innovation in large firms (Fagerberg 2005, p. 6). For a long time, Schumpeter's writings were ignored by mainstream economics. It took almost half a century before Schumpeter's ideas were rediscovered and cultivated in economics.

---

[15] Schumpeter's concept of methodological individualism is mainly focused on the relationship between prices and the behavior of individuals (Heertje 2004, p. 1453). See also Arrow (1994).

[16] For a review of Schumpeter's contributions to economic theory, see Fagerberg (2003) and Hanusch and Pyka (2007b).

[17] Marx originally used this phrase to describe his vision of capitalism's destructive potential (Elliott 1978).

2.3   Some General Reflections on the Neo-Schumpeterian Approach

From the mid-1970s to the mid-1980s, scholars began focusing on endogenous determinants of economic change and the economic consequences of technological change processes. It was at this time that the pioneering work of a few scholars laid the foundations for the neo-Schumpeterian school of thought.

To start with, Christopher Freeman published his first book *"The Economics of Industrial Innovation"* in the mid-1970s (Freeman 1974). He was among the first to revive Schumpeterian ideas by highlighting the role of innovation and technological change processes. In the same year, Nelson and Winter (1974) published an article in which they strongly criticized the traditional neoclassical paradigm. This paper laid the foundation for a new generation of endogenous growth models. Only a few years later, Nelson and Winter (1982) presented the *"Evolutionary Theory of Economic Change"* and introduced the concept of organizational routines. The book is still considered by many to be one of the most influential publications in economics and organizational science. Giovanni Dosi published at least two seminal papers in the 1980s. The first paper (Dosi 1982) starts with a critique of the traditional "technological push" theory of technological change and introduces the concept of technological paradigms and technological trajectories. In his second paper, Dosi (1988) addresses the sources and effects of innovative activities in market economies and explains the entire process from the recognition of opportunity to actual innovative effort and changes in the structure and performance of entire industries. Neo-Schumpeterian approaches explicitly considered path dependencies and lock-in effects (Page 2006; Arthur 1989; David 1985) when analysing economic change processes. In the same decade, Keith Pavitt significantly contributed to clarifying another key issue in innovation economics (Pavitt 1984). His analysis provided an empirically substantiated taxonomy that enabled two types of high-tech sectors to be clearly separated: "science-based" and "specialized suppliers". Pavitt showed that factors leading to successful innovation can differ significantly across industries and sectors (Fagerberg 2005, p. 16). In the 1970s, Nathan Rosenberg published several seminal articles on the relationship between science, invention, innovation and economic growth (Rosenberg 1973, 1974). Together with Stephen J. Kline, he developed an alternative to the linear innovation model, the so-called interactive chain-linked model of innovation (Kline 1985; Kline and Rosenberg 1986). Since then, technological change is no longer seen as a purely "technology-push" or a "demand-pull" but rather as intertwined forces that co-exist (Swann 2009, p. 15). Finally, Eric von Hippel prepared the groundwork for a better understanding of how the demand side, i.e. the consumer, may trigger and affect innovation processes (Von Hippel 1986, 1988). He demonstrated that the users' distinctive knowledge is an important impetus for user-initiated innovation (Klepper and Malerba 2010, p. 1516). This work has some important overlaps with the systemic innovation approach. Lundvall (1988), one of the pioneers of the national innovation system approach, emphasized the importance of user-producer interactions quite early on (Klepper and Malerba 2010, p. 1516). Since the late 1980s, many others have contributed to the field and some of these contributions will be discussed in more detail in subsequent chapters.

The overview above does not claim to be exhaustive; however, it gives an initial idea of the key topics that were on the research agenda of neo-Schumpeterian scholars at the time. The merits can hardly be summarized in a few points. Nonetheless, these early contributions broke new ground in economics in several ways. The first and maybe most important insight is that the explicit consideration of innovative activities is crucial for an in-depth understanding of economic change processes, economic outcomes and economic prosperity. Secondly, the early writings teach us that the assumptions of the classical-neoclassical paradigm are at least questionable. For instance, the notion of the very nature of information and knowledge changed significantly. Thirdly, the writings contributed to an in-depth understanding of how processes at the micro-level, as well as the determinants and mechanisms that trigger and fuel these processes, affect economic performance, growth and prosperity. They paved the way towards more dynamically oriented approaches and models in economics. Not only were the current interactions of firms on markets deemed important in explaining the emergence of technological change patterns at multiple analytical levels, the historical paths or trajectories were considered to be important as well. Fourthly, scholars recognized that innovation determinants, the process of generating novelty itself, and the outcomes of innovative efforts could differ significantly across industries. This implies that the insights from one industry do not necessarily hold true in other – even similar – settings. Finally, this early work teaches us that both supply-side factors as well as demand-side factors are important for a comprehensive understanding of innovation processes. Moreover, they point to the fact that the innovation process itself is neither linear in nature nor is it limited to the individual efforts of single economic entities. Instead, it is characterized by small incremental steps and accompanied by multiple feedback loops. Generating novelty is a highly uncertain and, in most cases, collective process which is characterized by the multiple interactions of independent, yet heterogeneous, economic actors with different knowledge endowments, capabilities, goals and strategies.

2.3.1   Clarification of Terms and Concepts

The clarification of some basic terms and concepts is essential for our discussion below. We start by taking a look at the very nature of information and knowledge. Then we proceed by disentangling the notions of creativity, invention and innovation. Finally, we address some early innovation models and provide a critical assessment of these concepts.

2.3.1.1 On the Nature of Information, Knowledge and Learning

The terms information and knowledge are usually used interchangeably in classical-neoclassical economics. Perfect competition assumes that all firms share the same information (Rutherford 2007, p. 118). Market theory usually makes an explicit assumption about "perfect knowledge" (Boulding 1966, p. 3). This assumption does not imply that firms have complete information about the past and the future (Rutherford 2007, p. 118), but that the information that the firms need in order to make their decisions is immediately available and symmetrically distributed over all economic agents.[18] "What this means in effect is that the acquisition of knowledge of prices or exchange opportunities in a perfect market is costless, so that knowledge is, as it were, a free good." (Boulding 1966, p. 3) The reality, however, looks quite different. Information is rarely complete or costless and economic agents have to make a considerable effort to search for, identify, gather, process and store information (Rutherford 2007, p. 118).

Neo-Schumpeterian scholars have explicitly addressed these issues. Malerba (2007, p. 16) argues that knowledge and learning are key building blocks of the neo-Schumpeterian approach. We began by drawing upon Cowan et al. (2000, p. 216) and Morroni (2006, pp. 26–27) to distinguish between data, information and knowledge.

Data is derived from the senses and consists of various senses that reach the brain from the outside world, whereas information is a structured and organized set of data. Neither data nor information is self-interpreting. Instead, it is knowledge which provides a cognitive dimension. The cognitive context gives a meaning to information and allows it to be interpreted. Morroni (2006, p. 26) sums it up by saying: "Knowledge is acquired by elaborating bits of information, and derives from the ability to search, select, memorize, store, retrieve, structure, compute, embody and use bits of relevant information within a *cognitive system*." By drawing upon the seminal work of Polanyi (1958, 1967), scholars have frequently argued that knowledge can be distinguished by a tacit dimension (non-codified knowledge) and an explicit dimension (codified knowledge). Cowan et al. (2000, p. 212) argued that this distinction can be misleading, at least in some sense. In a similar vein, Witt et al. (2012) suggested reconsidering the frequently used distinctions of overt versus tacit knowledge and encoded versus non-encoded knowledge. Instead, they propose distinguishing between at least three types of knowledge: **(I)** encoded knowledge that can be considered overt knowledge, **(II)** non-encoded knowledge that can be articulated and encoded in principle, and **(III)** non-articulable and, therefore, inherently non-codifiable knowledge that is considered to be, in fact, tacit knowledge (Witt et al. 2012). Their main argument is that the technical terms of storing, accessing, and transferring knowledge affect the key characteristics of knowledge regarding its public good character (ibid).

---

[18] Not only neo-Schumpeterian scholars but also other approaches in economics, such as agency theory, have questioned this problematic assumption (Ackerlof 1970; Spence 1976; Spence 2002).

Another key issue that is frequently addressed by neo-Schumpeterian scholars is an organization's ability to store and generate new stocks of knowledge.[19] It is well-recognized that both a firm's knowledge stock and its ability to generate new stocks of knowledge play a crucial role in understanding innovation and technological change. By referring to the concept of organizational routines, originally developed by Nelson and Winter (1982), several refinements have been proposed to explain a firm's ability to access and generate knowledge.[20] The sources and channels that can be used by a firm to access and generate knowledge are subject to discussion later on in this chapter.

2.3.1.2 Disentangling Creativity, Inventions and Innovation

Understanding creativity requires that we look at the personal and interpersonal level. Swann (2009, p. 25) claims that creativity is a rather chaotic activity or process that follows no specific rules or well-defined algorithms. Creative individuals are either intrinsically or extrinsically motivated, they need a certain degree of autonomy to develop their new ideas, and they have to be both introverted and extroverted at the same time (Swann 2009, pp. 119–127). It has been argued that networks play a key role due to the combinatory nature of the creativity process itself (Swann 2009, p. 128). In this context, Uzzi et al. (2007, p. 448) have argued that creativity is the consequence of a social system of actors that amplify or stifle one another's creativity. However, creativity and invention should be carefully separated. According to Swann (2009, p. 25), an invention is the result of a more or less structured research process or other form of creative activity and appears in the form of an idea, sketches or models for a potentially commercializable new product or process. In a nutshell, an invention is considered to be the result of a creative process (Swann 2009, p. 25).

The terms invention and innovation are closely linked but they are clearly not the same. This distinction is anything but new in economics (cf. Sect. 2.1). According to Pyka and Scharnhorst (2009, p. 9) an innovation can be defined as "[...] the implementation of a new or significantly improved idea, good, service process or practice which is intended to be useful or practical in the sense that either efficiency gains or new returns are generated". An innovation is the first attempt at carrying out an invention in practice in the form of a commercializable application (Swann

---

[19] Knowledge generation is closely related to organizational learning models. According to Bierly et al. (2000) learning can be defined in an organizational context as the process of linking, expanding, and improving data, information, knowledge and wisdom. For an overview of conceptual organizational learning models, see Bierly et al. (2000). For the intellectual roots of the adaptive learning approach in organization theory, see Sect. 2.2.

[20] Several other concepts have been proposed in subsequent years to understand how firms generate knowledge. These are, most notably, the concept of higher level routines, such as organizational capabilities (cf. Winter 2003) and the concept of dynamic capabilities (cf. Zollo and Winter 2002). For an overview, see Becker (2004) and Easterby-Smith et al. (2009).

2009, p. 25; Fagerberg 2005, p. 4).[21] In this context it is important to note that the capabilities of an inventor differ significantly in many respects from those of an innovator (Fagerberg 2005, p. 5). Inventors are the creative minds behind the innovations. Innovators and entrepreneurs need to have other skills that are important for commercializing new ideas.[22] Finally, an innovation is not a single event but rather the result of a process involving many interrelated innovations (Fagerberg 2005, p. 5). This recognition leads us to take a brief look at early innovation models and their limitations.

2.3.1.3 Early Innovation Models and the Emergence of Systemic Thinking

The early innovation model considers innovation to be a linear sequence of activities (Tidd 2006, p. 3). The model was originally proposed by Vannevar Bush (1945). It assumes that a firm has to traverse a well-defined set of stages – basic research, applied research, development, production, marketing and distribution – and at the end of this sequence there will be an innovation. The model is based on the assumption that innovation is basically nothing more than the result of applied science (Fagerberg 2005, p. 8). In early innovation models it was either the new opportunities stemming from research activities – "technology push" – or the needs of the market – "demand pull" – which triggered the sequence that resulted in an innovation (Tidd 2006, p. 3). The linear model of innovation is limited in several ways. Firstly, in practice, innovation is a coupling and matching process, where interaction is the critical element (Tidd 2006, p. 3). Secondly, the model ignores feedback loops that can occur between different stages of the model (Fagerberg 2005, p. 8). Finally, it fails to explain all of the innovations that occur within firm and new ventures (ibid). In response to these limitations, Stephen J. Kline and Nathan Rosenberg (Kline 1985; Kline and Rosenberg 1986) proposed the interactive chain-linked model of innovation. Since then, several generations of innovation models[23] have been proposed and discussed in the literature.[24] The latest generation of innovation models in Rothwell's (1994) framework points to the systemic

---

[21] For further details on the classification of innovations, see the "Oslo Manual" (OECD 2005).

[22] From Schumpeter's perspective, the entrepreneur is primarily an innovator (Swann 2009, p. 131). In his "Theory of Economic Development", Schumpeter identified the following main types of entrepreneurial behavior (cf. Schumpeter 1934 cited according to: (Swedberg 2000; Goss 2005, p. 206): **(I)** the introduction of new goods; **(II)** the introduction of new production methods; **(III)** the struggle into new markets; **(IV)** finding new sources of raw materials; and **(V)** reorganizing an industry in a new way.

[23] According to Rothwell (1994) there are five generations of innovation models: **(a)** the technology push model, **(b)** the market push model, **(c)** feedback loop or coupling models, **(d)** parallel-line models **(d)** systemic or network models.

[24] A review of the development stages of innovation models is not the subject of this investigation. For an overview and discussion, see Tidd et al. (2005) and Tidd (2006).

perspective which is characterized by extensive networking activities among the actors involved, flexible and customized responses and continuous innovation (Tidd 2006, p. 3). This brings us to the systemic perspective in innovation research. The original approach, the so-called "national innovation system" concept, was developed in the early 1980s by neo-Schumpeterian scholars (Freeman 1988; Lundvall 1988, 1992; Nelson 1992).[25]

2.3.2 *Understanding the Nature of Economic Change*

The neo-Schumpeterian approach (cf. Hanusch & Pyka 2007a; Winter 2006) provides an appropriate theoretical framework for analyzing the determinants and economic consequences of technological change processes. The aim of this study is to contribute to an in-depth understanding of how innovations are created and what factors affect the process of novelty generation in a given industrial setting. More precisely, we seek to shed some light onto collective innovation processes in complex interorganizational innovation networks. To accomplish this task, we need to take a closer look at the theoretical cornerstones of the neo-Schumpeterian approach.

In a recent article, Hanusch and Pyka (2007a) provide an overview and discussion on the intellectual roots of the neo-Schumpeterian approach. They identify five historical channels of influence (cf. Hanusch and Pyka 2007a, pp. 277–279): **(I)** the seminal writings of Schumpeter[26] on the role of innovation and entrepreneurial activity on economic change and prosperity; **(II)** the contributions in the broader field of evolutionary economics[27]; **(III)** the field of complexity economics[28]; **(IV)** approaches that are dedicated to *change and development*[29] and, last but not least, **(V)** *systemic approaches*[30] in economics and related disciplines. All these research fields have left their mark and significantly contributed to the emergence and solidification of the neo-Schumpeterian approach in economics. Today, most neo-Schumpeterian scholars would agree that the following characteristic features form the very heart of the approach.[31]

---

[25] This concept and further developments are subject to discussion in Sect. 2.3.

[26] For the most influential contributions, see Schumpeter (1912, 1939, 1942).

[27] For an overview of the field of evolutionary economics, see Witt (2003, 2008a, b).

[28] For an excellent overview of contemporary research in the field of complexity economics, see Antonelli (2011). For groundbreaking work on complexity economics, see Kirman (1989, 1993), Foster (2005) and Arthur (2007).

[29] For an overview of evolutionary thinking in economics and related disciplines, see Sect. 2.2.

[30] Systemic research has its roots in general system theory (Bertalanffy 1951, 1968; Boulding 1956). For an overview of innovation system approaches, see Sect. 2.3.3.

[31] This discussion is guided by Dosi and Nelson (1994) and Hanusch and Pyka (2007a, c).

2.3.2.1   The Notion of Individuals and Firms

As already outlined above, the notion of individual economic actors in classical-neoclassical and evolutionary theories differs considerably. The neo-Schumpeterian approach considers individuals to be bounded-rational economic actors (Simon 1955, 1991) who are continuously searching for new opportunities in uncertain and permanently changing environments (Hanusch and Pyka 2007a, p. 278). Learning and the cognition of the economic actors are central to the neo-Schumpeterian approach (ibid).[32] This notion of individual behavior is not compatible with the traditional "homo oeconomicus" concept in classical-neoclassical economics. Alternative concepts were developed and applied by neo-Schumpeterian scholars to capture the behavior of inventors and innovators in complex economic systems. The "homo creativus" can be characterized as an individual that is searching for opportunities in order to meet the challenges of unpredictable qualitative change in uncertain economic environments (Foster 1987; Dopfer 2011). The notion of firms in neo-Schumpeterian economics is closely related to these considerations. In traditional microeconomic theory, the firm is treated as a "black box" which can be, in principle, fully described by a simple production function. Firms are homogenous economic entities that produce homogenous goods and services which can be consumed by fully rationally-behaving economic agents in perfect markets. In contrast, from a neo-Schumpeterian perspective, firms are considered to be heterogeneous economic entities. They have very specific resource endowments and they follow their own strategies to accomplish their individual goals. The neo-Schumpeterian approach acknowledges that "[. . .] firms often no longer compete in a price dimension only, as competition in innovation has taken the dominant role" (Hanusch and Pyka 2007a, p. 281).

2.3.2.2   Qualitative Change, Punctuated Equilibria and Pattern Formation

The neo-Schumpeterian approach applies a dynamic perspective. The approach is concerned with all facets of open and uncertain developments in socioeconomic systems (Hanusch and Pyka 2007a, p. 276). The models and theoretical explanations focus on procedural issues and not on static snap-shots of economic phenomena. Static and competitive static models are considered to be necessary but not sufficient in understanding the dynamic processes that generate economic outcomes at the micro or macro-level. Evolutionary theories seek to explain the movement of something over time or they explain reasons for why something is what it is at a given point in time (Dosi and Nelson 1994, p. 154). In this context, Hanusch and Pyka (2007a, p. 276) argue that the future developmental potential of socioeconomic systems, triggered and fueled by the underlying innovation processes,

---

[32] Both inventors and entrepreneurs receive due attention in neo-Schumpeterian economics.

has to be regarded from a normative perspective. They identify the following three constitutive elements or normative principles of neo-Schumpeterian economics (Hanusch and Pyka 2007a, pp. 276–277): **(I)** qualitative change is a multi-level phenomenon that has to be analyzed against the backdrop of the existing constrains and changing circumstances, **(II)** qualitative change can be described as a process characterized by punctuated equilibria patterns with regular and smooth development phases and with transitional or disruptive phases, and **(III)** qualitative change inherently shows strong non-linearities and multiple feedback effects which are reflected in pattern formation and other forms of spontaneous structuring. The last point, in particular, appears in most evolutionary models and theories. These models involve both random elements, which generate or renew some variation, and mechanisms that systematically address variation patterns (Dosi and Nelson 1994, p. 154).

2.3.2.3 Path Dependencies and Irreversibilities

The analysis of change processes is at the very heart of evolutionary economics in general and neo-Schumpeterian economics in particular. Any kind of economic change process or dynamic development which takes place in historical time is accompanied by path dependencies and irreversibilities (Hanusch and Pyka 2007a, p. 277). David (1985, p. 332) defines path dependencies as follows: "A path-dependent sequence of economic changes is one of which important influences upon the eventual outcome can be exerted by temporally remote events, including happenings dominated by chance elements rather than systematic forces." David (1985) demonstrates that path-dependent processes can result in a "lock-in" effect. The basic idea behind the concept is straightforward. A course already adopted in the past can produce considerable switching costs at a later point in time. This can result in a situation in which second-best solutions prevail. At the same time, it is important to note that path-dependent processes do not imply that the future is in any way closed (Araujo and Harrison 2002, p. 6). Hanusch and Pyka (2007a, p. 278) note in this context that the outcomes of evolutionary processes are not determined ex ante but are rather the result of true uncertainty underlying all processes of novelty generation.

2.3.3 A Closer Look at Selected Key Concepts

We will now look at three concepts that are essential for the purpose of this book: the proximity concept, the innovation system concept and the network concept.

2.3.3.1 The Proximity Concept

The proximity concept has increasingly raised attention, not only among neo-Schumpeterian scholars but also in the areas of sociology, geography and management science. Both theoretical (e.g. Boschma 2005a, b; Torre and Rallet 2005; Visser 2009; Boschma and Frenken 2010) and empirical studies (e.g. Oerlemans et al. 2001; Oerlemans and Meeus 2005; Owen-Smith and Powell 2004; Whittington et al. 2009) have improved our understanding of how proximity relates to firm innovation. In the most general sense, proximity can be defined as "[...] being close to something measured on a certain dimension" (Knoben and Oerlemans 2006, pp. 71–72). The concept acknowledges that firms are simultaneously exposed to a variety of proximity dimensions such as institutional proximity, organizational proximity, cultural proximity, technological proximity, network proximity and geographical proximity (cf. Knoben and Oerlemans 2006, p. 71). However, most proximity concepts have some considerable limitations. According to Knoben and Oerlemans (2006, p. 71), one of the most notable issues is that previous research failed to provide a clear separation between proximity dimensions. This is still reflected in conceptual overlaps across many proximity dimensions (ibid).

Only a few studies (cf. Boschma 2005a, b; Boschma and Frenken 2010) have explicitly addressed this conceptual limitation. The proximity concept originally proposed by Boschma (2005b) can be regarded as an integral part and extension of the evolutionary economic approach (Boschma and Frenken 2010).[33] The concept encompasses five proximity dimensions: **(I)** cognitive proximity, **(II)** organizational proximity, **(III)** institutional proximity, **(IV)** geographical proximity, and **(V)** social proximity. If not otherwise stated, the following discussion on proximity dimensions draws upon the work of Boschma (2005b, pp. 63–71) and Boschma and Frenken (2010, pp. 122–124).

The cognitive proximity dimension points to the fact that successful knowledge transfer and learning processes among firms require, at least to a certain extent, a common cognitive basis among the actors involved. A high level of cognitive proximity facilitates effective communication. Firms with similar knowledge stocks and expertise are more likely to learn from one another than firms with entirely different cognitive backgrounds. The authors draw upon the absorptive capacity concept originally proposed by Cohen and Levinthal (1989, 1990) to substantiate their line of argument. However, it is important to note that the positive effects of cognitive proximity can have repercussions. Following Nooteboom (2008), the authors put forward the argument that too much cognitive proximity can also be detrimental to interactive learning. The reasoning behind the argument is straightforward: there probably remains little scope for mutual knowledge

---

[33] For a discussion on the intersections between evolutionary economics and approaches in economic geography, see Boschma and Frenken (2006).

transfer and interactive learning processes if the knowledge stocks, competencies and skills are highly similar.

Organizational proximity refers to the extent to which relations are shared within an organization or between organizations. The basic argument of the authors is that interactive learning and knowledge creation depends, among many other factors, on a capacity to coordinate the exchange of complementary knowledge within and between organizations. Organizational arrangements are considered to be the arenas in which knowledge transfer and learning can take place. A sufficiently high level of organizational proximity is assumed to reduce uncertainty and opportunism. This, in turn, is likely to facilitate interactive learning and knowledge creation. Too much organizational proximity can cause a lack of flexibility, thereby generating exactly the opposite effect.[34]

Institutional proximity addresses both the formal and the informal institutions at the macro-level. These institutions affect the extent to and way in which organizations interact. Formal institutions can be laws, norms, rules or codified codes of conduct. Informal institutions can be cultural norms, values, habits, or other conventions. In this context, institutions are considered to be enabling mechanisms that stabilize the environment in which knowledge transfer and interactive learning processes take place. An above average level of institutional proximity can prevent mutual knowledge transfer and learning processes due to institutional lock-in effects.[35] Other proximity conceptualizations have acknowledged the latter type of institutional proximity by introducing a separate "cultural proximity" dimension (Knoben and Oerlemans 2006, p. 71).

Geographical proximity focuses on the spatial distribution of economic actors. This proximity dimension addresses the physical distance between economic actors. Firms can benefit from geographical proximity in many ways. For instance, being located close to others is assumed to facilitate face-to-face contact. Short distances simplify the exchange of information and enable interactive learning processes. This, in turn, is likely to enhance the innovative performance at firm level. The main drawback of too much spatial proximity is that firms can become inward looking and isolate themselves from the outside word. Thus, after exceeding a certain degree of geographical clustering, the benefits outlined above may start to have the opposite effect.[36]

Social proximity is the last of the five proximity dimensions proposed by Ron Boschma and further developed by himself and Koen Frenken. Social proximity refers to the relationship between economic actors at the micro-level.[37] This type of

---

[34] The underlying theoretical arguments used in this context draw upon governance and transaction cost issues in market, hybrid and hierarchical organizational forms (cf. Sect. 2.5.2).

[35] To substantiate this line of argument, Boschma (2005b) points to the fact that too much institutional proximity can cause structural inertia.

[36] For an in-depth discussion, see (cf. Sect. 12.2).

[37] The social proximity concept is strongly influenced by the social capital and embeddedness literature (cf. Sect. 2.5.4).

proximity is frequently referred to as relational proximity (Coenen et al. 2004). The occupation of advantageous network positions can enhance access to external knowledge sources (Grant and Baden-Fuller 2004; Buckley et al. 2009), facilitate interorganizational learning processes (Hamel 1991; Schoenmakers and Duysters 2006; Nooteboom 2008) and is positively related to firm innovativeness (Ahuja 2000; Stuart 2000). The benefits of social proximity can have the reverse effect if firms become too densely embedded (cf. "overembeddedness" phenomenon) or missembedded due to structural inertia (cf. "network inertia" phenomenon).[38]

The proximity framework proposed by Boschma (2005b) and further developed by Boschma and Frenken (2010) is characterized by the following four features. Firstly, the framework provides a clear definition and separation of the proximity dimensions outlined above. The proximity dimensions are independent of each other. Secondly, this implies that one can reduce as well as extend the list of relevant proximity dimensions without changing the meaning of each dimension (Boschma and Frenken 2010, p. 124). Thirdly, the framework lays the ground for analyzing each dimension separately and, at the same time, it allows the interplay between selected proximity dimensions to be explored. Finally, the proximity framework applies a process-oriented perspective and explicitly addresses both the positive and the negative impact of proximity on knowledge transfer, interactive learning and firm-level innovation outcomes. In summary, the concept lays the foundations for a more dynamic understanding of proximity in all its facets.

2.3.3.2 The Innovation System Concept

Quite early on, neo-Schumpeterian scholars explicitly addressed the collective nature of innovation processes and introduced the "systemic innovation approach" (Freeman 1988; Lundvall 1988, 1992; Nelson 1992). According to this approach, innovations were considered to be the outcome of multiple interactions between elements in an integrated system (Lundvall 1992). The first conceptualization was the "national innovation system approach" (NIS). It emphasizes the role of knowledge generation through the process of interactive learning in an evolving social environment which is determined by national level institutions and increasingly challenged by the process of internalization and globalization (cf. Lundvall 1992, pp. 2–3). Since then, several refinements have been proposed. An innovation system can be defined not only from a national dimension but also along several

---

[38] The overembeddedness concept was originally introduced by Uzzi (1996). The structural network inertia concept (cf. Kim et al. 2006) is strongly influenced by organizational ecology research (cf. Sect. 2.2).

other dimensions: regional dimension (RIS),[39] sectoral dimension (SIS),[40] or technological dimension[41] (Carlsson et al. 2002, p. 233). The common denominator of all these concepts is that they all involve creation, diffusion and use of knowledge and each of them can be fully described by a set of components, relationships among these components and their attributes (ibid).[42]

In addition, systemic approaches have three other common features. Firstly, innovations are not the result of linear processes but rather the outcome of repeated knowledge exchange and learning processes between various types of actors in these socio-economic systems. An innovation system is characterized by multiple interactions and feedbacks and it allows for the reproduction of individual or collective knowledge (Lundvall 1992, p. 2). Secondly, innovation systems are dynamic rather than static entities because the elements and relations in the systems are subject to change. Carlsson et al. (2002, p. 234) summarize this as follows: "Another dimension is that of time. In a system with built-in feedback mechanisms, the configuration of components, attributes, and relationships is constantly changing." Thus, past events determine current actions and affect future innovation outcomes. Thirdly, the structural characteristics of innovation systems – such as the actors, types of relationships, system boundaries and the broader environments in which the system is embedded – affect the interactions of actors and subsequent innovation outcomes (Carlsson et al. 2002).

2.3.3.3 The Network Concept

Knowledge is considered to be the key factor in determining the competitiveness of firms and the economic growth and prosperity of nations (Saviotti 2011, p. 141). An in-depth understanding of knowledge generation and diffusion processes is a central theme in neo-Schumpeterian economics because these processes underlie the generation of novelty in terms of inventions and innovations (Hanusch and Pyka (2007c, p. 3). In general, firms can gain access to new knowledge via internal or

---

[39] The regional innovation system approach (RIS) is strongly influenced by the idea that innovation is the outcome of spatially or territorially determined learning processes between the actors in the system (Cooke 2001).

[40] The sectoral innovation system's approach (SIS) emphasizes the cognitive dimension by arguing that interactive learning processes and subsequent innovation outcomes are fostered by the technological and contextual relatedness of the actors in the system (Malerba 2002).

[41] The technological innovation system approach (TIS) focuses on generic technologies with general applications over many industries (Carlsson et al. 2002).

[42] Carlsson et al (2002, pp. 234–235) define components, relationships and attributes as follows: Firstly, components are the basic elements or operating parts of a system. They can be individuals, organizations, businesses, banks, universities, research institutes and public policy agencies (or parts or groups of each). Secondly, relationships involve all kinds of market and non-market links between the components of the system. Finally, attributes are considered to be the properties of the components and the relationships between them and they specify the very nature or type of system.

external channels of knowledge (Malerba 1992). The internal perspective refers to learning and knowledge generating processes within the boundaries of the firm, whereas the external perspective highlights the importance of collaborative partnerships. We focus here on the latter perspective. By now it is well recognized that cooperation and network embeddedness play a key role in a firm's effort to innovate. Various types and aspects of networks have attracted a great amount of attention in economics and related disciplines over the past few years. These are, for instance, "strategic networks" (Jarillo 1988; Gulati et al. 2000), "alliance networks" (Koka and Presscott 2008; Phelps 2010), "knowledge networks" (Saviotti 2011; Ozman 2013; Wang et al. 2014), and "innovation networks" (Pyka 2002; Cantner and Graf 2011; Leven et al. 2014).[43] In order to gain a comprehensive understanding of collective innovation processes, one needs to understand the very nature of innovation networks. For the purpose of this book, three aspects of innovation networks are of primary importance: **(I)** network structure, **(II)** network positioning **(III)** network evolution.

Cooperation activities are, at first glance, micro-level phenomena. However, it is important to recognize that each tie formation and tie termination process shapes the structural configuration of the industry's innovation network. Thus, all of the collaborative ties among a well-defined set of heterogeneous economic actors spans the structure of the overall interorganizational network at the macro-level. Firms follow individual cooperation strategies that are guided by firm-specific goals and motives. At the same time, each cooperating firm occupies, whether consciously or not, a position within the industry's innovation network. Contemporary research provides sound evidence for the relatedness between a firm's strategic network position and its subsequent innovation outcomes. However, the type of positioning that matters for a firm in its efforts to innovate can significantly differ from industry to industry. In addition, the structural positioning of a firm in an industry network is by no means static. The firm's own tie formations and tie termination processes, as well as the cooperation activities of other network actors, are reflected in a continuous change in its network position over time. This brings us to the last and maybe most important point. The structure of an innovation network is subject to change due to node entries and exits as well as tie formations and tie terminations. The implications are straightforward: a comprehensive analysis of how a firm's network position affects its innovative performance requires a dynamic setting and a solid understanding of the industry's characteristics.

To conclude, the overlaps and intersections of the three concepts are obvious. The proximity concept explicitly acknowledges the importance of networks by integrating a relational proximity dimension. The systemic approach can be seen as a broader concept that inherently entails innovation networks.

---

[43] Note that these types of networks are not mutually exclusive. They can be defined within firms ("intra-organizational network", cf. Rank et al. 2010), between firms ("interfirm networks", cf. Schilling and Phelps 2007), or between various types of organizations ("interorganizational networks", cf. Broekel and Hartog 2013).

2.4   Why Do Some Firms Outperform Others?

So far we have introduced and discussed the general theoretical framework that constitutes the basis for this study. Now we turn our attention to the micro-level. The guiding question of this section is why some firms outperform others. In search of an answer to this question we take a brief look at different schools of thought in economics and management science that have analyzed the determinants of a firm's performance. After this short overview, we will turn our attention to the knowledge-based theory of the firm. This approach explains a firm's competitive advantage by drawing upon its ability to access knowledge and generate new stocks of knowledge which are key drivers behind commercial success.

2.4.1   The Structure-Conduct-Performance Approach

Industrial economists were among the first to address the question raised above. In the early 1950s the "structure-conduct-performance" approach (SCP) was the accepted scientific doctrine. The approach was strongly influenced by the contributions made by Mason (1939, 1957) and Bain (1950, 1951). One of the key concerns of industrial economists at that time was to analyze a firm's behavior and strategic response to imperfect market conditions. Retrospectively, this was quite a controversial approach against the backdrop of the omnipresent neoclassical perfect competition paradigm. The core idea of the traditional SCP model is based on the notion that the structural characteristics of a firm's industrial environment affect its conduct and its performance outcomes (Porter 1981, p. 610). The industrial dimension is defined as a relatively stable economic and technological context in which a firm competes against other firms (Schmalensee 1988, p. 644; Porter 1981, p. 611) and is usually specified by characteristics such as number of sellers and buyers, entry barriers, level of vertical integration or product differentiation etc. (Conner 1991, p. 124). The conduct dimension reflects the firm's choice of strategic decisions such as pricing and advertising decisions (Porter 1981, p. 611) and is assumed to be determined by the industry's structure (Conner 1991, p. 124). Finally, the performance dimension is quite broadly defined in the traditional SCP frameworks and encompasses not only profitability or cost minimization but also a firm's innovative performance (Porter 1981, p. 611). The main implications of this approach are strongly connected with firm size and monopoly power arguments. Accordingly, vertical integration, advertising and, product differentiation are considered to be strategic options to control prices, erect entry barriers and to sustain or increase monopoly power (Conner 1991, p. 125). Several instances of criticism[44] were raised against the traditional SCP paradigm which led to many refinements in subsequent years.[45]

2.4.2   The Resource-Based View

The resource-based view (RBV) has its intellectual roots in the seminal contribution of Edith T. Penrose (1959). It is interesting to note that the basic ideas of the RBV emerged at nearly the same time as the SCP paradigm but it took nearly three decades to attract the attention of scholars in the field. Proponents of the RBV (Wernerfelt 1984; Barney 1991; Peteraf 1993) proposed a very different line of argument to explain a firm's above-average rents. They argue that firm-specific factors – not market conditions – are the key drivers behind a firm's competitive advantage.[46] More precisely, the RBV suggests that a firm's ability to achieve and maintain a profitable market position and outperform competitors mainly depends on its ability to generate, exploit und utilize firm-specific resources (Barney 1991). The concept of idiosyncratic, immobile resources is the crucial element of the resource-based approach. Barney (1991, p. 101) defines resources as all kinds of "[. . .] assets, capabilities, organizational processes, firm attributes, information, knowledge [. . .]" that are under the control of the firm and "[. . .] that enable the firm to conceive of and implement strategies that improve its efficiency and effectiveness". The explanandum is the firm's sustainable competitive advantage (Barney 1991, p. 102) which simply can be defined as a competitive advantage that lasts for a long period of time (Porter 1985) or a competitive advantage that continues to exist even after efforts have been made to duplicate it (Lippman and Rumelt 1982). A firm resource must fulfill four basic conditions to enable a firm to realize a sustainable competitive advantage (cf. Barney 1991, pp. 105–112): Firstly, it must be valuable by opening up new opportunities and by neutralizing threats; secondly, the resource has to be rare; thirdly, it must be difficult for other firms to imitate; finally, there must be no strategically equivalent substitute that is valuable

---

[44] The static nature of the structural dimension within this framework has been particularly criticized. For instance, Schumpeter claimed that a static view of competition (i.e. price competition over existing products) fails to see that creating or adapting innovations is a much more effective way to compete because it makes the rivals' positions obsolete (Schumpeter 1950, p. 84: cited by Conner 1991, p. 127). For a discussion on further limitations of the SCP paradigm, see Porter (1981, pp. 11–14).

[45] For excellent overviews of the traditional SCP paradigm and its refinements, see Schmalensee (1988) and Conner (1991).

[46] It is important to note that the emergence of the RBV can be seen as a critical response to the black box view of the firm in the SCP paradigm (Foss and Ishikawa 2007, p. 750). The RBV highlighted asymmetric information in factor markets and heterogeneous economic actors to explain differential rents (ibid). However, the RBV supplements rather than replaces the SCP paradigm (Kraaijenbrink et al. 2010, p. 350).

but neither rare nor difficult to imitate.[47] In summary, the RBV emphasized above-average returns as rents that stem from costly-to-copy productive assets which are inherent to the firm (Conner 1991, p. 144). Criticism has addressed several of the RBV's limitations.[48] The three most salient issues will be viewed in more detail below.

Firstly, Priem and Buttler (2001) argued that the RBV is tautological in a sense that main assertions are true by definition. Barney (2001, pp. 41–46) responded to this critique by showing in detail how several elementary building blocks of the RBV can be parameterized and thus be empirically tested. Secondly, it has been argued that the VRIN/O criterion is neither a necessary nor sufficient condition for realizing a sustainable competitive advantage (Kraaijenbrink et al. 2010, p. 351). This critique is fundamental as it addresses the core of the RVB.[49] Finally, the conceptualization of the sustainable competitive advantage concept itself was subject to criticism. According to Foss and Ishikawa (2007, p. 750) one central shortcoming of the RBV is its implicit reliance on the competitive equilibrium in a sense that resources are mainly used by firms to generate a superior rent equilibrium. A careful look at Barney's seminal contribution reveals this issue. "In a sense, this definition of sustainable competitive advantage is an equilibrium definition" (Barney 1991, p. 102). In response to these limitations, Foss and Ishikawa (2007) draw upon Austrian theories of capital and entrepreneurship and propose a dynamic resource-based view. The limitations of the RBV outlined above led to numerous refinements and extensions. A focus on knowledge as a firm's most decisive source of superior rents resulted, in particular, in the emergence of an entirely new paradigm in the mid-1990s.

2.4.3   The Knowledge-Based View

The knowledge-based view (KBV) is based on the notion that a firm's ability to assess and generate knowledge is decisive in outperforming competitors (Kogut

---

[47] The so-called originally proposed VRIN criterion has been expanded into the VRIN/O criterion by Barney (2002). Accordingly, firms must acquire and control valuable, rare, inimitable, and non-substitutable resources and capabilities, plus have the organization (O) in place that can absorb and apply them (Kraaijenbrink et al. 2010, p. 350).

[48] For an overview and discussion see Priem and Buttler (2001), Foss and Ishikawa (2007) and Kraaijenbrink et al. (2010). See Barney (2001) for a response to Priem and Butter's critique.

[49] This critique is centered around two lines of argument (cf. (Kraaijenbrink et al. 2010, pp. 355–356). On the one hand, it has been argued that uncertainty and immobility are the true basic conditions for achieving a sustainable competitive advantage whereas other conditions are simply additional to these. On the other hand, it has been argued that the RBV does not sufficiently recognize the role of the individual actors (i.e. entrepreneurs or managers). Thus, to generate a sustainable competitive advantage a firm needs: (a) a bundle of resources rather than only a single resource, and (b) the managerial capabilities to recognize and exploit the opportunities that are inherent to these resource bundles.

and Zander 1992; Spender and Grant 1996; Grant 1996). According to the KBV, the primary rationale for a firm's existence is the creation, transfer and application of knowledge (Decarolis and Deeds 1999, p. 954). The KBV has some characteristic features that clearly distinguish it from all other theories of the firm described above.[50]

Firstly, the KBV focuses on knowledge-accessing and on knowledge-acquiring processes. This concomitantly implies that a firm's knowledge stock is subject to change. It is not a firm's currently existing stock of knowledge but rather a firm's ability to access, recombine and create new stocks of knowledge out of existing firm-specific capabilities which is vital for gaining a competitive advantage (Nonaka et al. 2000). As a consequence, the KBV clearly paves the way towards a more dynamic and process-oriented understanding of how firms create above-normal returns. Secondly, the KBV accentuates the role of individuals and decision makers during the process of creating new knowledge out of existing knowledge stocks. Knowledge is regarded as context specific, relational, dynamic and essentially related to human action (Nonaka et al. 2000, p. 2). In other words, the creation of new stocks of knowledge always requires a certain degree of interaction at the interpersonal, inter-unit or interorganizational level. Thirdly, the KBV explicitly considers resource bundles instead of focusing on single idiosyncratic resources. Dierickx and Cool (1989, p. 1504) point to a fact that is often overlooked "[. . .] resource bundles need to be deployed to achieve or protect such privileged product market positions." In other words, it is not single resources but rather the synergies that stem from the interdependencies between multiple resources that are the sources of a firm's competitive advantage. In a similar vein, Teece (2007) argues that asset complementarities are important sources of a firm's above-normal returns. Last but not least, the KBV is based on the notion that knowledge and skills are the major source of a firm's competitive advantage because "[. . .] it is through this set of knowledge and skills that a firm is able to innovate new products/processes/services, or improve the existing ones more efficiently and/or effectively." (Nonaka et al. 2000, p. 2). In other words, the KBV clearly establishes a link between a firm's ability to access new stocks of knowledge, generate novelty in terms of innovations and the market success it will subsequently have.

---

[50] It is important to note that the KBV is not a closed theoretical paradigm but rather a set of theoretical ideas and concepts that are strongly influenced by related disciplines such as psychology and cognitive science (e.g. models of individual learning), organizational science (e.g. models of single-loop learning and double-loop learning) and evolutionary economics (e.g. routines, capabilities or dynamic capabilities). See Dodgson (1993) for information on organizational learning approaches and Winter (2003) for further insights on dynamic capabilities.

2.5   An Interdisciplinary View on Alliance and Network Research

Last but not least, we move on to theoretical concepts that substantiate various facets of alliances and networks. Due to the multifaceted und complex nature of hybrid organizational forms[51] there is no single theoretical paradigm that can respond sufficiently to the broad variety of research questions in this research domain. Instead, we face a diverse array of theoretical approaches and methodological concepts. In an excellent review article, Osborn and Hagedoorn (1997) describe the field of organizational alliance and network research at that time to be rather chaotic. Nonetheless the authors come to the conclusion that the research field is characterized by a rich set of theoretical contributions from economics, international business, management science and sociology (ibid).

Since then little has changed. Alliance and network research is still a highly interdisciplinary field (Ozman 2009; Bergenholtz and Waldstrom 2011). As a consequence, in this section we draw upon theoretical concepts from various scientific disciplines instead of restricting our line of argument to one school of thought in order to accomplish the tasks at hand. We define different types of hybrid organizational forms and introduce the theoretical concepts that allow us to explain the existence of hybrids. Then, we discuss the rationales and motives for cooperating. Finally, we introduce theoretical concepts that substantiate and explain the relationship between network embeddedness and firm performance.

2.5.1   Defining Interorganizational Relations, Alliances and Networks

This study focuses on the analysis of interorganizational innovation networks in the German laser industry. Interorganizational relations are the ties or linkages that connect the nodes or actors of a network.[52] Oliver (1990, p. 241) defines interorganizational relations as "[...] relatively enduring transactions, flows, and linkages that occur among or between an organization and one or more organizations in its environment". The organizations can be private, public or non-profit (Cropper et al. 2008, p. 4). The linkages themselves can connect two or more partner organizations. In the first case we refer to these interorganizational relations as dyadic and in the second case as multiplicitous hybrid organizational forms

---

[51] The term "hybrid organizational form" has been used by organizational scholars (Powell 1987; Williamson 1991) to subsume all kinds of mainly formal cooperative arrangements in an interorganizational context, such as alliances and networks.

[52] This term comprises all kinds of strategic alliances or other types of collaborative partnerships.

(Cropper et al. 2008, p. 4).[53] Another essential distinguishing feature of interorganizational relations is the degree of formality. In principle, interorganizational relations can be both informal as well as formal in nature (Pyka 1997, p. 210). The latter case encompasses a broad variety of contractual agreements such as strategic alliances or strategic technology partnerships.

A general definition of strategic alliances has been proposed by Inkpen (2009, p. 389). He defines strategic alliances as "[. . .] collaborative organizational agreements that use resources and/or governance structures from more than one existing organization". In a similar vein, Gulati (1998, p. 293) defines strategic alliances as "[. . .] voluntary arrangements between firms involving exchange, sharing, or co-development of products, technologies, or services. They can occur as a result of a wide range of motives and goals, take a variety of forms, and occur across vertical and horizontal boundaries". A third, more technology oriented definition of strategic partnership has been proposed by Hagedoorn and Schakenraad (1994, p. 291). They define strategic technology alliances as "[. . .] the establishment of cooperative agreements aimed at joint innovative efforts or technology transfer that can have a lasting effect on the product-market positioning of participating companies".

The preceding definitions have some important implications and allow us to identify and summarize four characteristic features of strategic alliances. Firstly, strategic alliances are formalized voluntary agreements between mutual independent companies, firms or other types of organizations. Secondly, even though the motives for entering a strategic alliance can be quite different, the organizations have to agree upon a common goal (e.g. joint development of products, technologies, or services). Thirdly, the achievement of common goals requires a certain degree of resource exchange (e.g. knowledge or technology transfer) among the parties involved. In other words, all of the parties involved have to agree upon more or less formalized rights and obligations. Finally, the definitions above acknowledge the existence of heterogeneous structural alliance forms ranging from short-term supply contracts, licensing and franchise agreements and consultancy contracts, to consortia, long-term partnerships, joint ventures and shared product development projects (Gulati and Singh 1998; Podolny and Page 1998; Brass et al. 2004; Inkpen 2009). In this study we focus on one particular type of formal knowledge-related interorganizational relation: i.e. publicly funded R&D cooperation projects. Medium-term contractual R&D partnerships clearly fall within the spectrum of "arm-length" contracts on the one hand and "equity joint ventures" on the other (Contractor and Lorange 2002, p. 487) irrespective of whether the project partners received public funding or not.

---

[53] The term "multi-partner alliance" is widely used, especially in management science, to address this type of hybrid organizational form. It is important to note that a multi-partner alliance is not a collection of independent dyadic alliances, nor can it be considered a network of partners that maintain direct ties to a single focal firm; it is rather a cooperation setting that entails multilateral interaction among the partners involved (Lavie et al. 2007, p. 578).

Empirical investigations indicate a clear increase in strategic alliances over the past decades (Duysters et al. 1999; Hagedoorn 2002). As a consequence, companies increasingly face the challenge of managing more than one alliance simultaneously (Lavie 2007). Several concepts have been introduced in the literature to capture firm-specific portfolios of collaborative relationships. These services range from concepts such as alliance constellations (Das and Teng 2002), alliance networks (Goerzen 2005), alliance groups (Duysters and Lemmens 2004), alliance portfolios (Hoffmann 2007; Wassmer 2010) and ego networks (Hite and Hesterly 2001). The two latter concepts are most frequently to be found in the literature. An ego network or alliance portfolio is defined from the focal actor's perspective and consists of a set of direct, dyadic ties between the focal actor and the alters and indirect ties between the alters (Ahuja 2000). Ego networks do not include second-tier ties or second-step ties to which the focal actor is not directly connected (Hite and Hesterly 2001, p. 277).

So far, we have focused on firm-centered cooperation and ego network concepts. Now we change the perspective and turn our attention to the overall network level. The following quote nicely shows that researchers have realized quite early the importance of indirect linkages "[...] we need to understand not only which organizations will become partners but also which part(s) of each partner will belong to the hybrid" (Borys and Jemison 1989, p. 236). But what exactly do we mean in terms of networks?

In a most basic sense, any kind of network consists of two basic elements: nodes and ties between these nodes (Wasserman and Faust 1994). In accordance with this quite general notion of network, Brass and colleagues (2004, p. 795) define a network "[...] as a set of nodes and the set of ties representing some relationship, or lack of relationship, between the nodes." These two definitions have some important implications. Firstly, the network perspective emphasizes the interconnectedness of a well-defined population of actors. Secondly, not only realized linkages among embedded network actors but also missing linkages or potentially realizable linkages are important for an in-depth understanding of the network's structural configuration. Thirdly, the network actors can be linked by many types of usually non-hierarchical connections and flows, such as information, materials, financial resources, services, and social support (Provan and Kenis 2007, p. 482). For each of these dimensions of interconnectedness, all ties put together form a particular network structure (Borgatti and Halgin 2011, p. 1169) which affects the embedded network actors in multiple ways. In this context, Gulati and colleagues (2000, p. 203) emphasize the strategic dimension of an interorganizational network and argue that "[...] networks of relationships in which firms are embedded profoundly influence their conduct and performance". However, firm performance has many facets ranging from productivity to revenue and other dimensions of performance such as firm survival. Due to the objective of this study we are primarily interested in the innovative performance of an organization. This leads to the notion of innovation networks. Freeman (1991, p. 501) argues that "[...] the problem of innovation is to process and convert information from diverse sources into useful knowledge about designing, making and selling new products and processes. Networks were shown to be essential both in the acquisition and in the processing of information inputs". By now it is well recognized that innovation is increasingly based on interaction between a variety of actors through formal agreements like R&D cooperation (Brenner et al. 2011, p. 1). Thus, the very purpose of an innovation network is to exchange already existing information, knowledge and expertise among cooperating organizations in order to commonly generate new knowledge which can be embodied in new products, services or processes (Cantner and Graf 2011, p. 373).

By drawing upon the considerations above, we define an innovation network for the purpose of this study as follows: an innovation network consists of **(I)** a well-defined set of independent economic actors (i.e. private and public organizational entities) and **(II)** connections that allow for a unilateral, bilateral or multilateral exchange of ideas, information, knowledge and expertise (i.e. formal and informal knowledge-related interorganizational relations) between directly and/or indirectly connected organizational entities; **(III)** it is embedded in a broader socio-economic environment and **(IV)** it has a strategic dimension in the sense that the organizations involved cooperate to recombine and generate new knowledge enclosed in goods or services in order to respond to continuously changing market demands and customer needs.

2.5.2   Explaining the Existence of Hybrid Organizational Forms

This section concentrates on two broad theoretical streams in the field of alliance and network research that seek to explain the existence of hybrid organizational structures: economic theories and sociological theories.

Economists (Thorelli 1986; Jarillo 1988; Williamson 1991) utilize predominantly transaction cost based arguments[54] and state that hybrids are an organizational form positioned intermediately between markets and hierarchies. For instance Thorelli (1986) argues that a network has to be interpreted as part of the markets. He puts forward the argument that a "[. . .] network is the one intermediary between the single firm and the market, i.e. two or more firms which, due to the intensity of their interaction, constitute a subset of one (or several) market(s)" (Thorelli 1986, p. 38). In a similar vein, Jarillo (1988) studied the economic conditions for the existence of networks by referring to transaction cost arguments. With reference to prior work by Ouchi (1980), who suggested breaking down hierarchies into two distinct categories (i.e. bureaucracies and clans), Jarillo

---

[54] The dichotomy of markets and hierarchies can be ascribed to the seminal work of Coase (1937) on the nature of the firm. He was the first to use transaction cost arguments to explain why economic transactions are processed most efficiently by means of hierarchical organizational forms. These ideas were applied and developed further, in particular by Williamson (1975, 1985).

(1988) developed this idea further and proposed doing the same with markets by splitting this category into two sub-categories (i.e. classic markets and strategic networks). He ended up with a scheme that illustrates four distinct modes of organizing economic activity (Jarillo 1988, p. 34).

One of the most advanced transaction cost-based explanations of hybrid organizational forms was proposed by Williamson (1991). According to this line of argument, the key distinguishing feature of hybrids compared to other forms of governance is a flexible contracting mechanism that facilitates continuity and efficient adaptation (Lee 1992, p. 2). Williamson (1991) conducted a discrete structural analysis in order to compare three generic forms of economic organization – markets, hybrids and hierarchies – in terms of governance cost efficiency. His analysis is based on three key dimensions that allow the very nature of an economic transaction to be specified i.e. the level of uncertainty, frequency of disturbance and asset specificity. One of the key findings from his comparative-static analysis shows that transactions characterized by an intermediate level of asset specificity are most efficiently processed by hybrid organizational forms (Williamson 1991, p. 284).

Sociologists have emphasized the relevance of social embeddedness of economic activities (Granovetter 1985; Uzzi 1996, 1997).[55] Embeddedness literature is based on the notion that all economic interactions are embedded in a larger socio-economic context (Borgatti and Foster 2003, p. 994). Proponents of this school of thought proposed an alternative explanation for the emergence, existence and profanation of hybrid organizational forms. They argued that hybrids are not intermediates between markets and hierarchies. Instead they have to be seen as unique organizational structures and thus should be considered an organizational form in their own right (Powell 1987, 1990; Podolny and Page 1998). The main argument is that the concept of market and hierarchies fails to see and explain the enormous variety of forms that cooperative arrangements can take (Powell 1987, p. 67). Various types of cooperative agreements have started to emerge in an unprecedented fashion, especially in high-tech industries such as microelectronics, telecommunication, and biotechnology (Powell 1987 pp. 71–72). Powell's argumentation is closely related to a Schumpeterian notion of economic change. He argues (Powell 1987, p. 77) that the proliferation of hybrid organizational forms is closely connected to the rapidly transforming economic environments that are moving away from an older set of industries (characterized by mature well-established firms) towards a new set of industries (in which most firms are at a youthful stage) because hybrids provide a better fit with these new market and technology demands. According to this line of argument there are several main aspects that explain the emergence, existence and proliferation of hybrid organizational forms. Powell (1987, pp. 77–82) identifies four specific factors: (**I**) hybrid organizational forms allow greater flexibility and adaptability to rapidly changing environments which are characterized by a shift away from vertical integration and mass production towards more flexible forms of production and greater emphasis

---

[55] For an in-depth discussion, see Sect. 2.5.4.

on innovative products or services; **(II)** hybrids allow large organizations, which are usually considered to be structurally inert and thus resistant to change,[56] to overcome, at least to some extent, these limitations; **(III)** hybrids provide fast and flexible access to information and knowledge located outside the firm's boundaries; **(IV)** hybrids have to be understood as a variant or application of the "generalized reciprocity concept" (i.e. individual units do not exist in isolation but rather in relation to other units, cf. (Podolny and Page 1998) that creates legitimacy, reputation and mutual trust, and thus generates an efficient and reliable environment for exchange and transfer of information.

In summary, both perspectives provide valuable insights into the very nature and existence of hybrid organizational forms. The merits of the first school of thought are obvious. Even critics (e.g. Podolny and Page 1998, p. 58; Powell 1990, p. 296) acknowledged the contributions made by transaction cost theory. On the one hand, Williamson (1973, 1975, 1985) and other protagonists of the transaction cost school of thought (e.g. Klein et al. 1978; Jones and Hill 1988) contributed significantly to an in-depth understanding of how organizational structures affect economic outcomes. Mainstream economics usually apply a very simplified production function concept to describe firms in the economy. In contrast, transaction cost economics opened up the firm as a "black box" and brought economics and other related fields such as organization theory, law and business history much closer together (Powell 1990, p. 296). On the other hand, transaction cost scholars (Thorelli 1986; Jarillo 1988; Williamson 1991) explicitly acknowledged the existence of hybrids as an additional form of economic organization by integrating these intermediates into their originally dichotomous conceptualized market-hierarchy models.

Nonetheless the transaction cost perspective has several limitations. According to Ozman (2009, pp. 43–44) the criticisms of transaction cost economics can be grouped into three broad categories: **(I)** strong emphasis on opportunistic economic agents and a lack of recognition of mutual trust, **(II)** the static and cost-centered analysis perspective ignoring the dynamic nature of networks in rapidly changing environments, **(III)** negligence of social processes that are assumed to affect economic activities.

The sociological view has addressed most of these issues and significantly widened our understanding of networks by arguing that hybrid organizational forms represent an entirely new organizational form characterized by unique logics of exchange (Powell 1987, 1990; Borgatti and Foster 2003, p. 995). The sociological view has contributed considerably to moving away from a purely governance cost-centered network explanation towards a more comprehensive understanding of hybrid organizational forms. Moreover, the sociological notion of networks has paved the way for integrating a rich array of theoretical concepts (Granovetter 1973, 1985; Coleman 1988; Burt 1992) and methodological concepts (Wasserman

---

[56] There are at least two perspectives on the very nature of evolutionary change processes: the adaptation perspective and the selection perspective. This assumption is advocated by proponents of the latter perspective. For a discussion, see Sect. 2.2.

and Faust 1994; Carrington et al. 2005; Borgatti et al. 2013) into contemporary organizational network theory (Parkhe et al. 2006; Borgatti and Halgin 2011). In their comprehensive review, Borgatti and Foster (2003, p. 995) came to the conclusion that there is some evidence that the latter stream of research seems to have since then prevailed.

2.5.3   Addressing the Rationales and Motives Behind Cooperation

The aim of this section is to elaborate on the rationale behind cooperation and to discuss what motivates a firm to cooperate in its effort to innovate.

The rationale behind cooperating is most easily illustrated by using a simple game theory model (cf. Parkhe 1993; Kudic and Banaszak 2009; Faulkner 2006).[57] As shown by Parkhe (1993), strategic interactions in collaborative constellations can be characterized by a typical prisoner dilemma game structure.[58] We start the line of argument with the following initial situation. There are two parties (i.e. Organization A, Organization B) both of which have, in principle, two strategic options (i.e. cooperation and defection). Moreover, we introduce a basic assumption that determines the payoff structure for a basic "prisoner dilemma" game (cf. Eq. 2.1). The payment ε represents the exploitation gains for unilateral defection, λ stands for net payments in the case of mutual cooperation, τ typifies the net payments for mutual defection of both actors and δ captures the detriment of unilateral cooperation. The following inequation determines the initial game structure.

$$\varepsilon > \lambda > \tau > \delta \tag{2.1}$$

Figure 2.1 illustrates the initial situation (cf. Fig. 2.1, left), a non-cooperative solution (cf. Fig. 2.1, center) and a cooperative solution (cf. Fig. 2.1, right). We start the analysis with a single-shot game[59] by assuming there is an absence of mutual trust, reputation or any kind of sanctioning mechanism (cf. Fig. 2.1, center).[60] The game starts from Cell (I). Due to the assumed incentive structure, each of the two actors can increase their individual payoff by behaving opportunistically. In other words, the "Pareto optimal" payoff structure in Cell (I) is

---

[57] If not otherwise stated, this line of argument is guided by Kudic and Banaszak (2009) and Faulkner (2006).

[58] We do not discuss the basic assumption of the simple prisoner dilemma here. For an excellent introduction to game theory, see Dixit and Skeath (2004) or Saloner (1991).

[59] A single-shot game simply means that the game ends after one round.

[60] For an in-depth analysis of reputation effects on firm cooperation activities in an iterated prisoner dilemma game with exit option, see Arend (2009).

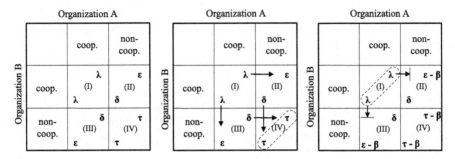

Fig. 2.1 The rationale to cooperate – cooperative versus non-cooperative games (Source: Author's own illustration, based on: Kudic and Banaszak (2009, p. 17) and Faulkner (2006, p. 612))

instable. This reflects a shift from Cell (I) to Cell (II) or Cell (III), respectively. Again, the situation is instable because at least one of the actors can increase its payoff by choosing a non-cooperative strategy. As a consequence both actors find themselves in a suboptimal equilibrium (cf. Fig. 2.1, center, Cell IV). In other words, we achieve a stable solution that is also known as the "Nash equilibrium". The moral of the story is straightforward: individually opportunistic behavior leads to suboptimal results for both actors. The main question now is what makes the two organizations cooperate?

In order to answer this question, we move away from a single-shot game to an iterated prisoner dilemma and introduce a sanction mechanism[61] to the model. "In a situation where the partners intend to work with each other over an indeterminate period" [...] "trust can be built" [...], "potential synergies from cooperation can be realized" [...], and "reputation comes into the equation" (Faulkner 2006, p. 613). Opportunistic organizations may find themselves in a situation where they cannot find an appropriate partner in later periods (ibid). This is because information about opportunistic behavior will spread rapidly throughout the entire network (Gulati et al. 2000, p. 209). As a consequence, the loss of a reputation[62] has been considered to be an important governance mechanism in strategic alliances (e.g. Robertson and Stuart 2007). Basically the same logic applies to a network setting.[63] The implementation of a sanction mechanism with a fixed sanction value β transforms the

---

[61] There is copious literature on different types of governance mechanism in an interorganizational context (White 2005; Oxley and Sampson 2004; Provan and Kenis 2007). Sanctions can make an appearance in the form of formal contracts, safeguards consisting of mutual hostages such as bilateral idiosyncratic tangible and intangible investments, quasi integration, joint decision-making or loss of reputation (for an overview and discussion, see Kudic and Banaszak 2009).

[62] Reputation concept is most easily explained by a simple example (cf. Weigel and Camerer 1988 p. 444): "[...] if a colleague always fulfills her promises, then you say she has a reputation for reliability" [...] "based on her past actions you infer that reliability is one of her attributes and she is a 'reliable person'."

[63] Gulati and colleagues (2000, p. 209) argue that networks can create strong disincentives for opportunistic behavior because building up a reputation is a long and difficult process, while destroying it quick.

non-cooperative game structure into a cooperative game. The sanction β must be at least so high that the alliance partners are indifferent about cooperative and non-cooperative behavior:

$$\lambda > \varepsilon - \beta > \tau - \beta > \delta \qquad (2.2)$$

Inequation 2.2 represents the new payoff structure which is illustrated by Fig. 2.1 (right). The right hand side of Fig. 2.1 shows that if Eq. 2.2 is fulfilled we retain the stable "Pareto optimum" solution in Cell (I). In other words, the incentives to defect disappear.

In a nutshell, integrating repeated interactions and implementing sanctions within the model leads to a situation in which the dominant strategy for both organizations is to cooperate (Faulkner 2006, p. 613).

Next we address a firm's motives for cooperating. Several early studies have provided us with interesting insights (e.g. (Alic 1990; Freeman 1991; Camagni 1993; Osborn and Hagedoorn 1997; Gulati 1998). Alongside them, Hagedoorn (1993, p. 371) was among the first who raised the question of "why companies cooperate in their efforts to innovate". The findings from both early and contemporary research allow us to assemble the broad variety of heterogonous and partially overlapping cooperation motives into six groups. These are: cost savings (Freeman 1991; Hagedoorn 1993, 2002), risk reduction (Hagedoorn 1993; Sivadas and Dwyer 2000), time savings (Hagedoorn 1993; Mowery et al. 1996), access to national and international markets (Perlmutter and Heenan 1986; Harrigan 1988; Johanson and Mattson 1988; Hagedoorn 1993), status and reputation building (Podolny 1994; Gulati 1998; Stuart et al. 1999), and last but not least, knowledge-related motives such as knowledge access (Grant and Baden-Fuller 2004; Rothaermel 2001; Buckley et al. 2009) and interorganizational learning (Hamel 1991; Khanna et al. 1998; Kale et al. 2000; Bessant et al. 2012).

The first two motives – cost and risk reduction – are closely related to a high level of uncertainty inherent to innovation processes (Hagedoorn 1993). At the bottom of the technological innovation processes is the search for new solutions for a given problem (Dosi 1988, p. 1126). There is no general algorithm that can be derived from the information about the problem (ibid). This implies that the search for novelty can generate enormous costs and it is by no means certain whether the problem can be solved at all. Joint research and development activities allow firms to share costs and reduce risks inherent to innovation processes (Hagedoorn 2002, p. 479). In addition, joint R&D efforts allow the period between invention and market introduction to be reduced (Hagedoorn 1993, p. 373). Even though the first two motives were named frequently in early studies, researchers realized that these cooperation motives were not as important for firms as initially expected. Freeman (1991, p. 507) concludes that especially "[...] cost-sharing and cost-minimizing appeared to play a relatively small role in comparison with strategic objectives relating to new technology and markets".

International business scholars presented other arguments to explain what motivates firms to cooperate. It has been argued that network relationships provide important ways for firms to get access to national and international markets (Johanson and Mattson 1988). Cooperation in highly interdisciplinary research areas can generate new ideas that go beyond the scope of the initial cooperation project. R&D cooperation allows firms to identify new and unexpected opportunities during the cooperation process. This can lead to an expansion of their product portfolios and open up new markets for them. Both opportunity seeking and market access are considered to be non-negligible cooperation motives (Hagedoorn 1993, p. 373).

Next we focus on a sociological line of argument and address the following two concepts: status and reputation. On the one hand, the status-based model of market competition acknowledges the dual nature status in a sense that a market actor's status affects the expectation of other market actors via two channels: its past demonstrations of quality and the status of its exchange partners (Podolny 1993, 1994; Benjamin and Podolny 1999, p. 563). On the other hand, reputation is an important asset of a firm (Fombrun and Shanley 1990) and firms spend time and considerable effort in building up their reputations. Reputation sends out a signal that allows others to judge an organization's reliability (Weigel and Camerer 1988, p. 444). Especially for young entrepreneurial firms there is often a lack of acceptance in the market. Thus, building up status and reputation can be an important motive for firms to cooperate in the area of R&D. In this context, Gulati (1998, p. 301) points out: "The status of an organization in the network affects its reputation and visibility in the system. The greater the reputation, the wider the organization's access to a variety of sources of knowledge, and the richer the collaborative experience, which makes it an attractive partner." Stuart and colleagues (1999) were able to empirically show that young companies endorsed by prominent cooperation partners perform better in term of survival and growth than comparable ventures without prominent partners.

Among the most pivotal motives for cooperating in R&D are knowledge access and interorganizational learning.[64] Solving technological problems during technological innovation processes involves the use of information drawn from previous experience, formal knowledge and various types of specific and uncodified capabilities (Dosi 1988, p. 1126). By now it is well-known that knowledge-related external linkages have to be considered as an [...] important ancillary and complementary source of scientific and technical information rather than a substitute for indigenous innovative activity" (Freeman 1991, p. 501). Two dominant streams have emerged in interdisciplinary alliance and network research over the past two decades: the "knowledge accessing approach" and the "knowledge acquiring approach" (Al-Laham and Kudic 2008).[65] The motives for accessing external

---

[64] These two categories subsume all kinds of knowledge-related cooperation motives that are directly linked to a firm's innovation process such as interorganizational learning, capturing tacit knowledge, technology transfer, technological leapfrogging etc. (cf. Hagedoorn 1993).

[65] This distinction refers to the underlying processes of knowledge generation (or "exploration") and knowledge application (or "exploitation") among partners in interorganizational relationships such as strategic alliances (Grant and Baden-Fuller 2004, p. 61).

technological knowledge are straightforward. Firms cooperate to gain access to multiple complementary stocks of knowledge (Grant and Baden-Fuller 2004). The access of external knowledge stocks does not necessarily require internalizing the cooperation partner's skills (Doz and Hamel 1997). According to the second approach, alliances can be regarded as "vehicles of learning" (Grant and Baden-Fuller 2004, p. 64) which allow a firm to share a particular part of its knowledge bases, exchange both explicit and implicit stocks of knowledge across firm boundaries and thus learn from one another (Hamel 1991).

2.5.4   Network Embeddedness and Firm Performance

This section draws upon social capital theory in sociology.[66] Several ideas and concepts in social capital literature have strongly influenced contemporary research on interorganizational networks.[67] For the purpose of this study we focus on the structural embeddedness perspective.[68] An in-depth understanding of the relationship between structural network embeddedness and firm performance requires a look at two concepts that have caused vigorous debate in the field of network research over the past two decades.

Traditionally, we can distinguish between the "closure concept" and the "brokerage concept" in network research. Both concepts are based on the idea that the social structure in which an economic actor is embedded affects its actions and performance outcomes in multiple ways. However, these two perspectives differ fundamentally with regard to the notion of what kind of structural topologies are beneficial and how the positioning of an economic actor within these structural patterns affects its subsequent economic outcomes. During the past two decades considerable efforts have been made to integrate these sociological concepts in interorganizational alliance and network research and to empirically test the extent to which various dimensions of firm performance are affected by structurally different network surroundings.[69]

---

[66] For an introduction and overview of concepts in social capital theory, see Lin (2002).

[67] For instance, a controversial discussion in social capital literature addressed the question of whether it was weak ties (Granovetter 1973; Levin and Cross 2004) or strong ties (Uzzi 1996; Krackhardt 1992) which affect the network actors' behavior and outcomes in social and economic networks. Originally, the "strength of ties" and a closely related concept i.e. "density of ties" were conflated, while later it was recognized that these two tie features have to be clearly distinguished and treated separately (Nooteboom 2008, p. 619).

[68] According to Gulati (1998, p. 296) one has to distinguish between at least two types of network embeddedness. On the one hand, relational embeddedness highlights the importance of direct cohesive ties. On the other hand, structural network embeddedness goes beyond direct ties and emphasizes the structural position of the actors.

[69] For an overview, see Ozman (2009, pp. 48–50) or Nooteboom (2008, pp. 618–620).

We start the discussion by taking a closer look at the closure concept. The basic idea of the "closure concept" goes back to Bourdieu (1986) and Coleman (1988). The concept is based on the notion that a network actor's positioning in a "cohesive" network (or "closed" network structure), characterized by a high degree of interconnectedness, is more advantageous than other more fragmented structural network patterns. The reasoning is straightforward. Cohesion facilitates "[. . .] the build-up of reputation, trust, social norms, and social control, for example by coalition building to constrain actions, which facilitates collaboration" (Nooteboom 2008, p. 619). This line of argument has far ranging implications for interorganizational innovation networks. The positioning within a cohesive network confers a competitive advantage to a firm due to improved coordination through repeated exchange with the same partners (Ozman 2009, p. 49). This enables and facilitates the successful transfer of tacit knowledge (ibid). As a consequence, a firm's involvement and positioning in these cohesive network structures is assumed to affect its subsequent performance outcomes in multiple ways.

This view is not shared by the proponents of the "structural hole" concept, of which Burt (1992) is one of the key protagonists. He argues that a network actor's position is most beneficial if it allows the actor to bridge the gap between previously unconnected cohesive subgroups of the network. Basically two main arguments substantiate the concept. Firstly, according to Burt (1992) much of the information that circulates in a system is redundant and the efficiency of the information structure can be increased by removing redundant ties and selectively retaining only those ties that bridge "structural holes" in the network (Nooteboom 2008, p. 618). Secondly, the well-known problem of overembeddedness (Uzzi 1997) can be avoided by pursuing an efficiency oriented cooperation strategy. Or to put it another way, carefully selecting ties can save time and energy that is needed to develop new non-redundant linkages (Nooteboom 2008, p. 618). Just like the "closure concept", the "structural hole" concept has also been transferred and applied to the field of interorganizational network research. It suggests that a firm's potential to achieve a competitive advantage rests on its ability to take a broker role and fill structural holes between densely connected subgroups of organizations in an innovation network (Ozman 2009, p. 48). In summary, according to this line of argument, those firms that have a broker position in the network are assumed to outperform others.

At first glance, these two concepts seem to be irreconcilable. However, quite recently we have been able to observe the emergence of integrative approaches that are in search of ways to combine these distinct theoretical standpoints (Burt 2000, 2005; Nooteboom 2008). Based on prior theoretical elaborations (Burt 2000) and empirical findings (Rowley et al. 2000) Nooteboom (2008, p. 619) argues that these two concepts are not incompatible. In doing so he provides the following two explanations (Nooteboom 2008, pp. 619–620): The first line of argument is based on a distinction between a competence dimension and a governance dimension. Accordingly, the competence dimension has to be interpreted "[. . .] in terms of access to new knowledge, the combination of complementary competencies, joint production of knowledge, and the creation of Schumpeterian 'novel combinations'

[...]" whereas the governance dimension addresses "[...] managing relational risks of opportunism and spillover or loss of appropriability of returns on innovation" (Nooteboom 2008, p. 619). In other words, the bridging of structural holes allows new opportunities for generating novelty to be explored whereas a cohesive network structure breeds trust and allows opportunistic behavior in networks to be overcome. The second line of argument is based on the distinction between "knowledge acquiring" and "knowledge accessing" in cooperative arrangements (cf. Al-Laham and Kudic 2008). In reference to prior research, Nooteboom (2008, p. 619) argues that strong ties in densely interconnected network areas promote the transfer of complex knowledge while weak ties that bridge structural holes transfer relatively simple knowledge.

In a nutshell, both theoretical as well as empirical findings substantiate the assumption that the "closure" and the "structural holes" concepts reflect different aspect of interorganizational cooperation activities. Accordingly, the concepts seem to be complementary and compatible with one another.

2.6 This Book in Light of the Preceding Discussion

For the purpose of this monograph we chose the neo-Schumpeterian approach as the general theoretical framework. The reasons for this are straightforward. The framework provides an appropriate setting for studying the consequences of innovation processes for economic growth and prosperity. At the same time, it provides a solid theoretical groundwork for investigating the determinants and mechanisms which are assumed to cause innovation processes and fuel technological change. We decided in favor of this theoretical framework because we are convinced that factors influencing the creation of novelty are best understood from a dynamic perspective. From a neo-Schumpeterian point of view, firms are considered to be heterogeneous economic entities (Hanusch and Pyka 2007c). Knowledge is no longer considered to be a pure public good and becomes a cornerstone of economic analysis (ibid). Moreover, it is important to note that the neo-Schumpeterian approach to economics explicitly acknowledges the collective nature of innovation process (Pyka 2002, 2007, 2009). This is reflected in at least three theoretical concepts: the "innovation system concept", the "proximity concept", and the "network concept". The theoretical principles or cornerstones that underpin the neo-Schumpeterian approach are inherent to all these concepts. To illustrate this, we briefly address the structure and structural evolution of interorganizational innovation networks. Firstly, evolutionary change of networks is a phenomenon that occurs at multiple levels simultaneously. Micro-level network change processes, i.e. tie formations & tie terminations and node entries & node exits affect the structural evolution at the meso (sub-group) level and the macro (overall network) level. Boschma and Frenken (2010, p. 129) argue that these processes of "creative destruction" are clearly Schumpeterian in nature. An in-depth understanding of micro-level network change processes provides the basis for explaining the evolution of networks (ibid). Secondly, a firm's past cooperation strategies and cooperation sequences affect its future cooperation options and decisions. As a consequence a network path can be observed over time for each firm in an industry network. This consideration is closely related to the next point. Early-stage cooperation decisions can cause lock-in effects that may restrict the scope of action at later points in time. Thirdly, a comprehensive understanding of structural network characteristics and structural network change process is a crucial prerequisite for an in-depth analysis of knowledge diffusion, knowledge exchange, and knowledge generating processes among heterogeneous economic actors. In short, the neo-Schumpeterian approach provides a solid basis for studying the nature of collective innovation processes and for analyzing micro-level innovation outcomes of network actors from various angles. Last but not least, the neo-Schumpeterian paradigm is open to other ideas and concepts from related disciplines. There are multiple conceptual overlaps with, for example, economic geography, economic sociology and management science.

References

Ackerlof GA (1970) The market for "lemons". Quality uncertainty and the market mechanism. Q J Econ 84(3):488–500

Agyris C, Schön DA (1978) Organizational learning: a theory of action perspective. Addison-Wesley, Reading

Ahuja G (2000) Collaboration networks, structural hole, and innovation: a longitudinal study. Adm Sci Q 45(3):425–455

Aldrich HE, Ruef M (2006) Organizations evolving, 2nd edn. Sage, London

Alic JA (1990) Cooperation in R&D. Technovation 10(5):319–332

Al-Laham A, Kudic M (2008) Strategische Allianzen. In: Corsten H, Goessinger R (eds) Lexikon der Betriebswirtschaftslehre, 5th edn. Oldenbourg Verlag, München, pp 39–41

Alter M (1982) Carl Menger and homo oeconomicus: some thoughts on Austrian theory and methodology. J Econ Issue XVI(1):149–160

Amburgey TL, Rao H (1996) Organizational ecology: past, present, and future directions. Acad Manag J 39(5):1265–1286

Amburgey TL, Al-Laham A, Tzabbar D, Aharonson BS (2008) The structural evolution of multiplex organizational networks: research and commerce in biotechnology. In: Baum JA, Rowley TJ (eds) Advances in strategic management – network strategy, vol 25. Emerald Publishing, Bingley, pp 171–212

Antonelli C (2009) The economics of innovation: from the classical legacies to the economics of complexity. Econ Innov New Technol 18(7):611–646

Antonelli C (2011) Handbook on the economic complexity of technological change. Edward Elgar, Cheltenham

Araujo L, Harrison D (2002) Path dependence, agency and technological evolution. Tech Anal Strat Manag 14(1):5–19

Arend RJ (2009) Reputation for cooperation: contingent benefits in alliance activity. Strateg Manag J 30(4):371–385

Arrow KJ (1994) Methodological individualism and social knowledge. Am Econ Rev 84(2):1–9

Arthur BW (1989) Competing technologies, increasing returns, and lock-in by historical events. Econ J 99(394):116–131

Arthur BW (2007) Complexity and the economy. In: Hanusch H, Pyka A (eds) Elgar companion to neo-Schumpeterian economics. Edward Elgar, Cheltenham, pp 1102–1110

Bain JS (1950) Workable competition in oligopoly: theoretical consideration and some empirical economics. Am Econ Rev 40(2):35–47

Bain JS (1951) Relation of profit rate to industry concentration: American manufacturing, 1936–1940. Q J Econ 65:293–324

Barney JB (1991) Firm resources and sustained competitive advantage. J Manag 17(1):99–120

Barney JB (2001) Is the resource-based "view" a useful perspective for strategic management research? Yes. Acad Manag Rev 26(1):41–56

Barney JB (2002) Gaining and sustaining competitive advantage. Prentice Hall, Upper Saddle River

Baum JA, Calabrese T, Silverman BS (2000) Don't go it alone: alliance network composition and startup's performance in Canadian biotechnology. Strateg Manag J 21(3):267–294

Becker MC (2004) Organizational routines: a review of the literature. Ind Corp Chang 13(4):643–678

Benjamin BA, Podolny JM (1999) Status, quality, and social order in the California wine industry. Adm Sci Q 44(3):563–589

Bergenholtz C, Waldstrom C (2011) Inter-organizational network studies – a literature review. Ind Innov 18(6):539–562

Bertalanffy LV (1951) Problems of general system theory. Hum Biol 23(4):302–312

Bertalanffy LV (1968) General system theory: foundations, development, applications. George Braziller, New York

Bessant J, Alexander A, Tsekouras G, Rush H, Lamming R (2012) Developing innovation capability through learning networks. J Econ Geogr 12(5):1087–1112

Betton J, Dess GG (1985) The application of population ecology models to the study of organizations. Acad Manag Rev 10(4):750–757

Bierly PE, Kessler EH, Christensen EW (2000) Organizational learning, knowledge and wisdom. J Organ Chang Manag 13(6):595–618

Borgatti SP, Foster PC (2003) The network paradigm in organizational research: a review and typology. J Manag 29(6):991–1013

Borgatti SP, Halgin DS (2011) On network theory. Organ Sci 22(5):1168–1182

Borgatti SP, Everett MG, Johnson JC (2013) Analyzing social networks. Sage, London

Borys B, Jemison DB (1989) Hybrid arrangements as strategic alliances: theoretical issues in organizational combinations. Acad Manag Rev 14(2):234–249

Boschma R (2005a) Role of proximity in interaction and performance: conceptual and empirical challenges. Reg Stud 39(1):41–45

Boschma R (2005b) Proximity and innovation: a critical assessment. Reg Stud 39(1):61–74

Boschma RA, Frenken K (2006) Why is economic geography not an evolutionary science? Towards an evolutionary economic geography. J Econ Geogr 6(3):273–302

Boschma R, Frenken K (2010) The spatial evolution of innovation networks: a proximity perspective. In: Boschma R, Martin R (eds) The handbook of evolutionary economic geography. Edward Elgar, Cheltenham, pp 120–135

Boulding KE (1956) General system theory – the skeleton of science. Manag Sci 2(3):197–208

Boulding KE (1966) The economics of knowledge and the knowledge of economics. Am Econ Rev 56(1):1–13

Bourdieu P (1986) The forms of capital. In: Richardson J (ed) Handbook of theory and research for the sociology of education. Greenwood, New York, pp 241–258

Brass DJ, Galaskiewicz J, Greve HR, Tsai W (2004) Taking stock of networks and organizations: a multilevel perspective. Acad Manag J 47(6):795–817

Brenner T, Cantner U, Graf H (2011) Innovation networks: measurement, performance and regional dimensions. Ind Innov 18(1):1–5

Brewer A (2010) The making of the classical theory of economic growth. Routledge, New York

Broekel T, Hartog M (2013) Explaining the structure of inter-organizational networks using Exponential Random Graph models. Ind Innov 20(3):277–295

Bruderer E, Singh JV (1996) Organizational evolution, learning, and selection: a genetic-algorithm-based model. Acad Manag J 39(5):1322–1349

Brüderl J, Preisendörfer P (1998) Network support and the success of newly founded businesses. Small Bus Econ 10(3):213–225

Buckley PJ, Glaister KW, Klijn E, Tan H (2009) Knowledge accession and knowledge acquisition in strategic alliances: the impact of supplementary and complementary dimensions. Br J Manag 20(4):598–609

Burt RS (1992) Structural holes: the social structure of competition. Harvard University Press, Cambridge

Burt RS (2000) The network structure of social capital. In: Staw BM, Sutton RI (eds) Research in organizational behavior, vol 22. JAI Press, Greenwich, pp 345–424

Burt RS (2005) Brokerage & closure – an introduction to social capital. Oxford University Press, New York

Bush V (1945) Science: the endless frontier – a report to the President on a program for postwar scientific research. US Government, Washington, DC

Camagni R (1993) Inter-firm industrial networks: the costs and benefits of cooperative behaviour. J Ind Stud 1(1):1–15

Campbell DT (1969) Variation and selective retention in sociocultural evolution. Gen Syst 14:69–85

Cantner U, Graf H (2011) Innovation networks: formation, performance and dynamics. In: Antonelli C (ed) Handbook on the economic complexity of technological change. Edward Elgar, Cheltenham, pp 366–394

Carlsson B, Jacobsson S, Holmen M, Rickne A (2002) Innovation systems: analytical and methodological issues. Res Policy 31(2):233–245

Carrington PJ, Scott J, Wasserman S (2005) Models and methods in social network analysis. Cambridge University Press, Cambridge

Carroll GR (1984) Organizational ecology. Annu Rev Sociol 10:71–93

Coase RH (1937) The nature of the firm. Economica (New Series) 4(16):386–405

Coenen L, Moodysson J, Asheim BT (2004) Nodes, networks and proximities: on the knowledge dynamics of the Medicon Valley Biotech Cluster. Eur Plan Stud 12(7):1003–1018

Cohen WM, Levinthal DA (1989) Innovation and learning: the two faces of R&D. Econ J 99 (397):569–596

Cohen WM, Levinthal DA (1990) Absorptive capacity: a new perspective on learning and innovation. Adm Sci Q 35(3):128–152

Coleman JS (1988) Social capital in the creation of human capital. Am J Sociol 94:95–120

Conner KR (1991) A historical comparison of resource-based theory and five schools of thought within industrial organization economics: do we have a new theory of the firm? J Manag 17 (1):121–154

Contractor FJ, Lorange P (2002) The growth of alliances in the knowledge-base economy. Int Bus Rev 11(4):485–502

Cooke P (2001) Regional innovation systems, clusters, and the knowledge economy. Ind Corp Chang 10(4):945–974

Corolleur F, Courlet C (2003) The Marshallian Industrial District, an organizational and institutional answer to uncertainty. Entrepren Reg Dev 15:299–307

Cowan R, David PA, Foray D (2000) The explicit economics of knowledge codification and tacitness. Ind Corp Chang 9(2):211–253

Cropper S, Ebers M, Huxham C, Ring PS (2008) Introducing inter-organizational relations. In: Cropper S, Ebers M, Huxham C, Ring PS (eds) Interorganizational relations. Oxford University Press, New York, pp 3–25

Cyert R, March JG (1963) Behavioral theory of the firm. Prentice-Hall, Englewood Cliffs

Das TK, Teng B-S (2002) Alliance constellations: a social exchange perspective. Acad Manag J 27 (3):445–456

David PA (1985) Clio and the economics of QWERTY. Am Econ Rev 75(2):332–337

Decarolis DM, Deeds DL (1999) The impact of stocks and flows of organizational knowledge on firm performance: an empirical investigation of the biotechnology industry. Strateg Manag J 20 (10):953–968

Dierickx I, Cool K (1989) Asset stock accumulation and sustainability of competitive advantage. Manag Sci 35(12):1504–1511

Dixit A, Skeath S (2004) Games of strategy, 2nd edn. W. W. Norton, New York

Dodgson M (1993) Organizational learning – a review of some literature. Organ Stud 14(3):375–394

Domar ED (1948) The problem of capital accumulation. Am Econ Rev 38:777–794

Dopfer K (2005) The evolutionary foundation of economics. Cambridge University Press, Cambridge

Dopfer K (2011) Economics in a cultural key: complexity and evolution revisited. In: Davis JB, Hand WD (eds) The Elgar companion to recent economic methodology. Edward Elgar, Cheltenham, pp 319–341

Dosi G (1982) Technological paradigms and technological trajectories. Res Policy 11(3):147–162

Dosi G (1988) Sources, procedures, and microeconomic effects of innovation. J Econ Lit 26 (3):1120–1171

Dosi G, Nelson RR (1994) An introduction to evolutionary theories in economics. J Evol Econ 4 (3):153–172

Doz Y, Hamel G (1997) The use of alliances in implementing technology strategies. In: Tushman MT, Anderson P (eds) Managing strategic innovation and change. Oxford University Press, New York, pp 556–580

Duysters G, Lemmens C (2004) Alliance group formation – enabling and constraining effects of embeddedness and social capital in strategic technology alliance networks. Int Stud Manag Org 33(2):49–68

Duysters G, De Man A-P, Wildeman L (1999) A network approach to alliance management. Eur Manag J 17(2):182–187

Easterby-Smith M, Lyles MA, Peteraf MA (2009) Dynamic capabilities: current debates and future directions. Br J Manag 20:1–8

Elliott JE (1978) Marx's "Grundrisse": vision of capitalism's creative destruction. J Post Keynes Econ 1(2):148–169

Fagerberg J (2003) Schumpeter and the revival of evolutionary economics: an appraisal of the literature. J Evol Econ 13(2):125–159

Fagerberg J (2005) Innovation: a guide to the literature. In: Fagerberg J, Mowery DC, Nelson RR (eds) The Oxford handbook of innovation. Oxford University Press, New York, pp 1–28

Faulkner D (2006) Cooperative strategy – strategic alliances and networks. In: Campbell A, Faulkner DO (eds) The Oxford handbook of strategy – a strategic overview and competitive strategy. Oxford University Press, New York, pp 610–648

Fombrun CJ, Shanley M (1990) What's in a name? Reputation building and corporate strategy. Acad Manag J 33(2):233–258

Foss NJ, Ishikawa I (2007) Towards a dynamic resource-based view: insights from Austrian capital and entrepreneurship theory. Organ Stud 28(5):749–772

Foster J (1987) Evolutionary macroeconomics. Unwin Hyman, London

Foster J (2005) From simplistic to complex adaptive systems in economics. Camb J Econ 29 (6):873–892

Freeman C (1974) The economics of industrial innovation. Penguin, Harmondsworth

Freeman C (1988) Japan: a new national system of innovation. In: Dosi G, Nelson RR, Silverberg G, Soete L (eds) Technical change and economic theory. Pinter, London, pp 330–348

Freeman C (1991) Networks of innovators: a synthesis of research issues. Res Policy 20(5):499–514

Goerzen A (2005) Managing alliance networks: emerging practices of multinational corporations. Acad Manag Exec 19(2):94–107

Goss D (2005) Schumpeter's legacy? Interaction and emotions in the sociology of entrepreneurship. Enterp Theory Pract 29(2):205–218

Granovetter MS (1973) The strength of weak ties. Am J Sociol 78(6):1360–1380

Granovetter MS (1985) Economic action and social structure: the problem of embeddedness. Am J Sociol 91(3):481–510

Grant RM (1996) Towards a knowledge based theory of the firm. Strateg Manag J 17(2):109–122

Grant RM, Baden-Fuller C (2004) A knowledge accessing theory of strategic alliances. J Manag Stud 41(1):61–84

Gulati R (1998) Alliances and networks. Strateg Manag J 19(4):293–317

Gulati R, Singh H (1998) The architecture of cooperation: managing coordination costs and appropriation concerns in strategic alliances. Adm Sci Q 43(4):781–814

Gulati R, Nohria N, Zaheer A (2000) Strategic networks. Strateg Manag J 21(3):203–215

Hagedoorn J (1993) Understanding the rational of strategic technology partnering – organizational modes of cooperation and sectoral differences. Strateg Manag J 14(5):371–385

Hagedoorn J (2002) Inter-firm R&D partnership: an overview of major trends and patterns since 1960. Res Policy 31(4):477–492

Hagedoorn J, Schakenraad J (1994) The effects of strategic technology alliances on company performance. Strateg Manag J 15(4):291–309

Hamel G (1991) Competition for competence and inter-partner learning within international strategic alliances. Strateg Manag J 12(1):83–103

Hannan MT, Freeman J (1977) The population ecology of organizations. Am J Sociol 82(5):929–964

Hannan MT, Freeman J (1984) Structural inertia and organizational change. Am Sociol Rev 49 (2):149–164

Hanusch H, Pyka A (2007a) Principles of neo-Schumpeterian economics. Camb J Econ 31 (2):275–289

Hanusch H, Pyka A (2007b) Schumpeter, Joseph Alois (1883–1950). In: Hanusch H, Pyka A (eds) Elgar companion on neo-Schumpeterian economics. Edward Elgar, Cheltenham, pp 19–27

Hanusch H, Pyka A (2007c) Elgar companion to neo-Schumpeterian economics. Edward Elgar, Cheltenham

Harrigan KR (1988) Joint ventures and competitive strategy. Strateg Manag J 9(2):141–158

Harrod RF (1948) Towards a dynamic economics. Macmillan, London

Heertje A (2004) Schumpeter and methodological individualism. J Evol Econ 14(2):153–156

Hite JM, Hesterly WS (2001) The evolution of firm networks: from emergence to early growth of the firm. Strateg Manag J 22(3):275–286

Hodgson GM (2006) Economics in the shadows of Darwin and Marx – essays on institutional and evolutionary economics. Edward Elgar, Cheltenham

Hoffmann WH (2007) Strategies for managing alliance portfolios. Strateg Manag J 28(8):827–856

Inkpen A (2009) Strategic alliances. In: Rugman A (ed) The Oxford handbook of international business. Oxford University Press, New York, pp 389–414

Jarillo CJ (1988) On strategic networks. Strateg Manag J 9(1):31–41

Jevons WS (1871) The theory of political economy (1888, 3rd edn. Macmillan, London

Johanson J, Mattson L-G (1988) Internationalization in industrial systems – a network approach. In: Hood N, Vahlen J-E (eds) Strategies in global competition. Croom Helm, New York, pp 287–331

Jones GR, Hill CW (1988) Transaction cost analysis of strategy-structure choice. Strateg Manag J 9(2):159–172

Kale P, Singh H, Perlmutter H (2000) Learning and protection of proprietary assets in strategic alliances: building relational capital. Strateg Manag J 21(3):217–237

Khanna T, Gulati R, Nohria N (1998) The dynamics of learning alliances: competition, cooperation, and relative scope. Strateg Manag J 19(3):193–210

Kim T-Y, Oh H, Swaminathan A (2006) Framing interorganizational network change: a network inertia perspective. Acad Manag Rev 31(3):704–720

Kirman A (1989) The intrinsic limits of modern economic theory: the emperor has no clothes. Econ J 99(395):126–139

Kirman A (1993) Ants, rationality, and recruitment. Q J Econ 108(1):137–156

Klein B, Crawford RG, Alchian AA (1978) Vertical integration, appropriable rents, and the competitive contracting process. J Law Econ 21(2):297–326

Klepper S, Malerba F (2010) Demand, innovation and industrial dynamics: an introduction. Ind Corp Chang 19(5):1515–1520

Kline SJ (1985) Innovation is not a linear process. Res Manag 28(4):36–45

Kline SJ, Rosenberg N (1986) An overview of innovation. In: Landau R, Rosenberg N (eds) The positive sum strategy: harnessing technology for economic growth. National Academy Press, Washington, DC, pp 275–304

Knoben J, Oerlemans LA (2006) Proximity and inter-organizational collaboration: a literature review. Int J Manag Rev 8(2):71–89

Kogut B, Zander U (1992) Knowledge of the firm, combinative capabilities, and the replication of technology. Organ Sci 3(3):383–397

Koka BR, Presscott JE (2008) Designing alliance networks: the influence of network position, environmental change, and strategy on firm performance. Strateg Manag J 29:639–661

Kraaijenbrink J, Spender JC, Groen AJ (2010) The resource-based view: a review and assessment of its critiques. J Manag 36(1):349–372

Krackhardt D (1992) The strength of strong ties – the importance of philos in organizations. In: Nohria N, Eccles RG (eds) Networks and organizations – structure, form, and action. Harvard Business Press, Boston, pp 216–239

Kudic M, Banaszak M (2009) The economic optimality of sanction mechanisms in interorganizational ego networks – a game theoretical analysis. In: 35th European International Business Academy conference, Valencia, pp 1–40

Lavie D (2007) Alliance portfolios and firm performance: a study of value creation and appropriation in the U.S. software industry. Strateg Manag J 28(12):1187–1212

Lavie D, Rosenkopf L (2006) Balancing exploration and exploitation in alliance formation. Acad Manag J 49(4):497–818

Lavie D, Lechner C, Singh H (2007) The performance implications of timing of entry and involvement in multipartner alliances. Acad Manag J 50(3):578–604

Lee V (1992) Organizational dynamics of market transition: hybrid forms, property rights, and mixed economy in China. Adm Sci Q 37(1):1–27

Leven P, Holmström J, Mathiassen L (2014) Managing research and innovation networks: evidence from a government sponsored cross-industry program. Res Policy 43(1):156–168

Levin DZ, Cross R (2004) The strength of weak ties you can trust: the mediating role of trust in effective knowledge transfer. Manag Sci 50(11):1477–1490

Levinthal DA (1991) Organizational adaptation and environmental selection-interrelated processes of change. Organ Sci 2(1):140–145

Levitt B, March JG (1988) Organizational learning. Annu Rev Sociol 14:319–340

Lin N (2002) Social capital: a theory of social structure and action. Cambridge University Press, Cambridge

Lippman S, Rumelt R (1982) Uncertain imitability: an analysis of interfirm differences in efficiency and competition. Bell J Econ 13:418–438

Lomi A, Negro G, Fonti F (2008) Evolutionary perspectives on inter-organizational relations. In: Cropper S, Ebers M, Huxham C, Ring SP (eds) The Oxford handbook of interorganizational relations. Oxford University Press, New York, pp 313–339

Lundvall B-A (1988) Innovation as an interactive process: from user-producer interaction to the national system of innovation. In: Dosi G, Freeman C, Nelson RR, Silverberg G, Soete L (eds) Technical change and economic theory. Pinter, London, pp 349–369

Lundvall B-A (1992) National systems of innovation – towards a theory of innovation and interactive learning. Pinter, London

Malerba F (1992) Learning by firms and incremental technical change. Econ J 102(413):845–859

Malerba F (2002) Sectoral systems of innovation and production. Res Policy 31(2):247–264

Malerba F (2007) Innovation and the evolution of industries. In: Cantner U, Malerba F (eds) Innovation, industrial dynamics and structural transformation. Springer, Heidelberg/New York, pp 7–29

Malthus RT (1798) An essay on the principle of population. Johnson, London

Malthus RT (1820) Principles of political economy. Murray, London

March JG (1991) Exploration and exploitation in organizational learning. Organ Sci 2(1):71–87

Marschall A (1890) Principles of economics – an introductory (1920, 8th edn. Macmillan, London

Marx K (1857) Grundrisse der Kritik der politischen Ökonomie (The Grundrisse – introduction to the critique of political economy,1973). Vintage, New York

Marx K (1867) Capital: a critical analysis of capitalist production, vol 1 (1974). Lawrence & Wishart, London

Mason E (1939) Price and production policies of large scale enterprise. Am Econ Rev 29:61–74

Mason E (1957) Economic concentration and the monopoly problem. Harvard University Press, Cambridge

Menger C (1871) Grundsätze der Volkswirthschaftslehre – Allgemeiner Teil (Principles of economics). Wilhelm Braumüller, Wien

Metcalfe SJ (2010) The open, evolving economy: Alfred Marshall on knowledge, management and innovation. In: Gaffard J-L, Salies E (eds) Innovation, economic growth and the firm – theory and evidence of industrial dynamics. Edward Elgar, Cheltenham, pp 3–30

Mill JS (1848) Principles of political economy. Longmans, London

Mill JS (1859) On liberty. Watts, London

Morroni M (2006) Knowledge, scale and transactions in the theory of the firm. Cambridge University Press, Cambridge

Mowery DC, Oxley JE, Silverman BS (1996) Strategic alliances and interfirm knowledge transfer. Strateg Manag J 17(2):77–92

Nelson RR (1992) National innovation systems: a retrospective on a study. Ind Corp Chang 1 (2):347–374

Nelson RR (2007) Understanding economic growth as the central task of economic analysis. In: Hanusch H, Pyka A (eds) Elgar companion to neo-Schumpeterian economics. Edward Elgar, Cheltenham, pp 840–853

Nelson RR, Winter SG (1974) Neoclassical vs. evolutionary theories of economic growth: critique and prospectus. Econ J 84(336):886–905

Nelson RR, Winter SG (1982) An evolutionary theory of economic change. Harvard University Press, Cambridge

Nonaka I, Toyama R, Nagata A (2000) A firm as a knowledge-creating entity: a new perspective on the theory of the firm. Ind Corp Chang 9(1):1–20

Nooteboom B (2008) Learning and innovation in inter-organizational relationships. In: Cropper S, Ebers M, Huxham C, Ring PS (eds) The Oxford handbook of interorganizational relations. Oxford University Press, New York, pp 607–634

OECD (2005) Oslo manual: guidelines for collecting and interpreting innovation data, 3rd edn. OECD Publishing, Paris

Oerlemans LA, Meeus MT (2005) Do organizational and spatial proximity impact on firm performance? Reg Stud 39(1):89–104

Oerlemans LA, Meeus MT, Boekema FW (2001) Firm clustering and innovation: determinants and effects. Pap Reg Sci 80(3):337–356

Oliver C (1990) Determinants of interorganizational relationships: integration and future directions. Acad Manag Rev 15(2):241–265

Osborn RN, Hagedoorn J (1997) The institutionalization and evolutionary dynamics of interorganizational alliances and networks. Acad Manag J 40(2):261–278

Ouchi WG (1980) Markets, bureaucracies, and clans. Adm Sci Q 25(1):129–141

Owen-Smith J, Powell WW (2004) Knowledge networks as channels and conduits: the effects of spillovers in the Boston biotechnology community. Organ Sci 15(1):5–21

Oxley JE, Sampson RC (2004) The scope and governance of international R&D alliances. Strateg Manag J 25(8):723–749

Ozman M (2009) Inter-firm networks and innovation: a survey of literature. Econ Innov New Technol 18(1):39–67

Ozman M (2013) Networks, irreversibility, and knowledge creation. J Evol Econ 23(2):431–453

Page SE (2006) Path dependence. Q J Polit Sci 1(1):87–115

Parkhe A (1993) Strategic alliance structuring: a game theoretic and transaction cost examination of interfirm cooperation. Acad Manag J 36(4):794–829

Parkhe A, Wasserman S, Ralston DA (2006) New frontiers in network theory development. Acad Manag Rev 31(3):560–568

Pavitt K (1984) Sectoral patterns of technical change: towards a taxonomy and a theory. Res Policy 13(6):343–373

Pavitt K (1998) Technologies, products and organization in the innovating firm: what Adam Smit tells us and Joseph Schumpeter doesn't. Ind Corp Chang 7(3):433–452

Penrose ET (1959) The theory of the growth of the firm. Wiley, New York

Perlmutter HV, Heenan DA (1986) Cooperate to compete globally. Harv Bus Rev 64(2):136–152

Peteraf MA (1993) The cornerstones of competitive advantage: a resource-based view. Strateg Manag J 14(3):179–191

Phelps C (2010) A longitudinal study of alliance network structure and composition firm explor-atory innovation. Acad Manag J 53(4):890–913

Podolny JM (1993) A status-based model of market competition. Am J Sociol 98(4):829–872

Podolny JM (1994) Market uncertainty and the social character of economic exchange. Adm Sci Q 39(3):458–483

Podolny JM, Page KL (1998) Network forms of organization. Annu Rev Sociol 24(1):57–76

Polanyi M (1958) Personal knowledge: towards a post-critical philosophy. University of Chicago Press, Chicago

Polanyi M (1967) The tacit dimension. Doubleday, New York

Porter ME (1981) The contributions of industrial organization to strategic management. Strateg Manag J 6(4):609–620

Porter ME (1985) Competitive advantage. Free Press, New York

Powell WW (1987) Hybrid organizational arrangements: new form of transitional development? Calif Manag Rev 30(1):67–87

Powell WW (1990) Neither market nor hierarchy: networks forms of organization. Res Organ Behav 12(1):295–336

Priem RL, Buttler JE (2001) Is the resource-based "view" a useful perspective for strategic management research? Acad Manag Rev 26(1):22–40

Provan KG, Kenis P (2007) Modes of network governance: structure, management, and effective-ness. Econ Innov New Technol 18(2):229–252

Pyka A (1997) Informal networking. Technovation 17(4):207–220

Pyka A (2002) Innovation networks in economics: from the incentive-based to the knowledge based approaches. Eur J Innov Manag 5(3):152–163

Pyka A (2007) Innovation networks. In: Hanusch H, Pyka A (eds) Elgar companion to neo-Schumpeterian economics. Edward Elgar, Cheltenham, pp 360–377

Pyka A, Scharnhorst A (2009) Network perspectives on innovations: innovative networks – network innovation. In: Pyka A, Scharnhorst A (eds) Innovation networks. Springer, Heidel-berg/New York, pp 1–16

Rank ON, Robins GL, Pattison PE (2010) Structural logic of intraorganizational networks. Organ Sci 21(3):745–764

Ricardo D (1817) On the principles of political economy and taxation. Dent & Sons, London

Robertson DT, Stuart TE (2007) Network effects in the governance of strategic alliances. J Law Econ Org 23(1):242–273

Rosenberg N (1973) Innovative responses to materials shortages. Am Econ Rev 63(2):111–118

Rosenberg N (1974) Science, invention and economic growth. Econ J 84(333):90–108

Rosenberg N (2011) Was Schumpeter a Marxist? Ind Corp Chang 20(4):1215–1222

Rothaermel FT (2001) Incumbent's advantage through exploiting complementary assets via interfirm cooperation. Strateg Manag J 22(6):687–699

Rothaermel FT, Deeds DL (2004) Exploration and exploitation alliances in biotechnology: a system of new product development. Strateg Manag J 25(3):201–221

Rothwell R (1994) Towards the fifth-generation innovation process. Int Mark Rev 11(1):7–31

Rowley TJ, Behrens D, Krackhardt D (2000) Redundant governance structures: an analysis of structural and relational embeddedness in the steel and semiconductor industries. Strateg Manag J 21(3):369–386

Rutherford D (2007) Economics – the key concepts. Routledge, New York

Saloner G (1991) Modelling, game theory and strategic management. Strateg Manag J 12(2):119–136

Sato R (1964) The Harrod-Domar model vs the Neo-Classical Growth model. Econ J 74 (294):380–387

Saviotti PP (2011) Knowledge, complexity and networks. In: Antonelli C (ed) Handbook on the economic complexity of technological change. Edward Elgar, Cheltenham, pp 120–141

Schilling MA, Phelps CC (2007) Interfirm collaboration networks: the impact of large-scale network structure on firm innovation. Manag Sci 53(7):1113–1126

Schmalensee R (1988) Industrial economics: an overview. Econ J 98:643–681

Schoenmakers W, Duysters G (2006) Learning in strategic technology alliances. Tech Anal Strat Manag 18(2):245–264

Schumpeter JA (1908) Das Wesen und der Hauptinhalt der theoretischen Nationalökonomie. Dunker & Humblot, Leipzig

Schumpeter JA (1912) Theorie der wirtschaftlichen Entwicklung (The theory of economic developmnt 1934). Duncker & Humblot, Berlin

Schumpeter JA (1939) Business cycles – a theoretical, historical and statistical analysis of the capitalism process. McGraw-Hill, New York

Schumpeter JA (1942) Kapitalismus, Sozialismus und Demokratie (Capitalism, socialism and democracy, 1950). Harper & Bros, New York

Silverberg G, Verspagen B (2005) Evolutionary theorizing on economic growth. In: Dopfer K (ed) The evolutionary foundation of economics. Cambridge University Press, Cambridge, pp 506–539

Simon HA (1955) A behavioral model of rational choice. Q J Econ 69(1):99–118

Simon HA (1991) Bounded rationality and organizational learning. Organ Sci 2(1):125–134

Sivadas E, Dwyer RF (2000) An examination of organizational factors influencing new product success in internal and alliance-based processes. J Mark 64(1):31–49

Smith A (1776) An inquiry into the nature and causes of the wealth of nations [1904, Edwin Cannan ed.]. Methuen, London

Solow RM (1956) A contribution to the theory of economic growth. Q J Econ 70(1):65–94

Solow RM (1957) Technical change and the aggregate production function. Rev Econ Stat 39 (3):312–320

Spence M (1976) Informational aspects of market structure: an introduction. Q J Econ 90(4):591–597

Spence M (2002) Signaling in retrospect and the informational structure of markets. Am Econ Rev 92(3):434–459

Spender JC, Grant RM (1996) Knowledge and the firm: overview. Strateg Manag J 17(2):5–10

Staber U (1998) Inter-firm co-operation and competition in industrial districts. Organ Stud 19 (4):701–724

Stuart TE (2000) Interorganizational alliances and the performance of firms: a study of growth and innovational rates in a high-technology industry. Strateg Manag J 21(8):791–811

Stuart TE, Hoang H, Hybles RC (1999) Interorganizational endorsements and the performance of entrepreneurial ventures. Adm Sci Q 44(2):315–349

Swan T (1956) Economic growth and capital accumulation. Econ Record 32(63):334–361

Swann PG (2009) The economics of innovation – an introduction. Edward Elgar, Cheltenham

Swedberg R (2000) Entrepreurship. In: Swedberg R (ed) The social science view of entrepreneur-
   ship. Oxford University Press, New York, pp 7–44
Teece DJ (2007) Explicating dynamic capabilities: the nature and microfoundations of (sustain-
   able) enterprise performance. Strateg Manag J 28(13):1319–1350
Thorelli HB (1986) Networks: between markets and hierarchies. Strateg Manag J 7(1):37–51
Tidd J (2006) A review of innovation models. Imperial College discussion paper series, 06/1, pp
   1–15
Tidd J, Bessant J, Pavitt K (2005) Managing innovation: integrating technological, market and
   organizational change, 3rd edn. Wiley, Chichester
Torre A, Rallet A (2005) Proximity and localization. Reg Stud 39(1):47–59
Tushman ML, Nelson RR (1990) Introduction: technology, organizations, and innovation. Adm
   Sci Q 35(1):1–8
Uzzi B (1996) The sources and consequences of embeddedness for the economic performance of
   organizations: the network effect. Am Sociol Rev 61(4):674–698
Uzzi B (1997) Social structure and competition in interfirm networks : the paradox of
   embeddedness. Adm Sci Q 42(1):35–67
Uzzi B, Amaral LA, Reed-Tsochas F (2007) Small-world networks and management science
   research: a review. Eur Manag Rev 4(2):77–91
Van De Ven AH, Poole MS (1995) Explaining development and change in organizations. Acad
   Manag Rev 20(3):510–540
Van Witteloostuijn A (2000) Organizational ecology has a bright future. Organ Stud 21(2):X–XIV
Veblen T (1898) Why is economics not an evolutionary science? Q J Econ 12(3):373–397
Visser E-J (2009) The complementary dynamic effects of clusters and networks. Ind Innov 16
   (2):167–195
Von Hippel E (1986) Lead users: a source of novel product concepts. Manag Sci 32(7):791–805
Von Hippel E (1988) The sources of innovation. Oxford University Press, New York
von Mises L (1969) The historical setting of the Austrian school of economics. Arlington House,
   New York
Walras L (1874) Elements of pure economics (1954). George Allen and Unwin, London
Wang C, Rodan S, Fruin M, Xu X (2014) Knowledge networks, collaboration networks, and
   exploratory innovation. Acad Manag J 57:484–514
Wasserman S, Faust K (1994) Social network analysis: methods and applications. Cambridge
   University Press, Cambridge
Wassmer U (2010) Alliance portfolios: a review and research agenda. J Manag 36(1):141–171
Weick KE (1979) The social psychology of organization. Addison-Wesley, Reading
Weigel K, Camerer C (1988) Reputation and corporate strategy: a review of recent theory and
   applications. Strateg Manag J 9(5):443–454
Wernerfelt B (1984) A resource based view of the firm. Strateg Manag J 5(2):171–180
White S (2005) Cooperation costs, governance choice and alliance evolution. J Manag Stud 42
   (7):1383–1413
Whittington KB, Owen-Smith J, Powell WW (2009) Networks, propinquity, and innovation in
   knowledge-intensive industries. Adm Sci Q 54(1):90–122
Williamson OE (1973) Organizational forms and internal efficiency – markets and hierarchies:
   some elementary considerations. Am Econ Rev 63(2):316–325
Williamson OE (1975) Markets and hierarchies: analysis and antitrust implications. Free Press,
   New York
Williamson OE (1985) The economic institutions of capitalism. Free Press, New York
Williamson OE (1991) Comparative economic organization: the analysis of discrete structural
   alternatives. Adm Sci Q 36(2):269–296
Winter SG (2003) Understanding dynamic capabilities. Strateg Manag J 24(10):991–995
Winter SG (2006) Toward a neo-Schumpeterian theory of the firm. Ind Corp Chang 15(1):125–141
Witt U (2003) Evolutionary economics and the extension of evolution to the economy. In: Witt U
   (ed) The evolving economy. Edward Elgar, Cheltenham, pp 3–37

Witt U (2008a) What is specific about evolutionary economics? J Evol Econ 18(5):547–575

Witt U (2008b) Recent developments in evolutionary economics. Edward Elgar, Cheltenham

Witt U, Broekel T, Brenner T (2012) Knowledge and its economic characteristics: a conceptual clarification. In: Arena R, Festre A, Lazaric N (eds) Handbook of economics and knowledge. Edward Elgar, Cheltenham

Yamakawa Y, Yang H, Lin Z (2011) Exploration versus exploitation in alliance portfolio: performance implications of organizational, strategic, and environmental fit. Res Policy 40 (2):287–296

Zollo M, Winter SG (2002) Deliberate learning and the evolution of dynamic capabilities. Organ Sci 13(3):339–351

Part II
Industry, Data and Methods

Chapter 3
Laser Technology and the German Laser Industry

I did realize that the laser had wider applications such as communications and cutting and welding, but I never envisaged the breadth of applications for which it is used today.

(Charles H. Townes 2010)

Abstract The quantitative analysis of innovation networks in the German laser industry requires a profound understanding of industry characteristics. This includes a solid understanding of the emergence and the very nature of the underlying technology itself. At the same time, a closer look at the laser industry value chain allows us to identify the core of the industry and provides the basis for analyzing knowledge exchange and interorganizational learning processes, which in turn are seen as a central prerequisite for an in-depth understanding of the collective nature of innovation processes. Chapter 3 is divided into five sections: Section 3.1 provides some brief information on the historical background of laser research. Section 3.2 gives a short introduction of basic laser operating principles. Section 3.3 outlines the most notable milestone developments over the past 50 years. Section 3.4 provides a closer look at the German laser industry by discussing the emergence of laser technology and specific industry characteristics in East and West Germany. Finally, in Sect. 3.5, attention is drawn to the configuration of the industry value chain.

3.1 Historical Background of Laser Research

In 2010, the laser celebrated its 50th anniversary. Lasers are artificial light sources that emit a coherent light beam characterized by some distinctive physical properties that make lasers useful for a broad range of technological applications (Buenstorf 2007, p. 182). The term *laser,* which stands for "Light Amplification by Stimulated Emission of Radiation", was originally coined by Gould (1959). He presented his ingenious idea for the construction of an optical maser at the "Ann

© Springer International Publishing Switzerland 2015
M. Kudic, *Innovation Networks in the German Laser Industry*,
Economic Complexity and Evolution, DOI 10.1007/978-3-319-07935-6_3

Arbor Conference on Optical Pumping" in 1959 and is still considered by many to be the actual inventor of the laser (Hecht 2005, p. 46).

Only one year after Gould's seminal article, Maiman (1960) put the first stable laser device into operation. Maiman and his colleagues demonstrated a flash-lamp-pumped ruby laser capable of delivering coherent red radiation (TSB 2010, p. 21). In these early days, the scope and the potential of this groundbreaking invention were by no means fully recognized. Irnee D'Haenens, a physicist who worked closely with Maiman, jokingly called the laser "a solution looking for a problem" (Hecht 2010, p. 21). Nowadays, laser applications have become part of everyday life (Hecht 2010) and the invention of the laser instigated the new field of quantum electronics in physics that spans the disciplines of theoretical quantum physics and electrical engineering (Slusher 1999, p. 472). However, the intellectual roots of modern laser research are even older and extend as far back as the end of the nineteenth century.

The detection of electromagnetic microwaves by James C. Maxwell in 1873 and Heinrich Hertz in 1888 (Bertolotti 2005, pp. 115–116) represent important milestones leading up to the invention of the *maser*,[1] a precursor to the laser, in the early 1950s by Jim P. Gordon, Herbert J. Zeiger and Charles H. Townes at Columbia University (1954, 1955). However, a crucial prerequisite for the invention of the laser was the breakthrough by two Soviet scientists, Basov and Prokhorov (1950) in the field of microwave research. They found a way to produce positive feedback of stimulated radiation based on a resonant circuit by showing that the circuit losses were smaller than the gain in energy afforded the wave by stimulated molecular transitions (Townes 1964, p. 61).

The second crucial event was the discovery of quantum theory which was at the forefront of physics at the beginning of the twentieth century. The fundamental concepts of classical physics began to crumble when Max Planck's seminal article *"On the law of distribution of energy in the normal spectrum"* in 1900 provided an explanation for "blackbody radiation"[2] by assuming that atoms emit and absorb energy (E) which is the product of a frequency (ν) and a fundamental constant (h), later named Planck's constant (Phillips 2003, p. 1). Only a few years later, Einstein (1905) published his seminal article *"On a heuristic viewpoint concerning the production and transformation of light"* which was strongly influenced by Planck's theory on blackbody radiation. In his article, he provided the theoretical foundation for what he called "energy quanta" and gave an explanation for the photoelectric

---

[1] The abbreviation MASER stands for "Microwave Amplification by Stimulated Emission of Radiation" (Schawlow and Townes 1958). Microwaves are the same as light electro-magnetic radiation but with a wider wavelength and a lower frequency compared to radiation in the visible spectrum.

[2] A blackbody is a theoretical object which absorbs emitted radiation. The first experimental setup for measuring blackbody radiation was designed at the end of the nineteenth century by the German Imperial Institute of Physics and Technology. It consisted of a long tube with a small hole which provided a nearly perfect absorbing installation, a so-called "blackbody" (Carson 2000, p. 8; Bertolotti 2005, p. 61).

effect (Carson 2000, p. 10) according to which electromagnetic radiation with very short wavelengths ejects electrons from the surface of metal when radiation frequency exceeds a threshold value (Phillips 2003, p. 18). The frequency or wavelength allows electromagnetic radiation to be classified on a spectrum ranging from radio magnetic radiation, microwaves and visible or invisible light, to x-ray radiation and gamma-ray radiation (ibid). Einstein illustrated this phenomenon using light as an example. This ultimately resulted in the recognition of the wave-particle duality of light.[3] Since then, light has no longer been recognized as a particle or as a wave.[4] Instead, physicists regard light as electromagnetic radiation at any wavelength which consists of photons, elementary particles of light, and which exhibits properties of both waves and particles (Carson 2000, p. 17).

In 1913, Niels Bohr developed a general theory of atom structure which paved the way for the development of quantum mechanics (Bertolotti 2005, p. 139). Einstein (1917) realized that an atom's energy level can be affected by the emission or the absorption of photons in two different ways: spontaneous emission or stimulated emission (Hecht 2005, p. 10). Spontaneous emission of radiation occurs when an atom emits energy without any external causation (Prokhorov 1964, p. 110). Most of the light we see in day-to-day life, such as daylight or artificial light, is largely spontaneous emission (Hecht 2005, p. 10). In contrast, the phenomenon of stimulated emission can be observed when an atom is forced to release energy due to interaction with an external field (Prokhorov 1964, p. 110). Initially, stimulated emission only seemed to be of theoretical interest (Hecht 2005, p. 10). However, it turned out that the principle of stimulated emission provided the theoretical foundation for the invention of the laser.

3.2   Basic Operating Principles of Lasers

Since the early days of laser research, several new types of lasers have been invented. Despite the rapid developments in laser technology, most lasers still operate on the basis of the same basic operating principles.

In the most basic sense, a laser is a unit that amplifies a light wave by simulated emission and emits a cascade of photons (Townes 1999, p. 12). The amplification of light by stimulated emission of radiation is achieved "[...] by exciting the electronic, vibrational, rotational, or cooperative modes of a material into a non-equilibrium state so that photons propagating through the system are amplified coherently by stimulated emission" (Slusher 1999, p. 471). As a result we can

---

[3] In 1924 DeBroglie proposed that all particles, not only photons, are associated with waves. The so-called wave-particle duality of light constitutes a central pillar of modern quantum mechanics (Phillips 2003).

[4] For a long time classical physics was dominated by two theoretical explanations about the nature of light: the "wave theory of light" and the "particle theory of light" (Bertolotti 2005, p. 13).

observe a coherent amplification "[...] that is, amplification of a wave at exactly the same frequency and phase" (Townes 1999, p. 13). The basic operating principle of a laser is illustrated in Fig. 3.1.

According to Fischer (2010, p. 54), the basic components of a laser are the "gain medium", "pumping system" and "optical resonator". These components can be made of several modular subunits. We will start by looking at the so-called "gain medium" (cf. Fig. 3.1, VI). To a large extent it determines the frequency and wavelength under which the laser will ultimately operate. The "gain medium" is also known as the "active medium" or "active material". This material is characterized by very special properties in a sense that "[...] molecules or atoms have been put in an abnormal condition, with more molecules in excited states than in ground, or lower states" (Townes 1999, p. 14). It can be of a solid, gaseous, liquid or plasma state and provides the basis for raising electrons to a higher energy level by absorbing the emitted pumping energy. In most, but not all, cases, laser types are named after the gain medium used. For instance, Maiman's first laser operated on the basis of a ruby which is a crystal of aluminum oxide with chromium atoms (Bertolotti 2005, p. 228). The "pumping system" provides energy in the active medium (cf. Fig. 3.1, II) which is needed to lift some electrons into an excited state. The excitation of the gain medium can be accomplished by using optical radiation, electrical current and discharges or on the basis of chemical reactions (Slusher 1999, p. 471). In his experimental setup, Maiman found a rather simple solution; he used standard photography equipment consisting of some very powerful helical flash-lamps (Bertolotti 2005, p. 231). The next important issue is the energy efficiency of lasers. The discharged energy cannot be entirely utilized during the pumping process. Thus, a notable amount of waste heat is produced (cf. Fig. 3.1, V). Finally, a so-called "optical resonator", consisting of one full reflective (cf. Fig. 3.1, I) and one semi-transparent (cf. Fig. 3.1, III) mirror, reflects a beam of coherent light between the two mirrors. In general, the resonator system is a very significant element of a quantum oscillator (Prokhorov 1964, p. 110). Spontaneous emission becomes aligned by the mirrors in the optical resonator and generates a "standing wave" at a particular frequency and wavelength (Bertolotti 2005, p. 231). The pumping energy and the optical resonator amplify the simulated radiation until the discharged energy reaches a critical level and a laser beam is emitted (cf. Fig. 3.1, VII). The coherent light beam is emitted out of the resonator by a semi-transparent

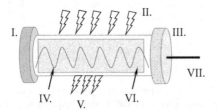

	I.	Fully reflective mirror
	II.	Pumping energy
	III.	Semi-transparent mirror
	IV.	Standing wave
	V.	Lost thermal energy
	VI	Gain medium
	VII	Laser beam

Fig. 3.1 Functioning principles of a laser (Source: Author's own illustration, based on: Bertolotti (2005, p. 231) and Fischer (2010, pp. 54 & 79))

mirror (Slusher 1999, p. 471). In Maiman's experiment the capacitor bank discharged several 1000 V until the threshold energy level of about 0.7–1.0 J was suddenly reached (Bertolotti 2005, p. 228). Ruby lasers are normally only operated in pulses because of the comparably high amount of power required to reach the threshold (Townes 1964, p. 73). In general, lasers can generate a continuous constant-amplitude output, known as continuous wave, or a pulsed output, by using a specific pulsing technique. The most commonly used techniques are q-switching and mode-locking, which will be explained in detail later.

3.3 From an Ingenious Idea to the Emergence of a New Technology

The early 1960s were characterized by several technological refinements of the initially presented laser device. Since then, remarkable advances have been made with regard to enhanced pumping systems which operate on the basis of chemical or nuclear reactions. The utilization of various types of gain media began quite early and resulted in entirely new types of lasers, such as solid-state lasers, gas lasers, semiconductor lasers, and organic dye lasers. In the recent past, researchers started working on entirely new laser operating principles. Figure 3.2 illustrates milestone developments in laser research over the past 50 years and provides the basis for the following chronological discussion.[5]

The most important types of lasers had already been developed by the late 1960s. Solid-state lasers operate on the basis of glass or crystalline gain materials which are doped with impurities – e.g. neodymium (Nd), chromium (Cr), erbium (Er), etc. – which affect the electrical and optical properties of the material. Only a few months after Maiman's ruby laser, Sorokin and Stevenson (1960, 1961) presented the uranium-doped calcium florid solid-state laser developed at the IBM laboratories in Yorktown Heights, NY, USA. The first solid-state nd:glass laser was invented by Snitzer (1961) at American Optical Corporation in Southbridge and the first nd:YAG laser was put into operation by Geusic et al. (1964) at Bell Labs in New Jersey, USA.

At almost the same time, Javan et al. (1961) developed the first gas laser at Bell Labs. They generated a continuous laser beam based on neon-helium gas. A wide variety of gases can be used as a gain medium. These include carbon monoxide (CO), carbon dioxide (CO_2) and argon (Ar). Only three years later, the CO_2-laser was invented by Patel (1964), again at Bell Labs. Carbon dioxide lasers are quite efficient and still one of the most powerful types of laser. In the same year, Bridges

---

[5] Unless otherwise stated, the following overview and discussion of milestone developments in laser research are based on Hecht (2005, 2010), Bertolotti (2005) and guided by a chronological overview of laser research provided by photonics.com (http://www.photonics.com/LinearCharts/Default. aspx?ChartID = 2, accessed: November 2011).

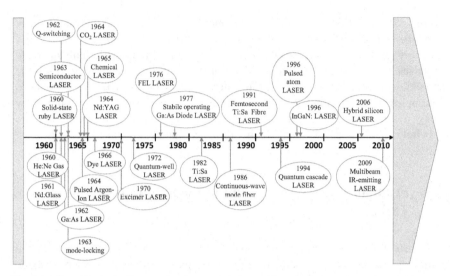

Fig. 3.2 Chronological milestone developments in laser research (Source: Author's own illustration, based on: www.photonics.com (Accessed: November 2011))

(1964) invented another gas laser – the argon-ion laser – at Hughes Aircraft Company in California. In the following years, Sorokin and Lankard (1966) invented a novel type of laser at the IBM research laboratories in New York, United States which operates on the basis of organic dyes. Even though dye lasers usually work on the basis of a liquid gain material, new developments proved that solid-state dyes could also be used as an active medium.

Beyond the invention of new laser types, several further advances were made in the first decade of laser research. For instance, new pumping techniques were tested, which led to the discovery of the chemical laser by Kasper and Pimentel (1965). Additionally, different pulsing techniques were developed in the 1960s. Hellwarth and McClung (1962, 1963) increased the peak power of ordinary ruby lasers 100-fold by applying the q-switching method, a pulsing technique which allowed a high-energy short-pulsed beam to be generated. Only one year later, Hargrove et al. (1964) successfully demonstrated the implementation of the mode-locking technique in a helium-neon laser. Mode-locking allows the emission of light for extremely short periods of time – so-called ultra-short pulses – ranging from tens of picoseconds (10^{-12} s.) down to a few femtoseconds (10^{-15} s.).

After the initial turbulent years of laser research further breakthroughs were achieved, particularly with respect to maximizing average and peak emissions, power efficiency, charging techniques and minimizing pulse durations. The most remarkable achievements in the 1970s were the invention of the excimer laser, quantum-well laser, free-electron laser and the gallium-arsenide diode laser which many consider to be the first stably operating semiconductor laser. In the early 1970s, Soviet scientists Basov et al. (1970) developed the first excimer laser at the Lebedev Physical Institute in Moscow. Excimer lasers operate on the basis of a short lived molecule – a so-called exited dimer – which consists of a noble gas – e.g. argon (Ar) or krypton (Kr) – and a reactive gas – e.g. fluorine (F) or chlorine (Cl).

The first quantum-well laser was invented between 1972 and 1974 by Charles W. Henry and his colleagues at Bell Labs (Dingle et al. 1974). Quantum-well lasers require a significantly lower threshold energy level to emit a laser beam compared to conventional semiconductor lasers. The first free-electron laser (FEL) was developed in the 1970s by John Madey and his colleagues at Stanford University (Elias et al. 1976). Free-electron lasers exhibit the same functionality as conventional lasers but they work on the basis of completely new operating principles. These types of laser use an electron beam as a gain medium which is accelerated to almost the speed of light. The beam moves through a periodic transitive magnetic field and this results in the sudden release of coherent radiation. Free-electron lasers have the widest frequency spectrum of all lasers and allow for extremely high peak power. Even though the first semiconductor laser was invented in the early 1960s, it took more than 15 years to present a stable system that operates on a constant basis at room temperature (Hecht 2010, p. 23). This laser type was a gallium-arsenide diode laser developed in 1977 at Bell Labs (ibid).

The 1980s were characterized by the development of the titanium-sapphire laser by Peter F. Moulton at the MIT Lincoln Laboratory in Massachusetts, USA. This solid-state laser uses a sapphire crystal – i.e. aluminum oxide (Al_2O_3) – doped with titanium (Ti) ions as a gain medium. The major advantage of this type of laser is the generation of ultra-short pulses and the possibility to adjust the wave-length over a considerably wide bandwidth. The latest generation of Ti:sapphire lasers is advancing into the realm of ultra-short lasting no more than a few attoseconds (10^{-18} s.). This allows for countless applications in fundamental research in natural science fields such as biology, chemistry and physics (Klein and Kafka 2010, p. 289). The second remarkable advancement in the 1980s was the invention of the fiber laser made possible by the pioneering work of Kao and Hockham (1966) on light signal transmission properties of optical glass fibers. Thus, in the 1980s, David N. Payne and his colleagues developed the first single-mode continuous-wave erbium-doped fiber laser at Southampton University in the United Kingdom (Mears et al. 1986). The laser was able to operate at room temperature at the all-important telecommunications wavelength of 1.55 μm (Nature 2010, p. 281).

At least three up-and-coming discoveries were made in the field of laser research in the 1990s – the quantum cascade laser, the pulsed atom laser and the InGaN laser. Based on the groundbreaking theories of Kazarinov and Suris (1971), it took more than 20 years to demonstrate the first stably operating quantum cascade laser (QCL). In the mid-1990s, the first successfully operating quantum cascade laser was presented by Faist and his colleagues (1994) at Bell Labs. These specific types of semiconductor lasers work on the principle of inter-subband transition and emit over the whole mid-infrared range (Mueller and Faist 2010, p. 291). Since then, QCLs have been used for numerous applications in environmental science, process control and medical diagnostics (Mueller and Faist 2010, p. 291). In the same decade, the first pulsed atom laser was demonstrated by Ketterle and Misner (1997) at MIT, Massachusetts. Atom lasers use matter instead of light to generate a coherent beam that behaves like a wave. The first stably operating atom laser was presented by Bloch et al. (1999) at the Max Planck Institute for Quantum Optics in Munich, Germany. Nakamura and his colleagues (1996) demonstrated the first blue laser diode at the Nichia Corporation in Tokyo, Japan. This specific type of multi-quantum well laser was electronically pumped and operated on the basis of indium-gallium nitride (InGaN). Compared to conventional GaN semiconductor lasers, the blue diode laser showed stimulated emission through current injection which was an important breakthrough at the time (Nature 2010, p. 281).

The invention of the hybrid silicon laser and the demonstration of the multi-beam infrared laser are considered to be some of the most important technological developments in laser research in the first decade of the twenty-first century. In 2006, researchers at the University of California, Santa Barbara and Intel Corporation announced the first electrically powered continuous wave hybrid AlGaInAs-silicon laser (Fang et al. 2006). These novel hybrid silicon lasers combine the light-emitting capabilities of binary semiconductors, such as indium phosphide (InP), with the light-routing and cost advantages of silicon and provide the potential for the implementation of optical data pipes inside of computers operating on a terabit level (Paniccia et al. 2006, p. 2). In 2009, the first multibeam-multiwave infrared emitting laser was presented by an international team of scientists (Yu et al. 2009). The multi-beam abilities of these novel types of semiconductor lasers qualify them for a broad range of applications in climate monitoring and communications and in many other applications that shape our everyday life.

3.4 The Onset of Laser Research in East and West Germany

In the early 1960s both public science and industry quickly initiated their own efforts towards laser construction and commercialization in Germany (Buenstorf 2007, p. 185). The emergence of laser technology in Germany is quite a unique and interesting case because of the breakup of Germany into the German Democratic Republic (GDR) and the Federal Republic of Germany (FRG) after the Second World War. The separation into East and West Germany was accompanied by the emergence of two largely detached sectoral innovation systems in the early years of laser research (Albrecht 1997).

We will start by taking a closer look at the beginnings of laser research in the GDR. After the Second World War the Soviet authorities established the "German Academy of Sciences" (GAS) in Berlin-Adlershof. This research center was later renamed "Academy of Sciences of the German Democratic Republic" and, until German reunification, was one of the most influential research facilities in the GDR. The "Institute for Optics and Spectroscopy" (IOS), located in Berlin-Adlershof, was actively involved in the research activities that paved the way for the first GDR laser (Albrecht 2010b, p. 177). The origin of laser research in the GDR in the early 1960s was mainly located in the federal state of Thuringia, specifically in and around Jena (Albrecht 1997, 2010b). The "Institute for Applied Optics" (IAO) and the "Physical Institute" (PI) at Friedrich Schiller University (FSU) in Jena were deeply involved in GDR laser research activities at the time (Albrecht 2010b, p. 182). In 1962, the "Laser Commission" was founded to coordinate the research activities between these research institutes and industry. The first GDR ruby laser was officially demonstrated in Berlin on August 8, 1962 (TSB 2010, p. 22). VEB Carl Zeiss[6] was the most obvious choice for the industrial production of lasers due to its expertise in optics and electrical engineering. The following year, VEB Carl Zeiss announced the launch of its first commercial laser device which was developed together with physicists from FSU (Albrecht 2010b, p. 176). In 1967, the company started to produce the first generation of stably operating gas lasers and by 1974 had successively extended its product range to a total of six different types of neon-helium gas lasers (Albrecht 2010b, p. 178). At nearly the same time, the development of the argon laser and the CW CO_2 laser began at the Central Institute of Optics and Spectroscopy (ZOS) in Berlin (TSB 2010, p. 22). In 1971, the CW CO_2 laser was further developed at the Center for Scientific Instruments (ZWG) located at GAS in Berlin and finally transferred to the VEB FEHA in Halle in 1975 (ibid).

This early success story in the field of laser research was primarily due to a few very well-connected GDR scientists.[7] However, the headway that was made was disrupted due to several factors. Firstly, the fields of laser research were pre-defined and directed by the GDR's Central Committee. For instance, the research activities of FSU had to be aligned with the centrally-planned medium-term production plans of VEB Carl Zeiss (Albrecht 2010b, p. 190). This led to a significant impairment in the freedom of research. Secondly, both Schramm (2005) and Albrecht (2010b, p. 198) conclude that there was a climate of mistrust between industry and science due to the progressive infiltration of informants in key positions during the late 1960s. Finally, the politically motivated change in the strategic orientation of VEB Carl Zeiss from precision opto-electronics to mass production dramatically hampered the firm's research potential (Albrecht 2010b).

---

[6] The company was founded in 1846 and over the years maintained its position as one of the leading manufacturers of microscopes, cameras, optical measuring instruments and other optical devices in Germany. After World War II, US forces occupied Thuringia for a short period and relocated engineers and managers to the federal state of Baden-Wurttemberg to build up Carl Zeiss GmbH in Oberkochen. The remaining sections of Carl Zeiss were taken over by the Soviet Military Administration and integrated into the GDR as the state-owned company VEB CARL ZEISS Jena. For more details, see: http://www.zeiss.com/corporate/en_de/history/company%20history/at-a-glance.html, accessed: August 2014.

[7] According to Albrecht (2010b, pp. 179–182), the most notable actors in the initial interpersonal laser research network in the GDR were: Wilhelm Schuetz (FSU), Konrad Kuehne (FSU), Bruno Elschner (FSU), Gerhard Wiederhold (FSU), Rudolf Ritschl (GAS), Paul Goerlich (VEB CARL ZEISS) as well as some international partners, such as Alfred Kastler, a physicist who received the Nobel Prize for his work on optical pumping techniques.

The development of the first laser greatly attracted the attention of physicists in the Federal Republic of Germany (FRG). In the early 1960s, two physicists – Hans Boersch at the Technical University of Berlin and Hermann Haken at the Technical University of Stuttgart – worked on typical laser subjects (Albrecht 2010a, p. 173). Some of the components for the first ruby laser at the TU Berlin in 1962 were made by several small companies in Berlin and the flash bulb was supplied by Osram (TSB 2010, p. 21). Only a few years later, research into microwaves and lasers became the subject of several projects funded by the "German Research Foundation" (GRF) (Albrecht 2010a, p. 161). Within the GRF's research priority program on "high-frequency physics", a total of 245 research applications in the field of maser and laser research were registered between 1958 and 1967 (Albrecht 2010a, p. 162). It is remarkable that more than half a dozen research facilities, which received a grant from the GRF research program on "high-frequency physics" in 1962, were located in the federal state of Baden-Wurttemberg (Albrecht 2010a, p. 174). A total of 27 laser research projects were awarded in the period between 1962 and 1966 (ibid). In the early 1980s Gerd Herziger, a former doctoral student of Hans Boersch, took over the formation and direction of the Fraunhofer Institute for Laser Technology in Aachen (TSB 2010, p. 21). In the late 1980s, the Solid-state Laser Institute (FLI) was founded as an affiliated institute of the Free University Berlin and TU Berlin, respectively (TSB 2010, p. 23).

The emergence of industrial laser research in West Germany was closely related to the microwave research activities of the Siemens Group in the late 1950s. In 1959, the first ruby maser was demonstrated in the research laboratories of Siemens and Halske AG in Munich (Albrecht 1997, p. 46). Industry data from the LASSSIE project (Buenstorf 2007) reveals that during the 1960s about a dozen firms entered the scene in West Germany. Apart from a few exceptions nearly all of these firms were located in the federal states of Baden-Wurttemberg and Bavaria.[8] In a very short period of time Siemens became the market leader in industrial laser technology research. Another dominant actor at that time was Carl Zeiss GmbH in Oberkochen which became one of the leading optical companies in West Germany after the Second World War (Albrecht 1997, p. 106).

In a nutshell, the onset of industrial laser research in the GDR was strongly influenced by the company VEB Carl Zeiss and Friedrich Schiller University, both located in Thuringia. In addition, the "Institute for Optics and Spectroscopy" (IOS), which was located at the "German Academy of Sciences" in Berlin, was vitally important for early laser research activities in the GDR. In contrast, the start of industrial laser research in West Germany was strongly affected by at least two dominant players, Siemens & Halske AG in Munich, who built the first German laser, and Carl Zeiss GmbH in Oberkochen with its long-standing tradition of

---

[8] Albrecht (1997, pp. 96–97) has identified the following companies active in the field of laser on the basis of patent and bibliometric data. Siemens in Munich and Erlangen, Telefunken in Ulm, Standard Elektrik Lorenz in Stuttgart, Carl Zeiss in Oberkochen, Osram in Augsburg, Sylvania-Vakuumtechnik in Erlangen, Impulsphysik in Hamburg, Jenaer Glaswerke Schott in Mainz, Quarzlampen Gesellschaft in Hanau, and Atlas Meß- & Analysentechnik in Bremen.

optical engineering. Scientific laser research in the FRG in the early 1960s was, with a few exceptions (most notable: TU Berlin), geographically concentrated in the southern part of Germany.

3.5   The German Laser Industry Value Chain

Figure 3.3 illustrates the value chain[9] of the laser industry and its links to the supply and market as well as to the contact points of technology and commercial partners. The laser industry value chain itself consists of the four main elements: "materials", "components", "laser beam sources & periphery" and "laser systems" accompanied by cross-sectional services that provide specific technical and commercial advice to these four elements.

To start with, we look at the market dimension (cf. Fig. 3.3, right). In the 1970s and 1980s the broad spectrum of potential laser applications soon led to a diffusion of the technology. Lasers started to play an important role in numerous application fields and industries, such as medical and biotech, military and security, ICT and

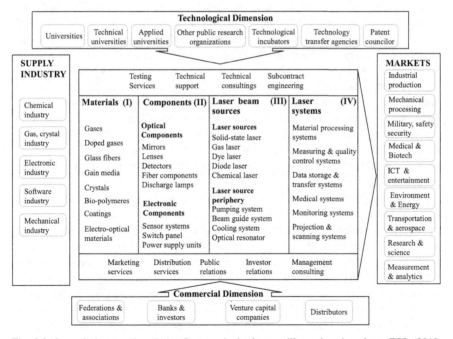

Fig. 3.3 Laser industry value chain (Source: Author's own illustration, based on: TSB (2010, p. 18))

---

[9] The following discussion is inspired by the supply chain concept reported in TSB (2010, p. 18).

entertainment, environment and energy, transportation and aerospace, research and science, measurement and analytics, and environment and energy.

Since the 1980s, lasers have been increasingly used in the fields of industrial production and material processing (Poprawe 2010). These two fields of application have played an important role in Germany due to the traditionally strong presence of machine building and machine tool companies in the German Mittelstand. The use of lasers in industrial production revealed the laser's potential of being an enabler of several subsequent innovations in large-scale production processes (Poprawe 2010, p. 31). Professor Boersch, one of Germany's laser technology pioneers, already recognized that the future of laser technology was going to be in material processing (TSB 2010, p. 21). Laser material processing systems can be divided into two broad categories: macro processing systems and micro processing systems. The first segment usually includes systems for cutting, welding, marking and other applications, such as rapid manufacturing and prototyping, ablation, and engraving of printing cylinders, whereas the second segment comprises systems used for the production of semiconductors, printed circuit boards, flat panel displays and lithography systems (Mayer 2006, p. 11). Today virtually every electronic entertainment device or computer has an optical drive that is practically inconceivable without the use of modern laser technology.

The left side of Fig. 3.3 illustrates the supply side of the industry. Not only the chemical, gas, and crystalline industries but also the software, electronic and mechanical industries occupy important positions on the supply side. The chemical industry plays a key role due to the diversity of chemical products which are usually used as gain media in lasers. The electronic and optical industries provide a broad range of ready-to-assemble components for lasers such as sensor systems, power supply units, optical resonator systems or discharge lamps. Even though these products are often highly specialized and focus on specific laser types, it is important to note that the core competencies of the component suppliers are clearly not in the field of laser design and production. Material and component suppliers constitute the first two elements of the industry value chain, in a narrower sense, illustrated at the center of Fig. 3.3 (I & II).

The third element (cf. Fig. 3.3, III) symbolizes laser source manufacturers that are considered to be the core of the industry. These firms are primarily concerned with the development, design and production of laser beam sources. In other words, their business activities are centered on the construction of laser beam units that constitute the key component of any kind of laser-based machine or system. Buenstorf (2007, p. 182) points to the fact that "[...] even though the laser as a product is unequivocally defined by its physical properties, there is no such thing as a general purpose laser." The different capabilities, origins and specialization patterns of German laser source manufacturers provide some good reasons to assume that the industry is characterized by a submarket structure in a sense that entrepreneurial firms enter as specialists in a single submarket but serve an increasing number of submarkets as they get older (ibid). The German laser industry is characterized by a high number of micro and small sized laser source

manufacturing firms (cf. Sect. 4.2.2). The majority of German laser source manufacturers operate in highly-specialized, narrow market segments. Exploring laser
industry journals (e.g. Laser und Optoelektronik) reveals that micro and small-sized
laser source manufactures often produce only a handful of laser types which are
designed for very specific applications. At the same time, rapid technological
developments in the field of laser research (cf. Sect. 3.3) steadily increase the
innovative pressure.

The fourth element of the laser industry value chain at the center of Fig. 3.3
(IV) highlights the importance of the industry's laser system providers. Buenstorf
(2007, p. 182) argues that lasers can significantly differ in economic dimensions,
such as production cost, energy efficiency, longevity, and ease of handling and
usually have to be customized to meet the user's needs. Laser system providers
purchase ready-to-assemble lasers from highly specialized laser source manufacturers and integrate these units into machines or systems which are assigned to a
particular application area. The core competencies of the laser system providers are
located in the application area of the laser system or the laser machine that they
produce and not in the field of laser source production. Nevertheless, due to their
strong market position these firms often trigger the development of novel laser
sources at upstream stages of the industry value chain without being directly
involved in the innovation process.

The technological dimension is displayed at the top of Fig. 3.3. At the onset of its
laser research, the German laser industry was perceived as lagging behind its
international competitors (Buenstorf 2007, p. 185). Substantial government efforts
were initialized to counteract this situation, for example, public funding initiatives
and the establishment of dedicated laser research institutes and technology transfer
agencies (ibid). The main reasons for this were straightforward: on the one hand,
political authorities had become aware of the economic potential of the technology.
On the other hand, there was clearly a demand for both basic and applied laser
research in Germany in order to catch up with international research in this field.
The technological dimension of the laser industry value chain includes all universities, applied universities and technical universities that are active in the field of
laser research. In addition, the German laser research landscape is characterized by
non-university research mainly conducted by public research institutes active in
both basic and applied research. Most of these institutes belong to one of the four
large German research associations: Fraunhofer, Helmholtz, Max Planck and Leibniz Society (cf. Sect. 7.2.4). Moreover, patent councilors and technological incubators are assigned to the technological dimension of the laser industry value chain.
Incubation organizations such as "business incubators", "science parks" or
"research parks" can be regarded from a theoretical perspective as policy-driven
instruments for counteracting the comparably high failure risk that newly
established ventures face in the very first years after their market entry (Schwartz
2013, p. 7).

The commercial dimension of the laser industry value chain (cf. Fig. 3.3, bottom) includes four groups of organizational entities: "federations & associations",
"banks & investors", "venture capital companies" and "distributors". Federations

and associations promote the firms' interests of generating support for laser-related technologies in politics and society. Banks, investors and venture capital companies secure financing for the firms, whereas distributors provide access to national and international markets.

To conclude the debate we turn to the most salient arguments that advocate the use of data from the German laser industry to conduct this research project. Firstly, the laser industry is a small but interesting part of the German optical technology industry. Laser technology requires knowledge from various academic disciplines, such as physics, optics and electrical engineering (Fritsch and Medrano 2015). It can clearly be characterized as a science-driven industry (Grupp 2000) in which a firm's ability to innovate is a key factor in its performance and success. The interdisciplinary and science-based nature of the industry is reflected in the high level of cooperation activities between German laser source manufacturers among themselves and with laser-related public research organizations (Kudic et al. 2011). Secondly, the economic potential of the industry is meanwhile well recognized by national and supra-national political authorities. Over the past few decades Germany has developed into a world market leader in many fields of laser technology. For instance, Mayer (2004) reports that 40 % of all laser beam sources purchased worldwide in 2003 were produced by German laser source manufacturers. The world market share for laser sources used in laser processing systems was even higher and amounted to 50 % that same year. Moreover, the potential of lasers to act as enablers of innovation in other industries has since become well recognized. Thirdly, the German laser industry provides quite a unique case due to the parallel technological development lines in East and West Germany before reunification in 1990. As we will show in more detail later, the influences of both lines of development are still clearly identifiable. This, however, qualifies the industry as being a relatively interesting candidate from a theoretical point of view. Last but not least, the majority of contemporary network studies on knowledge and innovation focus on the biotech industry (Fornahl et al. 2011; Owen-Smith and Powell 2004) and the semiconductor industry (Podolny et al. 1996; Stuart 2000). Findings, however, can diverge significantly due to inter-industry differences in terms of the industries' technological maturity, firm size distribution or industry life cycle stages. Thus, the quantitative analysis of interorganizational networks in other industries is clearly underrepresented[10] but urgently needed to check and validate previous empirical results.

---

[10] To the best of our knowledge there is only one study that explicitly analyzes interorganizational networks in the laser industry (Shimizu and Hirao 2009).

References

Albrecht H (1997) Eine vergleichende Studie zur Frühgeschichte von Laserforschung und Lasertechnik in der Bundesrepublik und der Deutschen Demokratischen Republik. Habilitation: Universität Stuttgart, Stuttgart

Albrecht H (2010a) The German Research Foundation and the early days of laser research at West German universities during the 1960s. In: Trischler H, Walker M (eds) Physics and politics – research and research support in twentieth century Germany in international perspective. Franz Steiner Verlag, Stuttgart, pp 161–196

Albrecht H (2010b) Innovationen im Zeichen von Planwirtschaft und SED-Diktatur – Die Anfänge der Entwicklung der Laser-Technologie in Jena in den 1969er Jahren. In: Dicke K, Cantner U, Ruffert M (eds) Die Rolle der Universität in Wirtschaft und Gesellschaft. IKS Garamond, Jena, pp 171–201

Basov NG, Prokhorov AM (1950) About possible methods for obtaining active molecules (English translation). Zh Eksp Teor Fiz 28(2):249–250

Basov NG, Danilychev VA, Popov YM, Khodkevich DD (1970) Laser operating in the vacuum region of the spectrum by excitation of liquid xenon with an electron beam (english translation). ZhETF Pis Red 12(10):473–474

Bertolotti M (2005) The history of the laser. Institute of Physics Publishing, Bristol

Bloch I, Haensch T, Esslinger T (1999) Atom laser with a cw output coupler. Phys Rev Lett 82 (15):3008–3011

Bridges WB (1964) Laser oscillation in single ionized argon in the visible spectrum. Appl Phys Lett 4(7):128–130

Buenstorf G (2007) Evolution on the shoulders of giants: entrepreneurship and firm survival in the German laser industry. Rev Ind Organ 30(3):179–202

Carson C (2000) The origins of the quantum theory. Beam Line (Stanf Linear Accel Center) 30 (2):6–19

Dingle R, Wiegemann W, Henry CH (1974) Quantum states of confined carriers in very thin AlxGa1-xAs-GaAs-AlxGa1-x as heterostructures. Phys Rev Lett 33(14):827–830

Einstein A (1905) On a heuristic viewpoint concerning the production and transformation of light (english translation). Ann Phys 17:132–148

Einstein A (1917) On the quantum mechanics of radiation (english translation). Physikalische Zeitschrift 18:121–128

Elias LR, Fairbank WM, Madey JM, Schwettman AH, Smith TI (1976) Observation of stimulated emission of radiation by relativistic electrons in a spatially periodic transverse magnetic field. Phys Rev Lett 36(13):717–720

Faist J, Capasso F, Sivco DL, Sirtori C, Hutchinson AL, Cho AY (1994) Quantum cascade laser. Science 264(5158):553–556

Fang AW, Park H, Jones R, Cohen O, Paniccia MJ, Bowers JE (2006) A continuous-wave hybrid AlGaInAs–silicon evanescent laser. IEEE Photon Technol Lett 18(10):1143–1145

Fischer EP (2010) Laser – Eine deutsche Erfolgsgeschichte von Einstein bis heute. Siedler Verlag, München

Fornahl D, Broeckel T, Boschma R (2011) What drives patent performance of German biotech firms? The impact of R&D subsidies, knowledge networks and their location. Pap Reg Sci 90 (2):395–418

Fritsch M, Medrano LF (2015) New technology in the region – agglomeration and absorptive capacity effects on laser technology research in West Germany, 1960–2005. Econ Innov New Technol 24 (forthcoming)

Geusic JE, Marcos HM, Van Uitert LG (1964) Laser oscillations in Nd-doped yttrium aluminium, yttrium gallium, and gadolinium garnet. Appl Phys Lett 4(10):182–184

Gordon JP, Zeiger HJ, Townes CH (1954) Molecular microwave oscillator and new hyperfine structure in the microwave spectrum of NH3. Phys Rev 95(1):282–284

Gordon JP, Zeiger HJ, Townes CH (1955) The maser – new type of microwave amplifier, frequency standard, and spectrometer. Phys Rev 99(4):1264–1274

Gould GR (1959) The laser: light amplification by stimulated emission of radiation. In: Ann Arbor conference on optical pumping, conference proceeding, 15–18 June 1959, pp 128–130

Grupp H (2000) Learning in a science driven market: the case of lasers. Ind Corp Chang 9 (1):143–172

Hargrove LE, Fork RL, Pollack MA (1964) Locking of He-Ne laser modes induced by synchronous intracavity modulation. Appl Phys Lett 5(4):4–5

Hecht J (2005) Beam – the race to make the laser. Oxford University Press, New York

Hecht J (2010) The first half-century of laser development – how a solution that once was looking for a problem has become part of everyday life. Laser Technik J 7(4):20–25

Hellwarth RW, McClung FJ (1962) Giant pulsations from ruby. J Appl Phys 33(3):838–841

Hellwarth RW, McClung FF (1963) Characteristics of giant optical pulsations from ruby. Proc IEEE 51(1):46–53

Javan A, Bennett WR, Herriott DR (1961) Population inversion and continuous optical maser oscillation in a gas discharge containing a He-Ne mixture. Phys Rev Lett 6(3):106–110

Kao CK, Hockham GA (1966) Dielectric-fibre surface waveguides for optical frequencies. IEE Proc 113(7):1151–1159

Kasper JV, Pimentel GC (1965) HCl chemical laser. Phys Rev Lett 14(10):352–354

Kazarinov RF, Suris RA (1971) Possibility of amplification of electromagnetic waves in a semiconductor with a superlattice (english translation). Sov Phys Semicond 5:707–709

Ketterle W, Misner H-J (1997) Coherence properties of Bose-Einstein condensates and atom lasers. Phys Rev A 56(4):3291–3293

Klein J, Kafka JD (2010) The Ti:Sapphire laser: the flexible research tool. Nat Photonics 4 (5):288–289

Kudic M, Guhr K, Bullmer I, Guenther J (2011) Kooperationsintensität und Kooperationsförderung in der deutschen Laserindustrie. Wirtschaft im Wandel 17(3):121–129

Maiman TH (1960) Stimulated optical radiation in ruby. Nature 187(4736):493–494

Mayer A (2004) Laser in der Materialbearbeitung – Eine Marktübersicht. Laser Technik J 1 (1):9–12

Mayer A (2006) Laser materials processing systems in 2005 – the world market reaches record volume. Laser Technik J 3(1):10–11

Mears RJ, Reekie L, Poole SB, Payne DN (1986) Low-threshold tunable CW and Q-switched fibre laser operating at 1.55 μm. Electron Lett 22(3):159–160

Mueller A, Faist J (2010) The quantum cascade laser: ready for take-off. Nat Photonics 4 (5):290–291

Nakamura S, Senoh M, Nagahama S-I, Iwasa N, Yamada T, Matsushita T et al (1996) InGaN-based multi-quantum-well-structure laser diodes. Jpn J Appl Phys 35(1b):74–76

Nature (2010) Technology focus – laser anniversary. Nat Photonics 4(5):278–295

Owen-Smith J, Powell WW (2004) Knowledge networks as channels and conduits: the effects of spillovers in the Boston biotechnology community. Organ Sci 15(1):5–21

Paniccia M, Krutul V, Jones R, Cohen O, Bowers J, Fang A et al. (2006) A hybrid silicon laser – silicon photonics technology for future tera-scale computing. Intel White Paper, pp 1–6

Patel KC (1964) Selective excitation through vibrational energy transfer and optical maser action in N2-CO2. Phys Rev Lett 13(21):617–619

Phillips AC (2003) Introduction to quantum mechanics. Wiley, Sussex

Podolny JM, Stuart TE, Hannan MT (1996) Networks, knowledge, and niches: competition in the worldwide semiconductor industry, 1984–1991. Am J Sociol 102(3):659–689

Poprawe R (2010) Part 1 – coherent light: an invention that once was searching for its applications became a versatile enabling tool. Laser Technik J 7(2):31–36

Prokhorov AM (1964) Quantum electronics. Nobel Lecture, pp 110–116

Schawlow AL, Townes CH (1958) Infrared and optical masers. Phys Rev 112(6):1940–1949

Schramm M (2005) Präzision als Leitbild? Carl Zeiss und die deutsche Innovationskultur in Ost und West 1945–1990. Technikgeschichte 72(1):35–49

Schwartz M (2013) A control group study of incubators' impact to promote firm survival. J Technol Transf 38(3):302–331

Shimizu H, Hirao T (2009) Inter-organizational collaborative research networks in semiconductor laser 1975–1994. Soc Sci J 46(2):233–251

Slusher RE (1999) Laser technology. Rev Modern Phys 71(2):471–479

Snitzer E (1961) Optical maser action of Nd+3 in barium crown glass. Phys Rev Lett 7(3):444–446

Sorokin PP, Lankard JR (1966) Stimulated emission observed from an organic dye, chloro-aluminum phtatocyanine. IBM J Res Dev 10(2):162–163

Sorokin PP, Stevenson MJ (1960) Stimulated infrared emission from trivalent uranium. Phys Rev Lett 5(12):557–559

Sorokin PP, Stevenson MJ (1961) Solid-state optical maser divalent samarium in calcium fluoride. IBM J Res Dev 5(1):56–58

Stuart TE (2000) Interorganizational alliances and the performance of firms: a study of growth and innovational rates in a high-technology industry. Strateg Manag J 21(8):791–811

Townes CH (1964) Production of coherent radiation by atoms and molecules. Nobel Lecture 12 (11):58–86

Townes CH (1999) How the laser happened – adventures of a scientist. Oxford University Press, New York

TSB (2010) Laser technology report – Berlin Brandenburg. TSB Innovationsagentur GmbH, Berlin

Yu N, Kats M, Pflügl C, Geiser M, Belkin MA, Capasso F et al (2009) Multi-beam multi-wavelength semiconductor lasers. Appl Phys Lett 95(16):1–3

Chapter 4
Methodological Reflections and Data Sources

> *You can use all the quantitative data you can get, but you still have to distrust it and use your own intelligence and judgment.*
>
> (Alvin Toffler).

Abstract In this chapter, we will start with some general methodological considerations about the design of longitudinal network databases. Then, we will look at the data sources and data collection methods that are required to construct the empirical basis of the subsequent investigations. Chapter 4 is divided into two sections: Section 4.1 focuses on methodological issues related to the construction of network datasets. Section 4.2 provides a description of the raw data sources and data collection methods that were used to construct a unique longitudinal laser industry database.

4.1 Initial Methodological Considerations

Each empirical network research project requires some fundamental a-priori considerations. Basically two types of variables can be included in a network dataset: structural or relational variables that are measured for pairs of actors, and composition or actor-attribute variables that are measured for each individual actor (Wasserman and Faust 1994, p. 29). Both types of variables can be calculated at different analytical levels. Network boundaries have to be defined at the very start. This requires an exact specification of the nodes and the ties of the network under investigation. The content and aim of the research project determine the very nature of the network. As already stated, the German laser industry provides a rich opportunity to study collective innovation processes.

The aim of the study is to analyze the innovation consequences of knowledge-transfer and interorganizational learning processes in the German laser industry innovation network by focusing on knowledge-related R&D linkages among a well-defined set of laser-related organizations. We concentrate below on laser source manufacturers (LSMs) and laser-related public research organizations

© Springer International Publishing Switzerland 2015

M. Kudic, *Innovation Networks in the German Laser Industry*,
Economic Complexity and Evolution, DOI 10.1007/978-3-319-07935-6_4

(PROs). The value chain discussion above (cf. Sect. 3.3) provides good reasons to assume that a substantial proportion of the innovation activities in the German laser industry occur within these two sets of actors and at the intersection of both actor sets. Other firms positioned at upstream and downstream stages on the industry value chain are of secondary importance for this study and were deliberately excluded from the network analysis in the strict sense. This concretization concurrently highlights three fundamental issues: **(I)** network boundary specification **(II)**, identification of all relevant laser-related organizations **(III)** and concretization of knowledge-related linkages.

To start with, we outline the boundary specification concept proposed by Laumann and colleagues (1989) which provides the following three distinct boundary specification strategies: positional, relational and event-based.[1] The positional strategy draws upon actor-specific attributes such as firm size, firm age, sector or industry affiliation, stock exchange listing and many other things to set the boundaries of the network. Knoke and Yang (2008, p. 16) point to the fact that membership lists are often outdated prompting the need to conduct one's own census to compile a complete membership list. The relational strategy is based on the assumption that a subset of all relevant network actors is initially known. These actors provide some kind of relational information that allows additional network actors to be identified and included. The relational approach comprises the following procedures according to Knoke and Yang (2008, p. 17): reputational method, snowball-sampling, fixed list sampling, expanding selection and the k-core method. The event-based strategy includes actors who participate in a previously defined set of activities that occur at a specific time and place (Knoke and Yang 2008, p. 20; Marsden 2005, pp. 9–10).

In this study a combination of positional and relational boundary specification procedures was employed to identify all LSMs and PROs throughout the entire observation period. We started with a positional strategy to identify all relevant laser firms using "laser industry affiliation" as an inclusion criterion. More precisely, all firms that were actively involved in the development, design and production of laser beam sources for at least one year between 1990 and 2010 were included in the sample. Since we employ a dynamic approach we had to identify a full set of LSMs for each year under observation. In a second step we applied the expanding selection procedure originally proposed by Doreian and Woodard (1992) to identify all laser-related public research organizations (PROs). The identification procedure starts with a "fixed list" (in our case the annually complied LSM lists) and adds all PROs that are linked to LSMs on our initial list to create an "extended list". In contrast to a simple snowball sampling method, Doreian and Woodard (1992) proposed including only those objects in the sample with several linkages to actors on the initial list (Marsden 2005, p. 10). This expanding selection method has both advantages and some notable limitations. For instance, the procedure ignores

---

[1] The discussion of these three strategies is guided by Marsden (2005, pp. 9–10) and Knoke and Yang (2008, pp. 16–18).

all PROs that were actively operating in the field of laser research but had no cooperation linkages to any LSMs between 1990 and 2010. Consequently we applied a complementary method based on bibliometric data to complete the PRO lists (cf. Sect. 4.2.3).

Finally, we had to concretize the types of linkages we looked at. Knowledge-related linkages can be both informal as well as formal in nature. According to Pyka (1997, p. 210) the former case includes "[. . .] any action that can contribute to disclosure, dissemination, transmission and communication of knowledge." The latter case encompasses a broad variety of structural forms ranging from short term contractual alliances and minority alliances characterized by an intermediate degree of hieratical control to long-term oriented equity alliances such as joint ventures (Gulati and Singh 1998). Common to all formalized partnerships is that all parties involved have to agree upon more or less formalized obligations, rights and common goals. Firms tend to announce the initialization of these partnerships in the press, through newsletters, on websites or through other communication channels making them much easier to identify than informal partnerships.

In this study we focus on one particular type of formal knowledge-related linkage i.e. publicly funded R&D cooperation projects. These partnerships are very well documented by official funding authorities. Other researchers have provided solid theoretical as well as methodological arguments for the use of nationally funded R&D cooperation project data (cf. Broekel and Graf 2011, p. 6; Fornahl et al. 2011) and supra-nationally funded R&D cooperation project data (cf. Scherngell and Barber 2009, 2011) for the construction of knowledge-related innovation networks. This will be discussed in more detail later (cf. Sect. 4.2.3).

4.2   Data Sources and Data Collection Methods

Based on the considerations above, a longitudinal database for the German laser industry was compiled that covers the time period between 1990 and 2010.[2]

Proprietary data as well as information sources that were free and subject to fees were tapped to stock the four main elements of the database (i.e. industry data, firm data, network data and innovation data). The database was used for the construction of two longitudinal datasets – (**I**) an event history dataset, (**II**) a panel dataset. Figure 4.1 illustrates the raw data sourced, the overall database structure, the two datasets that were employed to conduct the descriptive analysis (cf. Part III) and the econometric estimation in the main part of this study (cf. Part IV).

---

[2] Methodological and technical support for data processing was provided by the IWH department "Formal Methods and Databases" and data collection was supported by student assistants.

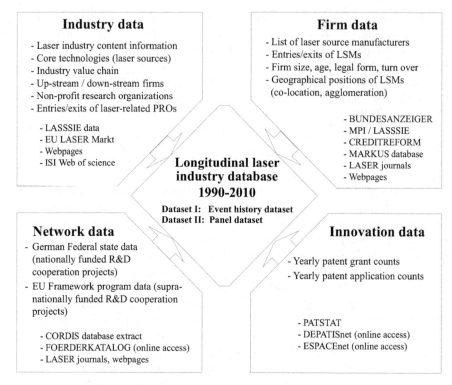

Fig. 4.1 Longitudinal laser industry database (Source: Author's own illustration)

4.2.1   Industry Data

Initial industry data[3] came from a proprietary dataset containing detailed information on firm entries and exits for the entire population of German laser source
manufacturers[4] between 1969 and 2005 (Buenstorf 2007). This dataset was originally designed to analyze industry dynamics in the German laser industry.
Buenstorf (2007, p. 186) points to the fact that "[...] studying the evolution of
industries requires the construction of original data [...]" because "[...] industry
classifications in official statistics are generally too broad [...]" and in-depth "[...]
firm-level information is not normally disclosed for the complete firm population
because of privacy considerations [...]". The same argument basically applies and
even increases in significance when studying the evolution of networks. Industrial

---

[3] This dataset was originally compiled by Guido Buenstorf, Max Planck Institute for Economics
in Jena.

[4] Corporations were declared to be laser source manufacturers based on their actual business
activities as reported in laser industry business registers and frequently published laser exhibition
catalogs.

sector classifications, like the NACE, SIC or the German WZ classification, group firms into coarse-meshed categories based on historically rooted industry development patterns. In short, laser source manufacturers cannot be clearly separated on the basis of these industry classification schemes. As outlined above, specifying network boundaries in a longitudinal research setting requires a detailed understanding of the configuration of the industry on an annual basis. Official classification schemes, however, are updated every 4–5 years at best.

The initial industry dataset has been modified in several ways to meet the requirements of an in-depth analysis of an innovation network in the German laser industry. Additional data sources were employed to gather supplementary information about firm entries and exits after 2005. In the first instance, we were given access to updated German laser industry data, again provided by Guido Buenstorf. Secondly, we used annually published laser industry business directories (i.e. *"Europäischer Laser Markt"*) provided by the *B-Quadrat* Publishing Company. This data source provided valuable information on the business activities of laser-related firms. Thirdly, data from Germany's official company register (i.e. *"Bundesanzeiger"*) and two additional data sources i.e. *MARKUS* database,[5] provided by Bureau van Dijk Publishing and the *Creditreform* archival database, provided by the Creditreform Company [6] were tapped to supplement our extended database in the contemporary part of the observation period. These data sources allowed us to complete industry data for the entire observation period.

For the purpose of this study, we selected the firm or business unit level. The reasons for this are straightforward: information on both R&D cooperation projects and patent applications or grants is commonly reported at the firm or business unit level. Thus, corporate level entities were decomposed and broken down into the business functions or market segments they serve. This allowed us to identify organizational entities within large corporations that were primarily concerned with the design, development or production of laser beam sources. Several archive data sources were evaluated to gather historical information and missing data in order to reconstruct complete life histories for all of the firms in the sample.[7] First, we included predecessors of currently existing firms in our sample. All changes to firm names and legal status over time were taken into consideration in order to ensure the full traceability of a firm's origin and development path. Firm exits due

---

[5] The *MARKUS* database contains information on 1.4 million officially registered companies in Germany, Austria and Luxembourg. Data on insolvent companies is usually excluded from this database. Data access was provided by the IWH department "Formal Methods and Databases".

[6] The Creditreform Company stores firm data on insolvent companies in an archive database.

[7] Quarterly published laser industry journals (i.e. "Laser & Optoelektronik", "LaserOpto" and "Photonik") provided by the AT Publishing Company, industry brochures (i.e. "Laser für die Materialbearbeitung" and "World of Laser") provided by the VDMA, freely accessible Internet wayback machines (http://www.archive.org/web/web.php, accessed: November 2011) and firm web pages were systematically screened and evaluated. These historical information sources provided access to in-depth industry and firm information over the entire observation period between 1990 and 2010.

to insolvencies, mergers or acquisitions, and several modes of population entries like, for instance, new company formations, spin-offs from existing firms or public research organizations were treated separately. We ended up with an industry dataset encompassing 233 laser source manufacturers over the entire observation period between 1990 and 2010.

Public research organizations (PROs) constitute the second set of organizations in the laser industry database. Two complementary methods were applied to obtain a complete list of all PROs that were actively operating in the field of laser research during the investigation period. We started with the "expanding selection method" according to Doreian and Woodard (1992), introduced in Sect. 4.1. Taking the initial list of 233 laser source manufacturers we screened our cooperation database and marked all laser-related research entities as long as these organizations established a link to at least one firm on our initial list. For each of these cases we checked whether the identified research entity was active in the field of laser research or not. Departments, research units or chairs were allocated to respective superordinate research institutes or universities. An extended membership list was created containing the full set of all identified PROs. Following the suggestion of Doreian and Woodard (1992) we adjusted this list by checking all PROs that were observed only once over the entire observation period. Next we excluded all non laser-related PROs from the list. All in all, at the end of the procedure 138 laser-related public research organizations remained in the sample.

As stated above, the expanding selection method is limited insofar as it completely ignores non-cooperating laser-related PROs. Thus, we applied a second methodological approach to solve this problem and complement our sample. Based on a bibliometric analysis we identified all German public research organizations which published laser papers, conference proceedings or articles in academic journals over the past two decades. Raw data for this analysis was provided by the LASSSIE project consortium (Albrecht et al. 2011). Data originally came from the INSPEC database.[8] This initial raw data source was taken and supplemented by an in-depth search for laser-related publications listed in the ISI Web of Knowledge database.[9] This enabled us to generate a comprehensive list of all PROs which have published at least one paper in the field of laser research. By comparing and consolidating the results of the expanding selection method and the bibliometric analysis we ended up with a final list of 145 laser-related PROs for the time span between 1990 and 2010. Finally, entry and exit dates and address data were retrieved for all PROs identified in the dataset.

---

[8] The INSPEC database contains over 11 million abstracts. The database includes journal articles, conference proceedings, technical reports and literature in the fields of physics, electronics and computing. For further details, see http://www.ovid.com/site/catalog/DataBase/107.jsp (Accessed: November 2011).

[9] The IWH library provides access to the ISI Web of Knowledge archive to the following extent: SCI 1995–2011, SSCI 1980–2011, AHCI 1995–2011. For detailed information on the database packages, their scope and contents see http://www.wokinfo.com (Accessed: September 2011).

In order to complement the picture of the German laser industry, a third group of laser related corporations – laser source providers (LSPs) – were included in our database. Raw data came from the LASSSIE project consortium (Albrecht et al. 2011). LSP data was made available on a higher regional aggregation level (i.e. "planning region") and in an anonymized form. We used information on LSP counts per year and per planning region to calculate some basic spatial measurements. More precisely, the data was used to calculate a geographical concentration index and to explore some basic descriptive industry change patterns at the overall industry level (cf. Sect. 7.1.1).

4.2.2   Firm Data

We gathered firm-level data for the entire population of 233 laser source manufacturers based on the same raw data sources that were used at the industry level. Data from Germany's official company register (i.e. "Bundesanzeiger") was used to reconstruct the firms' current addresses and address changes for the entire observation period. In addition we gathered information on the firms' legal status and changes in legal status. By drawing upon the ZIP code information we employed the ESRI ArcMap 10.0 Software package and a freely accessible geo-coding application[10] to gather GPS coordinates (latitudes & longitudes) on an annual basis for each firm in the sample.

Firm-level information on currently existing firms in our sample came from the MARKUS database. A typical company report provides a short company profile, some basic firm information (i.e. registration code, address data, founding date, ownership structure and management team etc.), a set of general financial figures (i.e. equity capital, market capitalization) and a set of time-variant indicators that are usually reported on an annual basis (i.e. number of employees, revenue etc.). Despite its high coverage, the MARKUS database also has some drawbacks. Most notably, companies that have closed down are usually removed from the database after a certain time. Consequently, additional raw data sources were needed to complete the missing data in our sample. The *Creditreform* archival database[11] was tapped to supplement missing data on all insolvent firms in our sample. All in all, we ordered data that was subject to fees on about 110 insolvent firms. As a result of an in-depth search[12] the majority of firm-level information was reconstructed. The response rate was 85%. Additionally, we had a relatively high coverage for time-variant variables throughout the lifespan of the firms. Finally we used the archive

---

[10] http://www.netzwelt.de/software/google-maps.html (Accessed: November 2011).

[11] The Creditreform Company supplies Bureau van Dijk Publishing Company – the provider of the MARKUS database – with current business data and has an extensive inventory and historical company data.

[12] I would like to thank Markus Bachmeyer from the Creditreform Company for his support.

data material described in the previous section to fill the remaining gaps in our firm-level dataset. For example, we evaluated our stock of historical laser industry journals that have been published quarterly and which contain company reports, background stories and interviews with founders or managers to extract some additional information. Moreover, the firms' actual web pages as well as expired web pages, accessed by using Internet wayback machines, provided us with rich information on the firms' histories, the milestones they achieved and the development paths they took.

4.2.3 Network Data

Up until now several generic approaches for collecting network data have been proposed such as survey methods, questionnaires, observations as a part of extended fieldwork and archive data methods (cf. Marsden 2005, pp. 10–21; Knoke and Yang 2008, pp. 21–32). For the purpose of this study we employed archival data sources to construct interorganizational networks. In general, archival data can be obtained from a broad range of archival records or documents such as journal articles, newspapers, patent citations, minutes from executive meetings, web pages, court records, annual reports and many other sources (Wasserman and Faust 1994, p. 50; Marsden 2005, p. 24). The use of archival data is especially suitable for the compilation of longitudinal network datasets. Archival records are "[. . .] relatively inexpensive, pose no burden on informant time and efforts, and may contain high-quality longitudinal information when data are maintained over time" (Knoke and Yang 2008, p. 28). However, Marsden (2005, pp. 24–25) points to the following issues that should be taken into consideration when using archive data. Firstly, archival network data should correspond to the conceptual ties of the research interest to avoid network misspecifications. Secondly, to ensure the validity of archive data, both conditions under which objects come to be included, as well as the conditions under which archives are constructed, should be carefully noted. Finally, when using electronically available archive data, technical problems, such as unexpected name changes potentially lead to errors that can be easily overlooked.

The use of archive data, which is collected, stored and issued by official authorities, mitigates at least some of these concerns. In this context, Knoke and Yang (2008, p. 30) point to the fact that "[. . .] government and economic organizations have been assembling massive amounts of information [. . .]" and that these archive data sources are still "[. . .] largely overlooked and scarcely tapped by organizational theorists". In this study we use data on nationally and supra-nationally funded R&D cooperation projects documented by the funding authorities. In doing so we draw upon two electronically available archive data sources: the *Foerderkatalog* database, provided by the German Federal Ministry of Education and Research (BMBF) and the *CORDIS* database, provided by the European Community Research and Development Information Service (*CORDIS*).

We are not the first to use these archive data sources to construct knowledge-related innovation networks (cf. Broekel and Graf 2011, p. 6; Fornahl et al. 2011; Scherngell and Barber 2009, 2011; Cassi et al. 2008). There are solid arguments that advocate for the use of these archive data sources for constructing innovation networks. Organizations that participate in R&D cooperation projects subsidized by the German federal state have to agree upon a number of regulations that facilitate mutual knowledge exchange and provide incentives to innovate (Broekel and Graf 2011, p. 6). In a similar vein, the EU has funded thousands of collaborative R&D projects in order to support transnational cooperation activities, increase mobility, strengthen the scientific and technological bases of industries and foster international competitiveness (Scherngell and Barber 2009, p. 534).

We chose a modular approach for compiling our network database. Each data source is assigned to an individual database segment. This design ensures expandability and enables additional types of cooperation, such as strategic alliances, to be included in our database in the future.

In the first instance, we exploited the *Foerderkatalog* database[13] to fill the first module. This raw data source encompasses information on a total of more than 110,000 completed or ongoing subsidized research projects and provides detailed information on the starting point, duration, funding, project description and some additional information on the project partners involved. Each registered project is equipped with a funding identification number and information on the department responsible in the subsidizing ministry. All in all, the publicly funded research projects that are listed in the *Foerderkatalog* database came from five German federal ministries.[14] We used the following data gathering procedure to extract the data needed. In a first step, we took the complete list of 233 laser source manufacturers and systematically searched for each company name in the *Foerderkatalog* database. The search mask offers two search options for the identification of organizational entities: *"Zuwendungsempfänger"* (i.e. grant recipient) & *"Ausführende Stelle"* (i.e. executing body). We chose the latter search field as we were interested in identifying the entities that were actually involved in the projects. To deal with spelling issues in the database search process we prepared a list containing various ways of spelling each firm's name. In order to separate collaborative projects we considered only those projects which were labeled as *"Verbundprojekte"* or *"Verbundvorhaben"*. This is in line with the data gathering procedure applied by Fornahl et al. (2011, p. 403). We ended up with a complete list

---

[13] http://foerderportal.bund.de/foekat/jsp/StartAction.do (Accessed: May–September 2011).

[14] These are the following ministries according to the database description available online. Federal Ministry of Education and Research (BMBF), Federal Ministry for the Environment, Nature Conservation and Nuclear Safety (BMU), Federal Ministry of Economics and Technology (BMWi), Federal Ministry of Food, Agriculture and Consumer Protection (BMELV), Federal Ministry of Transport, Building and Urban Development (BMVBS). For further information on the use of the *"Foerderkatalog"* data in the field of innovation research see Fornahl et al. (2011) and Broekel and Graf (2011). They analyze drivers of patent performance in the German biotech industry by using the same raw database that was used in this study.

of all publicly funded R&D cooperation projects for each of the 233 firms over the entire observation period.

The inherent problem with these cooperation project extracts was that they did not incorporate a listing of the cooperation partners involved. Thus, we had to employ the following search procedure in order to obtain a full list of cooperation partners at the project level. We made use of a special database particularity to accomplish this task. Cooperation projects are divided into subprojects ("Teilprojekte", "Teilvorhaben") which are executed by at least one of the project partners involved. The project designation follows a general scheme in a sense that certain parts of the "project title" of related subprojects are always exactly identical. Thus, we used the search field "Thema" (i.e. "project title") that is offered by the Foerderkatalog online interface and searched for specific parts of the project designation or project acronyms in order to identify all related subprojects. Following this procedure we ended up with a record for each of the 233 LSMs that contained a complete listing of R&D cooperation projects. This allowed us to amass detailed project level information on the subproject structure and all project partners involved. In total, we were able to identify 416 R&D projects with up to 33 project partners from various industry sectors, non-profit research organizations and universities.

Next we exploited the CORDIS database[15] to fill the second module of our network database. This CORDIS database encompasses all seven EU Framework Programs and covers a time span from 1983 to 2010. We only used the online interface of the database for consistency checks and not for data gathering. Instead we were provided with a database extract which includes a complete collection of R&D projects for all German companies which were funded by the European Commission. This database extract compromised all seven EU Framework program initiatives and covered a time span from 1983 to 2010.[16] In total, this database extract consisted of a project dataset with over 31,000 project files and an organization dataset with over 57,100 German organizations and roughly 194,000 international project partners. Each project data file is identified by a project ID and a record ID and provides information on project content, prime and secondary contractors, starting and ending date of the project, total costs and total funding, the framework program and some further information. Data files in the organization dataset are identified by an organization ID and a project ID and provide detailed information on an organization's address, ZIP code, country code, project status and the contact information of the project coordinators. The project ID allowed us to link these two datasets in order to identify a complete list of R&D projects for the

---

[15] The Community Research and Development Information Service provides a broad range of information and resources on European R&D funding activities. Data on publicly funded R&D cooperation projects can be accessed by tapping the following online interface: http://cordis.europa.eu/search/index.cfm?fuseaction=search.advanced (Accessed: May 2012).

[16] Additional programs and funding initiatives are included such as "Education & Training", "Energy & the Environment", "Health & Safety", "Information Society", "International Cooperation", "EURATOM Framework Program" as well as some regional programs.

entire sample of 233 LSMs. Based on this raw data, we identified 154 R&D projects with up to 53 project partners for the entire observation period.

Using information about publicly funded research projects to construct R&D networks potentially raises selectivity concerns.[17] In this case, empirical findings that higher innovativeness is related to larger networks might simply be caused by the inherent superiority of those actors who have been awarded more grants. In our case, these concerns seem to be of limited salience for the following reasons. The optical industry is regarded as one of the key technologies affecting the innovativeness and prosperity of the German economy as a whole (BMBF 2010). Prior work on German technology policy vis-à-vis the laser industry has shown that beginning in the mid-1980s, German policy makers identified lasers as a crucial technology for the future competitiveness of various German industries (Fabian 2011). As a consequence, substantial efforts were made to support the industry and funding of collaborative R&D projects was selected as a key policy instrument for this purpose. In other words, funding decisions were primarily motivated by the aim to make German actors more competitive than their international rivals; spurring on domestic competition through highly selective merit-based funding decisions appears to have been of secondary importance. Basically the same arguments apply with regard to European funding decisions. Scherngell and Barber (2009, p. 534) point out that one of the main EU Framework program objectives is to strengthen the scientific and technological bases of European industries and foster international competitiveness. The reasoning above is consistent with our data showing that not only medium and large sized firms but also a significantly high proportion of micro and small firms have received public funding for R&D cooperation (cf. Fig. 4.2). The diagram below gives an overview of funding received, either from a "*Foerderkatalog*" or "*CORDIS*" program, broken down by partner type and firm size. This is at least a first indication that funding decisions did not substantially vary across firms in our sample.

Finally, both cooperation data sources were used to construct interorganizational innovation networks on an annual basis based upon the following considerations. The decomposition of R&D cooperation projects with more than two partners requires a presumption in terms of the connectedness of the partners involved. In the simplest case, one can assume all project partners are directly linked to the project's lead partner but have no links among themselves. The structural implication of this "star-assumption" is illustrated in Fig. 4.3 (left). This assumption, however, seems quite unrealistic in the case of publicly funded R&D cooperation projects for two reasons. Firstly, project partners have to agree on three regulations that facilitate knowledge exchange: unrestricted use of the project's results, mutual cooperation to foster solution finding, and free-of-charge access to project-relevant know-how and intellectual property rights that existed before the project started (Fornahl et al. 2011, p. 403). Secondly, innovation incentives are incorporated into the project regulations in a sense that project partners that have made extraordinary contributions to an invention have to be explicitly acknowledged (ibid). In a similar

---

[17] This line of argument is taken from Kudic et al. (2011, p. 20).

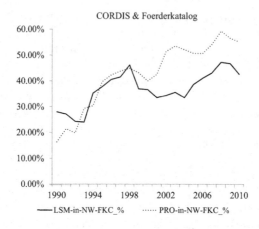

CORDIS & Foerderkatalog

	Micro LSMs	Small LSMs	Medium & large LSMs	Descriptive statistics	Max.	Min.	Avg.	Std. Dev.
2005	35.7%	35.1%	29.2%	LSM-in-NW-FKC	47.24%	24.05%	36.92%	6.90%
2000	33.8%	42.4%	23.7%	PRO-in-NW-FKC	59.31%	16.36%	42.74%	12.62%
1995	51.1%	21.1%	27.8%					

Fig. 4.2 Cooperation funding received – by partner type and firm size (Source: Authors own illustration)

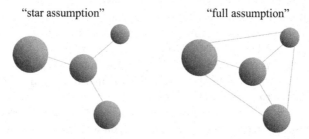

Fig. 4.3 Decomposition of R&D cooperation projects (Source: Authors own illustration)

vein, European Framework programs are explicitly designed to facilitate the circulation of information and knowledge among participating organizations and across national boundaries (Cassi et al. 2008, p. 664). These arguments, however, substantiate the assumption that all partners in nationally as well as supra-nationally funded R&D cooperation projects are mutually connected to one another. Consequently, for the purpose of this study, we stick to the "full assumption" which is illustrated in Fig. 4.3 (right).

The problem of converting multi-partner R&D cooperation projects into fully connected cliques was solved practically by programming a simple permutation tool in an MS Excel environment.[18] Networks consisting of fully connected cliques

---

[18] Support by the IWH department "Formal Methods and Databases", especially from Dr. Henry Dannenberg, is gratefully acknowledged.

are widespread and usually referred to as bipartite networks (Uzzi and Spiro 2005, p. 453). The specifics of bipartite networks are explicitly considered in the analytical part of this study.[19]

4.2.4   Innovation Data

The emergence of innovation is a complex phenomenon that is difficult to capture by a single measure (Ejermo 2009, p. 143). There is a long-standing discussion in the literature on the conceptual background of innovation measurement (cf. Smith 2005). Since then a broad range of innovation indicators have been proposed that can be grouped into the following broad categories: R&D-based measures, patent-based measures, market-based innovation count measures, bibliometric measures and survey-based measures (for an in-depth discussion see: Brenner and Broekel 2011; Smith 2005; Ejermo, 2009). A closer look at the contemporary literature on innovation measurement reveals that all of these measurements have certain advantages but also considerable disadvantages.

In order to measure the innovative performance at the firm level we decided in favor of using patent data. Despite some methodological constraints related to the use of patents to measure innovation performance (Patel and Pavitt 1995), patent indicators are commonly used in analyzing innovation processes (Jaffe 1989; Jaffe et al. 1993). The reasons are straightforward. Firstly, patents provide firms with a time-restricted monopoly on the use of their innovative products and services (Brenner and Broekel 2011, p. 12). They allow firms to protect their property rights for a certain period of time. Thus, firms have a natural incentive to secure their novel ideas against unauthorized use through patents. This argument substantiates the assumption that a notable proportion of a firm's innovative efforts are reflected in patent data. Secondly, patent data are official documents which are processed and stored by patent authorities. This ensures reliability and trustworthiness of the data. Thirdly, patent data is usually stored over long periods of time. Patent documents provide, among other things, application filing dates and publication dates which enable an exact time tracking of the event in interest. Furthermore, patent requirements have not changed remarkably over the past few years which enable patent data to be compared over time (Ejermo 2009, p. 145). Last but not least, patent data is frequently used in longitudinal settings simply because no better innovation indicators are available over long periods of time (Brenner and Broekel 2011, p. 13).

---

[19] Bipartite networks usually tend to show a higher overall connectedness than unipartite networks. We have generated annual network layers based on both the "star assumption" and the "full assumption" to examine the structural consequences of these assumptions. Subsequently, we calculated a set of overall network measures (i.e. overall density, clustering coefficient, degree centralization) on an annual basis and explored the results over time. A comparison of the results revealed nearly the same curve progressions but on a much different scale.

At the same time, patent-based indicators suffer from some notable limitations. Firstly, patents are an indicator of invention rather than innovation as they mark the emergence of a novelty but not the commercialization of the idea (Smith 2005, p. 160). In other words, the inventor's idea is not necessarily converted into a product that can be sold to a customer. Moreover, innovators have several other possibilities of appropriating the benefits of an invention which are not reflected in patent data (Fritsch and Slavtschev 2007, p. 204). For instance firms can secure their innovations by maintaining a high level of information security or by implementing other knowledge securing strategies (Liebeskind 1996).

Despite these limitations there is a widespread use of patent data for the construction of an innovation indicator at the firm level. In accordance with contemporary empirical network studies (Ahuja 2000; Stuart 2000; Whittington et al. 2009; Baum et al. 2000; Fornahl et al. 2011; Schilling and Phelps 2007), we decided to use annual patent counts as a proxy for firm innovativeness. Three patent data sources were tapped to gather the patent data needed. The European Patent Office's (EPO) database was used as the primary data source to generate a complete overview of the firms' patent activities.[20] Additionally, two patent data sources accessible online – *DEPATISnet* German Patent and Trade Mark Office database & *ESPACEnet* European Patent Office database – were tapped for data completion and to check results for integrity and consistency.

For the technical realization of the firm-specific patent search procedure we had to compile an SQL query.[21] This query was needed to tap the database and extract a comprehensive list of patents for each of the 233 LSMs in our sample. Each extracted patent datasheet included an extensive set of patent variables. We employed the following data gathering procedure to identify patent applications and patent grants which were needed for the purpose of this study. By drawing upon the initially complied list of 233 LSMs, we conducted a firm-specific search in order to identify and extract all patents which were assigned to the firms. A list of various ways to spell each firm's name was used to deal with spelling issues. In the case of micro firms (i.e. firms with less than 10 employees) we also searched for the founder's name. This allowed us to identify some additional patents that were otherwise overlooked. We used the 'like' function in the SQL query to identify the firm's name listed as at least one of the following two variables: "person_name" or "doc_std_name". Though this increased the likelihood of finding patents that would have otherwise been overlooked, it also led to the capturing of some irrelevant patent documents. Search results were entered into an MS Excel spreadsheet and processed using a two-stage approach. In the first stage, false positive results and potential double counts were identified and excluded from the datasheet. These records were removed manually from the data collection. In the second stage, patents were sorted into applications and grants. The EPO Patent database

---

[20] Data access was provided by the IWH department "Formal Methods and Databases".

[21] Support in generating the SQL query and for conducting the search by Katja Guhr, Martin Zenker and Dr. Iciar Dominguez Lacasa is gratefully acknowledged.

has a variable that indicates whether a patent was granted or not. Thus, we used the "patent first granted" flag in combination with the variable "publn_kind" to identify all granted patents.[22] Finally, the *DEPATISnet* database[23] and the *ESPACEnet* database[24] were employed to check the results. We ended up with a patent data file encompassing a comprehensive list of all patent applications and patent grants for each of the 233 LSMs over the entire observation period.

References

Ahuja G (2000) Collaboration networks, structural hole, and innovation: a longitudinal study. Adm Sci Q 45(3):425–455

Albrecht H, Buenstorf G, Fritsch M (2011) System? What system? The (co-) evolution of laser research and laser innovation in Germany since 1960. Working paper, pp 1–38

Baum JA, Calabrese T, Silverman BS (2000) Don't go it alone: alliance network composition and startup's performance in Canadian biotechnology. Strateg Manag J 21(3):267–294

BMBF (2010) Ideas, innovation, prosperity – high-tech strategy 2020 for Germany. Federal Ministry of Education and Research, Bonn

Brenner T, Broekel T (2011) Methodological issues in measuring innovation performance of spatial units. Ind Innov 18(1):7–37

Broekel T, Graf H (2011) Public research intensity and the structure of German R&D networks: a comparison of ten technologies. Econ Innov New Technol 21(4):345–372

Buenstorf G (2007) Evolution on the shoulders of giants: entrepreneurship and firm survival in the German laser industry. Rev Ind Organ 30(3):179–202

Cassi L, Corrocher N, Malerba F, Vonortas N (2008) Research networks as infrastructure for knowledge diffusion in European regions. Econ Innov New Technol 17(7):665–678

Doreian P, Woodard KL (1992) Fixed list versus snowball selection of social networks. Soc Networks 21(2):216–233

Ejermo O (2009) Regional innovation measured by patent data – does quality matter? Ind Innov 16 (2):141–165

Fabian C (2011) Technologieentwicklung im Spannungsfeld von Industrie, Wissenschaft und Staat: Zu den Anfängen des Innovationssystems der Materialbearbeitungslaser in der Bundesrepublik Deutschland 1960 bis 1997. Dissertation, TU Bergakademie Freiberg

Fornahl D, Broeckel T, Boschma R (2011) What drives patent performance of German biotech firms? The impact of R&D subsidies, knowledge networks and their location. Pap Reg Sci 90 (2):395–418

Fritsch M, Slavtschev V (2007) Universities and innovation in space. Ind Innov 14(2):201–218

Gulati R, Singh H (1998) The architecture of cooperation: managing coordination costs and appropriation concerns in strategic alliances. Adm Sci Q 43(4):781–814

Jaffe AB (1989) Real effects of academic research. Am Econ Rev 79(5):957–970

Jaffe AB, Trajtenberg M, Henderson R (1993) Geographic localization of knowledge spillovers as evidenced by patent citations. Q J Econ 108(3):577–598

---

[22] National and international patent classification schemes provided by the German Patent Office were used for the data classification: http://www.dpma.de/ponline/ipia/03_DPMA informativ_IPIA.pdf (Accessed: June–August 2011).

[23] Online access: www.depatisnet.de (Accessed: June–August 2011).

[24] Online access:www.espacenet.com (Accessed: June–August 2011).

Knoke D, Yang S (2008) Social network analysis. Sage, London

Kudic M, Buenstorf G, Guhr K (2011) Analyzing the relationship between cooperation events, ego-networks and firm innovativeness – empirical evidence from the German laser industry. In: Conference proceedings. The 5th international EMNet conference, Limassol, pp 1–42

Laumann EO, Marsden PV, Prensky D (1989) The boundary specification problem in network analysis. In: Freeman LC, White DR, Romney KA (eds) Research methods in social network analysis. George Mason University Press, Fairfax, pp 61–87

Liebeskind JP (1996) Knowledge, strategy and the theory of the firm. Strateg Manag J 17 (2):93–108

Marsden PV (2005) Recent developments in network measurement. In: Carrington PJ, Scott J, Wasserman S (eds) Models and methods in social network analysis. Cambridge University Press, Cambridge, pp 8–30

Patel P, Pavitt K (1995) Patterns of technological activity: their measurement and interpretation. In: Stoneman P (ed) Handbook of the economics of innovation and technological change. Blackwell, Oxford, UK, pp 14–51

Pyka A (1997) Informal networking. Technovation 17(4):207–220

Scherngell T, Barber MJ (2009) Spatial interaction modeling of cross-region R&D collaborations: empirical evidence from the 5th EU framework programme. Pap Reg Sci 88(3):531–546

Scherngell T, Barber MJ (2011) Distinct spatial characteristics of industrial and public research collaborations: evidence from the fifth EU framework programme. Ann Reg Sci 46 (2):247–266

Schilling MA, Phelps CC (2007) Interfirm collaboration networks: the impact of large-scale network structure on firm innovation. Manag Sci 53(7):1113–1126

Smith K (2005) Measuring innovation. In: Fagerberg J, Mowery DC, Nelson RR (eds) The Oxford handbook of innovation. Oxford University Press, New York, pp 148–177

Stuart TE (2000) Interorganizational alliances and the performance of firms: a study of growth and innovational rates in a high-technology industry. Strateg Manag J 21(8):791–811

Uzzi B, Spiro J (2005) Collaboration and creativity: the small world problem. Am J Sociol 111 (2):447–504

Wasserman S, Faust K (1994) Social network analysis: methods and applications. Cambridge University Press, Cambridge

Whittington KB, Owen-Smith J, Powell WW (2009) Networks, propinquity, and innovation in knowledge-intensive industries. Adm Sci Q 54(1):90–122

Chapter 5
Quantitative Concepts and Measures

Measure what is measurable, and make measurable what is not so.

(Galileo Galilei).

Abstract A broad range of concepts and measures are needed to provide a quantitative description of the industry and to analyze the initially raised research questions. Focus is on using applied methods in calculating geographical and network-related measures. Chapter 5 is divided into three sections: Section 5.1 presents some general graph theoretical concepts. Section 5.2 provides an overview of techniques and measures for the structural analysis of interorganizational networks. More precisely, we present most commonly used network measures at three analytical levels: actor level, subgroup level and overall network level. Finally, in Sect. 5.3 we outline a selection of spatial proximity and geographical concentration concepts that were applied in the analytical part of the study.

5.1 Graph-Theoretical Foundation and Basic Network Concepts

Various notional schemes, such as graph theoretical, sociometric and algebraic approaches, can be used to mathematically describe a network (Wasserman and Faust 1994, p. 69). The graph theoretical approach is suitable for defining and clarifying general network properties. From this perspective, a network is defined as a set of vertexes[1] joined by edges[2] (Newman 2010, p. 109). A graph (N, g) consists of a set of nodes $N = \{1...n\}$ and a real-valued matrix g ($= n \times n$), where g_{ij} represents the relation between the node i and the node j in the network (Jackson 2008, p. 21). The node-set consists of a finite number of nodes n which determine the size of the network. Depending on the specific type of network under

---

[1] Vertexes are also called "nodes", "actors", "agents", "players" and "entities".
[2] Edges are also called "ties", "links", "connections" and "relationships".

© Springer International Publishing Switzerland 2015
M. Kudic, *Innovation Networks in the German Laser Industry*,
Economic Complexity and Evolution, DOI 10.1007/978-3-319-07935-6_5

investigation the nodes can represent, for instance, individuals, organizations or nation-states. The matrix g is called the "adjacency matrix" and reflects, for the relation in question, all of the present or absent connections between the actors in the node-set (Wasserman and Faust 1994). Again, the type and the very nature of such a relation depends on the network under investigation and ranges, for instance, from friendship networks between individuals to strategic alliance networks between organizations and diplomatic relation networks between nation-states. Regardless of the object under investigation one can differentiate in general between weighted and unweighted networks. In unweighted networks the entities of g assume the value 1 to indicate the presence, and assume the value 0 to indicate the absence, of a connection between actors. Relations also can be weighted in order to track the intensity level of a relationship (Jackson 2008, p. 21). Furthermore, relations between network actors can be directed or undirected. A network is deemed to be directed when $g_{ij} \neq g_{ji}$; the network is considered to be undirected when $g_{ij} = g_{ji}$ (Jackson 2008, p. 21). The structure of the "adjacency matrix" is asymmetric in the first case and symmetric in the latter case. Directed graphs have higher information content than undirected graphs as they specify not only the presence or absence of a linkage but also who are the sending and who are the receiving entities in the network. Undirected graphs are used where the direction of relations does not make sense or must be, for logical reasons, always reciprocated (Borgatti et al. 2013). In summary, the graph (N, g), also known as an "adjacency matrix", provides a full description of all weighted or unweighted and directed or undirected mutual linkages for a finite number of nodes n in a node-set N.

The previous considerations provide the basis for the introduction of some fundamental network concepts. The first concept is the "dyad". A dyad D_{ij} is the most basic building block of a network which is defined as "a pair of actors and the (possible) tie(s) between them" (Wasserman and Faust 1994, p. 18). These building blocks constitute the smallest structural entity of a network. According to Wasserman and Faust (1994, p. 510) there are four states and three isomorphism[3] dyadic classes for the two actors i and j in a dyadic subgraph. These unique classes are the null dyad (i.e. $D_{ij} = \{0,0\}$, no tie between i and j), asymmetric dyad (i.e. $D_{ij} = \{1,0\}$, directed tie from i to j or $D_{ij} = \{0,1\}$, directed tie from j to i) and the mutual dyad (i.e. $D_{ij} = \{1,1\}$, undirected tie between i and j). In the case of symmetric networks only null dyads and mutual dyads need to be considered. The formation and termination of dyads have some fundamental implications for the structural features of networks in terms of size, density and fragmentation (Amburgey et al. 2008). The second concept is the triad. A triad T_{ijk} consists of a triple of actors (i, j, k), where $i \neq j \neq k$, and the (potential) tie(s) between these actors (Wasserman and Faust 1994, p. 559). A triadic subgraph has a much higher complexity level than a simple dyad. This becomes apparent when looking at the triadic states and isomorphism classes. For a triad T_{ijk}, 64 states and 16 unique

---

[3] Isomorphic means in this context that subgraphs are structurally indistinguishable from one another (Wasserman and Faust 1994, p. 560).

isomorphism classes can be identified whereas the accurate description of these classes requires a labeling scheme specifically developed for this purpose (Wasserman and Faust 1994). Holland and Leinhardt (1970, 1976) have proposed the so-called "M-A-N labeling scheme" which assigns four characters to each triad.[4] A basic understanding of triads and transitivity is important for the understanding of more complex concepts such as clustering coefficients. The third concept is the subgroup. A subgroup is defined as a subset of node set N and the (possible) linkages between them (Wasserman and Faust 1994, p. 19). There are several concepts to separate or identify subgroups in networks. The most basic subgroup concept is the component. A component is defined as a maximally connected subgraph with a path between all pairs of nodes within the subgraph but no connection to the nodes in other components (Wasserman and Faust 1994, p. 109). The components concept is of overriding importance for analyzing the fragmentation of a network. If a network contains more than one component, the network is considered disconnected. If there is only one component in a network we call this a connected network (Newman 2010, p. 142). There are other more-or-less restrictive subgroup concepts (i.e. cliques, cores and plexes etc.) which allow us to separate subsets of actors in networks based on some well-defined structural criteria (Newman 2010, pp. 193–197; Wasserman and Faust 1994, pp. 257–266). The fourth concept is the ego network concept. Ego networks consist of a focal actor – a so-called "ego" – and a set of directly connected actors – so-called "alters" – which can be connected among each other (Wasserman and Faust 1994, p. 42). This concept involves a shift in the analytical perspective from the overall network level to an actor-based perspective. Thus, the individual network structure of actors is highlighted and provides the basis for several ties as well as node-related structural measures (i.e. ego size, ego density, ego brokerage etc.). Ego network concepts are less restrictive with regard to data requirements but they also provide a limited analytical value since second tier ties and indirect structural effects are not considered. However, ego network concepts can also be applied in analyzing full network data (cf. Sect. 5.2.2). Depending on the underlying research question, these concepts can provide an additional analytical value. The last concept is the overall network or group concept.[5] In general a group can be defined as a finite set of actors that are clearly separable in a theoretical, conceptual or empirical manner and against which a certain type of relation can be measured (Wasserman and Faust 1994, p. 19). Thus, the group concept provides the graph theoretical foundation for all analytical concepts that addresses the overall network level.

---

[4] According to this scheme, the first character specifies the number of mutual dyads, the second character gives the number of asymmetric dyads, the third character displays the number of null-dyads, and the last character gives a further characterization of how the ties are directed at each other within these specific isomorphism classes by using the characters "D" (for down), "U" (for up), "T" (transitive), "C" (cyclic). For details, see Wasserman and Faust (1994, pp. 559–575).

[5] The graph theoretical terminology can be somewhat misleading in this context. Note that the term "group" refers to the overall graph. The term "subgroup" addresses subsets of actors in the overall network.

In addition to these fundamental network concepts there are three further aspects that have to be considered and clarified for the purpose of this study: contextual specification, time framing and level of analysis. Quantitative network analysis methods have been applied in a broad range of scientific fields. This analysis focuses on interorganizational innovation networks. Due to the aim of this study the actor set is restricted to a clearly defined subset of organizations – laser source manufacturing firms (LSMs) and laser-related public research organizations (PROs) in the German laser industry innovation system – and to a specific type of relationship between these actors – knowledge-related publicly funded R&D cooperation project linkages. Because of the very nature of these specific types of relations and the underlying mutual knowledge exchange and learning processes (Broekel and Graf 2011), we assume, for the purpose of this study, that there is a presence of undirected network ties. Secondly, networks can be analyzed from a static or dynamic perspective. Static means that all actor-specific attributes as well as all relational ties among these actors are measured at one given point in time or simply pooled together. In contrast, a dynamic perspective requires that the node attributes and all of the ties among these nodes are measured at repeated points in time. The crucial point in this context is that the node set of the underlying graph has to be adjusted for each time point as not only the ties but also the nodes can be subject to change over time. In other words, the entry and exit dynamics of both nodes and ties have to be measured and explicitly considered. Moreover, the identification of structural change patterns in interorganizational networks requires a sufficiently long observation period and an adequate separation of the observation period into distinct time intervals. The length of the intervals is determined by the underlying research questions and data availability issues. With regard to longitudinal research designs, Suitor et al. (1997, p. 6) point out: "Given the cost and complexity of network studies, time-series analysis will probably be the rarest form of study of network change". In this study we draw upon two longitudinal research designs to account for the change of the network over time: panel data design & event history data design. A panel data design enables a given sample to be analyzed across a discrete number of time points (Suitor et al. 1997, p. 6). Event history data provides an even more demanding type of network data due to the strict requirements regarding the exact time-tracking of relevant events. Thirdly, an in-depth analysis of network structures and embedded actors requires different analytical perspectives. According to Wasserman and Faust (1994, p. 25) there are at least three levels of analysis: actor level, subgroup level and overall network level. Quantitative network analysis methods on the actor level provide an analytical toolbox for identifying organization-specific properties such as functions, roles or positions. These micro-level measurements allow us to compare nodes with one another and to say something about how given nodes relate to the overall network structure (Jackson 2008, p. 37). The application of subgroup level methods requires the separation of a set of organizations that is homogenous with respect to specific network properties. Ego networks can be regarded as one specific type of subgroup as the focal actor's direct linkages clearly define the boundaries of the subgraph in question. The particularly interesting features of ego networks are that they can be defined for each actor in the actor set of a graph and the structural properties of these individualized subgroup networks can be directly ascribed to respective focal actors (Marsden 2005, p. 9). Finally, group level methods are required to analyze the structural features of the overall network structure. Measures such as size, density, reach, cliquishness, connectedness and many others allow us to get a fine-grained picture of structural features and properties of the network under investigation. Even though these types of measures are predominantly macro in nature (Jackson 2008, p. 37) most of these measures can also be applied at the subgroup level. The next sections provide – together with these three analytical levels – an overview and clarification of the network analysis methods which were applied for the purpose of this book.

5.2 Quantitative Network Analysis Methods

The predominant objectives of quantitative network analysis are to measure and represent the structural relations among a well-defined set of entities, and to explain both the occurrence and the consequences of structural patterns (Knoke and Yang 2008, p. 4). The analytical power of quantitative network analysis[6] methods lies, to a large extent, in the ability to analyze the connectedness of actors in a system (Wasserman and Faust 1994, p. 19). These methods provide general analytical tools which can be used for the structural analysis of complex socio-economic systems.[7] More precisely, these methods allow us to quantify overall network features along various aggregation levels and, at the same time, provide us with analytical instruments for the analysis of functions, roles and positions of actors in the network.

5.2.1 Methods for Calculating Network Positioning

In general, one can distinguish between two categories of location property measures. The first category – so-called "prestige measures" – requires directional relations whereas the second category – so-called "centrality measures" – is intended for the analysis of non-directional relations. Due to the conceptual orientation of this study, focus is placed upon the latter category of analytical methods.

---

[6] The historical roots of this concept are located in the field of sociological research. In this study the terms "social network analysis" and "quantitative network analysis" are used interchangeably.

[7] General system theory (Bertalanffy 1968) provides the general theoretical foundation for socio-economic and other systems by describing the general nature of a system by explicitly referring to system elements and some kind of relationships or forces between them.

The overall centrality concept can be defined very broadly as the extensive involvement of an actor with many other actors (Wasserman and Faust 1994, p. 173). However, since the development of the first centrality concepts (Bavelas 1948; Czepiel 1974) several remarkable methodological advances and refinements have subsequently been made (Freeman 1979). Today we can draw on a wide range of methods capturing various facets of centrality. According to Borgatti (2005) and Jackson (2008) there are basically four categories of centrality measures – degree centrality, closeness centrality, betweenness centrality, power centrality, and eigenvector centrality. What these centrality concepts share in common is that they seek to identify exposed actors within a network by referring to previously well-defined criteria.

Degree centrality is one of the first centrality concepts (Freeman 1979). The idea behind this network measure is straightforward as the concept defines actor centrality based on the connectedness to other actors in the network. In other words, degree centrality indicates how well a node is connected in terms of direct linkages (Jackson 2008, p. 38). Network actors with a high degree have greater opportunities because they have choices and this autonomy makes them less dependent on any other specific actor (Hanneman and Riddle 2005). According to Wasserman and Faust (1994, p. 179) the degree centrality C_D of a network actor n_i depends on the group size g with the maximum value $(g - 1)$ and is defined as:

$$C_D(n_i) = \frac{d(n_i)}{(g - 1)} \tag{5.1}$$

This measure ranges from 0 to 1 and can be compared across networks of different sizes (Wasserman and Faust 1994, p. 179). The measure allows for the identification of network actors which are highly connected at the micro-level and reveals, at the same time, an initial indication of highly active areas – so-called "hot spots" – in the overall networks at the macro-level. Densely connected actors are highly visible and well-recognized by others as major channels of relational information, whereas actors with a low degree centrality are clearly peripheral in the network (Wasserman and Faust 1994, p. 179). Nevertheless, the concept has some considerable drawbacks. For instance, second tier ties are not considered so that the concept fails to recognize the importance of brokers that have a low degree but who occupy a bridging position between otherwise unconnected network components (Jackson 2008, p. 38). Accordingly, Newman (2010, p. 188) illustrates this point by referring to a low-degree node which connects two subgraphs in a network and therefore occupies a position with a high betweenness centrality. In general, one can say that the degree centrality measure is very intuitive but relatively coarse measure of centrality (Borgatti et al. 2013, p. 168).

Closeness centrality is based on the distance of a particular actor to all other actors or a well-defined subgroup of actors in the graph. The concept basically goes back to Bavelas (1948), Beauchamp (1965) and Sabidussi (1966). The fundamental idea is that actors who are near many other actors can easily reach others in the network. One of the most commonly applied conceptualizations of the idea is the

Sabidussi Index[8] which measures actor closeness as a function of geodesic distances (Wasserman and Faust 1994, p. 184). The geodesic distance, also called the shortest distance, is simply defined as the shortest path between network actors (Newman 2010, p. 139). Thus, the shortest distance between network actors i and j is reflected by the distance function $d(n_i,n_j)$. Sabadussi's closeness index is defined as an inverse function of the total distance function d^T, which is the sum of all shortest distances from actor i to all other actors in a graph, whereas the sum is taken over all $j \neq i$ (Wasserman and Faust 1994, p. 184). Beauchamp (1965) proposes an actor-based closeness measure by simply multiplying Sabidussi's (1966) closeness index by $(g - 1)$ in order to get a standardized index ranging from 0 to 1 (Wasserman and Faust 1994, p. 185). Thus, according to Wasserman and Faust (1994, p. 185), the closeness centrality C_C of a network actor n_i depends on the group size g with the maximum value $(g - 1)$ as well as the total distance function d^T and is defined as:

$$C_C(n_i) = \frac{(g-1)}{\sum_{j=i}^{g} d(n_i, n_j)} = \frac{(g-1)}{d^T} \qquad (5.2)$$

Closeness measures were applied quite early to analyze the accessibility of information in communication networks (Leavitt 1951). This centrality conceptualization, however, offers further interesting application possibilities, for instance to measure the reachability of knowledge stocks dispersed among multiple actors in networks. Nevertheless, the concept has some notable limitations. According to the idea behind the geodesic distance concept, each node in a graph must be reachable by all other nodes in a graph. This however implies that a closeness centrality measure, as defined above, can only be applied for a connected graph (Wasserman and Faust 1994, p. 185).

The betweenness centrality concept provides the third network positioning measure. The logic behind the betweenness centrality concept is fairly different to the two previously discussed centrality concepts as it measures the extent to which an actor controls as many shortest paths as possible between other groups of network actors. Scholars recognized the basic idea of the concept quite early on (Bavelas 1948; Shimbel 1953) and pointed out the strategic importance of locations on geodesics for the actors themselves as well as for the characteristics of the overall network structure (Burt 1992). However, the first conceptionalization of the betweenness centrality, usually attributed to Anthonisse (1971) and Freeman (1977), quantifies how well a node is located in terms of the shortest paths it lies on (Jackson 2008, p. 39; Wasserman and Faust 1994, p. 189). The measurement of betweenness centrality requires the consideration of at least three actors i, j and k in

---

[8] Jackson (2008, p. 39) suggests that decay centrality is a richer way of measuring closeness. Instead of a simple distance function $d(n_i, n_j)$ a so-called decay parameter with $\delta^{d(n_i, n_j)}, 0 < \delta < 1$ is introduced. The specific feature of this measure is that distances get weighted by the decay parameter.

a graph whereas it is assumed the latter two actors j and k always choose the shortest path to connect one another. If more than two geodesics exist for these two actors, it is assumed that all geodesics are equally likely to be chosen (Wasserman and Faust 1994, p. 190). The number of geodesics is reflected by g_{jk} so that the probability of using one of these geodesics is $1/g_{jk}$ (Wasserman and Faust 1994, p. 190). Next, it has to be determined how many geodesics $g_{jk}(n_i)$ the actor i lies on between actors j and k (Jackson 2008, p. 39). Based on these considerations we can estimate how important the actor i is in terms of connecting j and k (Jackson 2008, p. 39) by calculating the probability $p_{jk} = g_{jk}(n_i)/g_{jk}$ as originally proposed by Freeman (1979). According to Wasserman and Faust (1994, p. 190) the betweenness centrality index for all n_i, defined as the sum of all p_{jk}, has to be standardized in order to obtain an actor centrality index ranging from 0 to 1 by using the maximum value of the index $(g-1)(g-2)/2$. This reflects the maximum possible number of pairs of actors not including n_i and leads to the actor betweenness centrality being defined as:

$$C_B(n_i) = \sum_{k \neq j} \frac{p_{jk}}{(n-1)(n-2)/2} \tag{5.3}$$

The actor betweenness measure can be calculated for connected as well as unconnected graphs and the standardization allows for comparisons across networks and relations (Wasserman and Faust 1994, p. 190). Betweenness centrality values are typically distributed over a wide range (Newman 2010, p. 189). More interesting, however, is the robustness of this measure against network changes over time. Although betweenness values may shift when the network structure changes due to node and tie entries and exits, the changes in centrality are relatively small compared to the large gaps between the leaders and other actors in terms of betweenness values. Thus the order of the centrality ranking list changes relatively infrequently (Newman 2010, p. 190). This is a quite interesting property, especially when analyzing evolutionary network change over time.

The category of power and eigenvector related measures are based on a more comprehensive understanding of centrality. These centrality concepts take two perspectives into account by measuring an actor's network position, i.e. the actor's own structural embeddedness, as well as the extent of interconnectedness of its neighbors. In other words, these measures are based on the premise that a node's importance is not simply determined by its own connectivity, closeness or brokerage position but also by its proximity to many other "important" nodes (Jackson 2008, p. 40). The term "important" refers in this context to a structurally exposed network position. One of the first conceptualizations of this idea was provided by (Katz 1953).[9] For the purpose of this study we concentrate on the eigenvector centrality measure originally proposed by Bonacich (1972, 1987). Eigenvector

---

[9] For an in-depth discussion on the Katz prestige measure, see Jackson (2008, pp. 40–41) or Newman (2010, pp. 172–175).

centrality can be regarded as a straightforward extension of the degree centrality concept that takes into account the fact that a network actor's direct neighbors are not all identical in terms of their own connectedness to other network actors (Newman 2010, p. 169):

$$C_E(n_i) = k_1^{-1} \sum_j A_{ij} n_j \qquad (5.4)$$

The eigenvector centrality C_E of a network actor n_i is proportional to the sum of centralities of its neighbors where A_{ij} represents an element of the adjacency matrix, x_j represents the neighbors' centralities and k_1 is the largest eigenvalue of the matrix A (Newman 2010, p. 170). The eigenvalue k_1 is nonnegative and constant (Jackson 2008, p. 41). An important property of the eigenvector centrality measure is, however, that an actor's centrality value can be large either because the actor itself has many direct relationships or because the actor has a few direct relationships to highly interconnected partners (Newman 2010, p. 170). The eigenvector centrality can be interpreted as a measure of popularity in a sense that an actor with a high eigenvector centrality is connected to other actors that are themselves well connected (Borgatti et al. 2013, p. 168).

5.2.2   Methods for Calculating Ego Network Characteristics

Now we focus on the measurement of firm-specific cooperation structures. This leads us to the concept of ego centered networks. An ego network consists of a focal actor called ego and a set of directly connected partners called alters, that can be connected among one another (Wasserman and Faust 1994, p. 42). Ego networks do not include second-tier ties or second-step ties to which the focal actor is not directly connected (Hite and Hesterly 2001, p. 277). The size of an ego network is determined by the ego itself and the number of partners the ego has. In other words, an ego size measure is simply defined as the number of all of ego's one-step-out neighbors, plus ego itself (Hanneman and Riddle 2005). Marsden (2002) proposed a reformulation of the classical centrality measures for ego networks. For the purpose of this study we concentrate exclusively on two structural measures: ego density and ego brokerage.

The first measure is the ego network density of a focal actor's individual cooperation portfolio. Borgatti et al. (2002) define ego network density E_D of a focal actor n_i as the number of currently existing ties T_i divided by the number of potentially possible ties, termed pairs P_i, times one hundred. The ego network density can be expressed as follows:

$$E_D(n_i) = \frac{T_i}{P_i} 100 \tag{5.5}$$

The maximum number of possible dyadic ties for a focal actor's ego network with a total number of N alter can be calculated as follows: $P_i = (N!)/(2!*(N-2)!)$. The ego network density measure can be extended and used for the structural analysis of ego networks that consist of directed relations, valued non-directed relations or valued directed relations (cf. Knoke and Yang 2008, pp. 53–54).

The second measure is ego network brokerage. Marsden (2002) demonstrated that brokerage measures for ego networks can differ substantially from complete network measures. The basic idea of this ego network brokerage concept is to capture the extent to which the focal actor is the "go-between" for other alters and establishes a linkage between otherwise unconnected nodes of the ego network (Hanneman and Riddle 2005). A straightforward quantification of the concept is provided by Marsden (2002, p. 410):

$$E_B(n_i) = \sum_{j \in A_i,\ j \neq i}^{N_i} \sum_{k \in A_i,\ k \neq i}^{j-1} \left[1 - a(p_j, p_k)\right] \tag{5.6}$$

The brokerage measure E_B for a focal actor n_i in an ego network setting based on an egocentric network A_i is defined by the number of pairs of nodes (excluding the focal actor) in the egocentric network that are not directly connected to one another and, therefore, are indirectly connected via a geodesic through the focal actor (Marsden 2002, p. 410). Up until now several refinements of this quite simple conceptualization have been proposed. For instance, the egocentric equivalent of Freeman's (1979) betweenness centrality measure takes into account that some pairs (p_j, p_k) may not only be connected via the focal actor, but also through other nodes in the egocentric network (Marsden 2002, p. 410). In a similar vein, Borgatti et al. (2002) provide us with a normalized brokerage measure. This normalized ego network measure is defined as the number of times the focal actor is located on the shortest path between two alters (i.e. the number of pairs of alters that are not directly connected) normalized by the potentially possible number of brokerage opportunities (i.e. a function of the focal actor's ego network size).[10]

5.2.3 Methods for Calculating Overall Network Measures

Now we will look at a selection of global network measures. First, we will briefly present three centralization indices. Then, we will lay the foundation for the

---

[10] For an in-depth discussion on further egocentric concepts and measures see: Marsden (2002) or Knoke and Yang (2008, pp. 53–56).

quantification of large-scale network properties by introducing clustering and average distance measures for complete networks.

In its most basic sense, network centralization can be defined as the extent to which a network is centralized around one or a few central actors (Freeman 1979). Centralization indices measure how heterogeneous the actors' centralities are in a well-defined population (Wasserman and Faust 1994, p. 176). Centralization measures provide an indicator of actor variability whereas other global network measures, such as average degree or graph density, provide quantifications of average actor tendencies (Wasserman and Faust 1994, p. 182). They draw upon actor-specific centrality measures. Freeman (1979) proposed three types of network centralization indices – closeness, degree and betweenness.

The closeness centralization index is based on the closeness centrality measure outlined above (cf. Bavelas 1948; Beauchamp 1965; Sabidussi 1966). This centralization index can be interpreted as an indication of the dispersion of closeness centralities over the entire population of actors (Knoke and Yang 2008, p. 67). Like the underlying closeness centrality measure C_C, the centralization index Z_C is only meaningfully defined for connected graphs (Wasserman and Faust 1994, pp. 184–187).[11] The numerator represents the sum of the differences between the largest closeness centrality $C_C(n^*)$ and the closeness centrality value to all other network actors $C_C(n_i)$. The denominator represents the maximum possible closeness centrality value for the graph under investigation. The index ranges from 0 to 1. Low index values indicate homogenous dispersion of closeness centrality values among the network actors and high values indicate that actor-specific centralities are unevenly dispersed:

$$Z_C = \frac{\sum_{i=1}^{g}[C_C(n^*) - C_C(n_i)]}{[(g-2)(g-1)/(2g-3)]} \tag{5.7}$$

The degree centralization index Z_D provides a measure that reflects how network actors vary in terms of their connectedness to other actors. In other words, the index draws upon the network actors' nodal degree. At the one extreme, the index equals zero if all nodes have the same degree centrality and at the other extreme the index reaches its maximum when degree centralities are extremely unevenly dispersed (Knoke and Yang 2008, p. 65). The calculation of the degree centralization index follows the same logic:

$$Z_D = \frac{\sum_{i=1}^{g}[C_D(n*) - C_D(n_i)]}{(g-1)(g-2)} \tag{5.8}$$

The betweenness centralization index Z_B captures a rather different network property. The index measures the extent to which each network actor controls or

---

[11] In the case of unconnected graphs, the index can be applied to at least the main component.

mediates the relationships between other actors in the network. The index equals
zero if all network actors have the same betweenness centrality and the index equals
one when a single actor mediates all geodesic paths in the network (Knoke and
Yang 2008, p. 69). We follow a procedure proposed by Wasserman and Faust
(1994, pp. 180–192) which allows us to quantify the dominator of the Freeman
Centralization Index (Wasserman and Faust 1994, p. 182):

$$Z_B = \frac{\sum_{i=1}^{g}[C_B(n*) - C_B(n_i)]}{[(g-1)(g-2)]/2} \tag{5.9}$$

All three indices were frequently applied in previous network studies to measure
network properties at the macro-level. However, the measures can also be used to
analyze subgroups. Finally, it is important to note that the latter two indices can be
calculated for both connected and unconnected graphs.

Now we take a look at two further measures that allow us to quantify the
clustering of a network and the average reachability among the network actors
involved.[12] We start with the clustering coefficient (cf. Watts 1999; Watts and
Strogatz 1998). This more indirect, tie-related concept captures the density of an
actor's surroundings and measures how many of its direct partners are
interconnected. In other words, the network is said to be highly clustered or cliquish
when many of the actor's contacts are connected to each other (Uzzi et al. 2007).
The overall network clustering coefficient is the average of all individual clustering
coefficients for the entire network. In contrast, the weighted overall clustering
coefficient is defined as the weighted mean of the clustering coefficient of all the
actors, each one weighted by its degree (Borgatti et al. 2002). The calculation of the
clustering coefficient is straightforward. The indicator simply measures the density
of triangles in a given network (Newman 2010, p. 264). Firstly, it is important to
consider that the percentage of closed triads is three times the total number of
closed triads (Uzzi et al. 2007, p. 79). Secondly, we have to quantify the number of
triangles (numerator) and the number of connected triples (denominator). This
leads to the following definition of the clustering coefficient (Uzzi et al. 2007,
p. 79):

$$CC = \frac{3\ x\ \textit{number of triangles}}{\textit{number of triples}} \tag{5.10}$$

The coefficient varies from 0 to 1 where a value of zero represents no clustering
and a value of one represents full clustering (Uzzi et al. 2007, p. 79).

Now we focus on the shortest paths and distances between the network actors. In
order to quantify the average reachablility among actors in a connected graph, we
have to take a closer look at the length of the geodesics between all pairs of actors.

---

[12] These measures are especially required for analyzing the emergence of large-scale properties at
the overall network level (cf. Sect. 8.3.2).

The shortest path between a pair of network actors is referred to as the geodesic distance (Wasserman and Faust 1994, p. 110). Paths between two actors can have different lengths in directed networks (Newman 2010, p. 242). In unconnected networks (i.e. networks with at least two components) the distance for at least one pair of actors can reach infinity (Wasserman and Faust 1994, p. 110). As most real world networks are not fully connected (Newman 2010, p. 237) this issue is usually tackled by focusing on the main component.[13] The average path length captures the reachability among all network actors in a connected graph or subgraph. The measure can be defined as "[...] the average number of intermediaries, that is, the degrees of separation, between any two actors in the network along their shortest path of intermediaries" (Uzzi et al. 2007, p. 78). Calculating the shortest path distance between pairs of nodes in a network is much harder than calculating the clustering coefficient and no exact expression for the mean distance has been found yet (Newman 2010, p. 560). As a consequence, we refer to the so-called average distance weighted reach concept (Borgatti et al. 2002; Schilling and Phelps 2007) to capture the reach of the network:

$$AR = \left[\sum_n \sum_j \frac{1}{d_{ij}} \right] / n \qquad (5.11)$$

The number of network nodes is given by n, and d_{ij} is defined as the number of smallest geodesic distances from actor i to a partner j; with $i \neq j$ (cf. (Schilling and Phelps 2007, p. 1118). The measure provides an important macro-level indicator by quantifying how far the distances between all pairs of network actors are on average.

5.3   Spatial Proximity and Concentration Measures

Most industries exhibit a pronounced tendency towards geographic concentration (Sorenson and Audia 2000, p. 424). As we will show in more detail later (cf. Sects. 7.1.2 and 7.1.3), the German laser industry is no exception in this regard. In this section we introduce two concepts that allow us to account for the geographical particularities at different levels of analysis.

---

[13] Newman (2010, p. 235) reports that the main component usually fills more than 90 % of the entire network in the majority of real world networks such as social networks, biological networks, information networks or technological networks. For the German laser industry network, we found that the main component fills 94.51 % of the network on average (cf. Sect. 8.3.3).

5.3.1   Methods for Calculating Spatial Proximity Measures

We start with a localized density measure LD[14] originally proposed by Sorenson and Audia (2000). The Euclidian distance can be used in the simplest case to calculate the geographical distance between two organizational entities. However using the shortest geographical distances can cause considerable errors in measuring because this measure ignores the curvature of the earth. As the organizations in our sample are spread over the entire federal republic, we decided to use the orthodromic distance. The shortest distance on a curved surface can be calculated by using the following formula (Sorenson and Audia 2000, p. 435):

$$gd_{ijt} = c\left\{\arccos\left[\sin\left(lat_i\right)\sin\left(lat_j\right) + \cos\left(lat_i\right)\cos\left(lat_j\right)\cos\left(\left|long_i - long_j\right|\right)\right]\right\}$$

(5.12)

The calculation of the shortest orthodromic distance between two organizations i and j requires information about the latitude and longitude of each firm. We proceeded as follows to collect all of the required information. First we included data on the organizations' addresses and address changes in our database. Next, we utilized ZIP code information to gather GPS coordinates on an annual basis. Latitudes and longitudes were measured in radians. Unlike Sorenson and Audia (2000, p. 435) we calculated the distances in kilometers by using the natural earth radius constant (c = 6378 km) and we split the overall population into two sub-populations, i.e. LSMs and PROs. Inspired by Whittington et al. (2009) we calculated the shortest distance on a curved surface not only for each LSM to all other LSMs but also for each LSM to all PROs in our sample. More precisely, in the case of the LSM-LSM distances, we generated a symmetric orthodromic distance matrix for each year under observation (t = 1...21). The number of rows and columns is determined by the number of actively operating firms in a given year (cf. Fig. 5.1, top). Likewise, in the case of LSM-PRO distances, we created asymmetric orthodromic distance matrices on an annual basis where the number of columns and rows in a given year is determined by the number of actively operating LSMs and PROs respectively (cf. Fig. 5.1, bottom). In some cases an even more detailed separation of the overall sample was needed.[15]

After these steps, both geographical co-location measures (LD$_{\text{LSM-LSM}}$ & LD$_{\text{LSM-PRO}}$) were calculated by using the following localized density measure

---

[14] If not otherwise stated, in this section we follow the methodological concept proposed by Sorenson and Audia (2000, pp. 433–435).

[15] To account for the heterogeneity of organizations in our PRO sample we put all universities and technical universities into one group, and all other public research organizations into another. These measures were predominantly used to check for consistency and robustness in our estimation results (for instance, an additional consistency check of estimation results in Chap. 12).

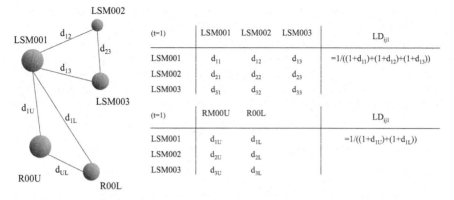

Fig. 5.1 Illustration of symmetric and asymmetric orthodromic distance matrices (Source: Author's own illustrations, based on: Sorenson and Audia (2000, p. 434))

proposed by Sorenson and Audia (2000, p. 434) and specified by Whittington et al. (2009)[16]:

$$LD_{it} = \sum_j \frac{1}{(1 + gd_{ijt})} \tag{5.13}$$

An LSM's localized distance measure at time t is defined as the sum of inverse distance functions to all other LSMs or PROs respectively. Thus, we end up with two types of localized density measures for each firm in the sample, each of which is calculated on an annual basis.

5.3.2   Methods for Calculating Geographical Concentration Measures

Now we turn our attention to the industry level. The Hirschman-Herfindahl Index and the Entropy measure are some of the most commonly used industry concentration measures (Acar and Sankaran 1999). Economists have used the Hirschman-Herfindahl Index (HHI) for quantifying the competiveness of markets and for analyzing the market shares held by firms in an industry. For the purpose of this study we draw upon the HHI index in order to set up a geographical concentration index at the industry level. To do so, we proceeded as follows: First, we used the planning region scheme, commonly used in Germany for the classification of territorial units for statistical purposes. This divides the territory into

---

[16] According to Whittington et al. (2009) the weighting factor x in the numerator of the originally proposed LD measure was taken to equal one.

97 geographical areas. Next, we generated a count variable for each type of laser-related organization in our database – LSMs, PROs and LSPs – that represents the number of organizations per planning region and year. Then we calculated the relative proportion of organizations on an annual basis for each planning region i (with $I = 1...97$). Finally, three concentration indices were established, one for each of the three organizational subsets in our database (i.e. LSMs, PROs or LSPs), by applying the following equation:

$$HHI_t = \sum_i p_{it}^2 = \frac{\sum_i^N a_{it}^2}{\left(\sum_i^N a_{it}\right)^2} \tag{5.14}$$

We ended up with an indicator that allows us to quantify the intensity of organizational crowding in the geographical space. The HHI moves towards zero if the organizations under observation are equally dispersed throughout the geographical space; the HHI has comparably large values if some organizations are widely dispersed whereas others show a pronounced tendency to crowd together.

References

Acar W, Sankaran K (1999) The myth of unique decomposability: specializing the Herfindahl and entropy measures. Strateg Manag J 20(1):969–975

Amburgey TL, Al-Laham A, Tzabbar D, Aharonson BS (2008) The structural evolution of multiplex organizational networks: research and commerce in biotechnology. In: Baum JA, Rowley TJ (eds) Advances in strategic management – network strategy, vol 25. Emerald Publishing, Bingley, pp 171–212

Anthonisse JM (1971) The rush in the directed graph – technical report BN 9/71. Stichting Mathematisch Centrum, Amsterdam

Bavelas A (1948) A mathematical model for group structure. Hum Organ 7(3):16–30

Beauchamp MA (1965) An improved index of centrality. Behav Sci 10(2):161–163

Bertalanffy LV (1968) General system theory: foundations, development, applications. George Braziller, New York

Bonacich P (1972) Factoring and weighting approaches to status scores and clique identification. J Math Sociol 2(1):113–120

Bonacich P (1987) Power and centrality: a family of measures. Am J Sociol 92(5):1170–1182

Borgatti SP (2005) Centrality and network flow. Soc Networks 27(1):55–71

Borgatti SP, Everett MG, Freeman LC (2002) Ucinet for windows: software for social network analysis. Analytic Technologies, Harvard

Borgatti SP, Everett MG, Johnson JC (2013) Analyzing social networks. Sage, London

Broekel T, Graf H (2011) Public research intensity and the structure of German R&D networks: a comparison of ten technologies. Econ Innov New Technol 21(4):345–372

Burt RS (1992) Structural holes: the social structure of competition. Harvard University Press, Cambridge

Czepiel JA (1974) Word of mouth processes in diffusion of a major technological innovation. J Mark Res 11:172–180

Freeman LC (1977) A set of measures of centrality based on betweenness. Sociometry 40 (1):35–41

Freeman LC (1979) Centrality in social networks: I. conceptual clarification. Soc Networks 1 (3):215–239

Hanneman RA, Riddle M (2005) Introduction to social network methods. University of California, Riverside

Hite JM, Hesterly WS (2001) The evolution of firm networks: from emergence to early growth of the firm. Strateg Manag J 22(3):275–286

Holland PW, Leinhardt S (1970) A method for detecting structure in sociometric data. Am J Sociol 76(3):492–513

Holland PW, Leinhardt S (1976) Local structure in social networks. Sociol Methodol 7:1–45

Jackson MO (2008) Social and economic networks. Princeton University Press, Princeton

Katz L (1953) A new status index derived from sociometric analysis. Psychometrika 18(1):39–43

Knoke D, Yang S (2008) Social network analysis. Sage, London

Leavitt HJ (1951) Some effects of communication patterns on group performance. J Abnorm Soc Psychol 46(1):38–50

Marsden PV (2002) Egocentric and sociocentric measures of network centrality. Soc Netw 24:407–422

Marsden PV (2005) Recent developments in network measurement. In: Carrington PJ, Scott J, Wasserman S (eds) Models and methods in social network analysis. Cambridge University Press, Cambridge, pp 8–30

Newman ME (2010) Networks – an introduction. Oxford University Press, New York

Sabidussi G (1966) The centrality index of graph. Psychmetrika 31(4):581–603

Schilling MA, Phelps CC (2007) Interfirm collaboration networks: the impact of large-scale network structure on firm innovation. Manag Sci 53(7):1113–1126

Shimbel A (1953) Structural parameters of communication networks. Bull Math Biophys 15 (4):501–507

Sorenson O, Audia PG (2000) The social structure of entrepreneurial activity: geographic concentration of footwear production in the Unites States, 1940–1989. Am J Sociol 106 (2):424–462

Suitor JJ, Wellman B, Morgan DL (1997) It's about time: how, why, and when networks change. Soc Networks 19(1):1–7

Uzzi B, Amaral LA, Reed-Tsochas F (2007) Small-world networks and management science research: a review. Eur Manag Rev 4(2):77–91

Wasserman S, Faust K (1994) Social network analysis: methods and applications. Cambridge University Press, Cambridge

Watts DJ (1999) Small worlds – the dynamics of networks between order and randomness. Princeton University Press, Princeton

Watts DJ, Strogatz SH (1998) Collective dynamics of 'small-world' networks. Nature 393 (6684):440–442

Whittington KB, Owen-Smith J, Powell WW (2009) Networks, propinquity, and innovation in knowledge-intensive industries. Adm Sci Q 54(1):90–122

Chapter 6
Dataset Design and Estimation Methods

All models are wrong, but some models are useful.
(George E.P. Box 1979)

Abstract The phenomenon under investigation guides the data collection process, the structural design of the dataset as well as the choice of empirical methods (Blossfeld et al. 2007, p. 4). The data collection process for the German laser industry database was described above. Now, we will turn our attention to the two latter points. Chapter 6 is divided into two sections: Section 6.1 presents the two compiled datasets. On the one hand, an event history dataset was constructed to analyze the propensity and timing of laser source manufacturers to cooperate and enter the German laser industry innovation network. On the other hand, a panel dataset was employed to analyze the determinants of firms in the German laser industry from various angles. Section 6.2 provides an overview and general discussion of estimation methods which were applied in Part IV of this book. We start with a brief discussion on non-parametric event history analysis models using continuous time, followed by an introduction of econometric models for panel count data.

6.1 Design and Scope of the Compiled Datasets

In essence, there are three types of dataset designs: cross-sectional datasets, panel datasets and event history datasets (Blossfeld et al. 2007, pp. 5–21). Due to the aim of this study focus is placed on the two latter dataset designs.

6.1.1 Dataset I: Event History Data Structure

The use of event history analysis methods requires a relatively demanding data design. Firstly, the compilation of an event-oriented longitudinal dataset requires the appropriate choice of time intervals and information on the origin state and destination state (Blossfeld et al. 2007, p. 42). Secondly, data has to be organized in

© Springer International Publishing Switzerland 2015   123
M. Kudic, *Innovation Networks in the German Laser Industry*,
Economic Complexity and Evolution, DOI 10.1007/978-3-319-07935-6_6

an event-oriented design in which each record is related to a particular duration in a predefined state (Blossfeld et al. 2007, p. 42). Finally, precise time-tracking of start and end dates is needed for the event under investigation to analyze the transitions from one state to another.

Some issues, however, require particular attention. The most notable is the setup of the analytical framework. Both dimensions of the analytical framework – "state space" and "time space" – have to be defined carefully to avoid misspecification (Blossfeld et al. 2007, p. 38). The choice of these two dimensions is driven by theoretical considerations and determines the choice of the empirical estimation models. In general, there are four types of models: the "single episode model", "multi state model", "multi episode model", and "multi state multi episode model" (Blossfeld et al. 2007, p. 39).[1] For the purpose of this study we employed a design that is based on the following considerations. On the one hand, we seek to understand what factors determine a firm's propensity to cooperate for the first time and enter the laser industry innovation network. On the other hand, we are interested in the factors that affect the length of time until the first cooperation occurs. Consequently, we constructed a single-episode event history dataset that provides the basis for conducting a non-parametric spell-duration analysis.

The single-episode event history dataset for the German laser industry is constructed and organized as follows: The time axis is defined on the basis of century months. All firm foundation dates as well as all start and end dates of cooperation events are given in century months. The unit of analysis is the firm. In cases where the number of censored observation units is small, it is acceptable to simply exclude them (Allison 1984, p. 11). Thus, firms founded before 1990 were excluded from the dataset to avoid left truncation and left censoring problems (Blossfeld and Rohwer 2002, pp. 39–41). Starting from the full population of 233 LSMs in our sample we identified 39 firms which were founded before 1990. Thus, a total of 194 firms were potentially at risk for conducting the first cooperation event. Out of this population we ended up with a total of 112 firms with at least one cooperation event during the observation period. The event of interest is the first cooperation for all laser manufacturing firms which are at risk in the time period between 1990 and 2010. The dataset allows us to analyze the transition from the origin state ("no cooperation") to the destination state ("first cooperation"). These two states allow us to define the risk set. At the same time the initial cooperation event marks the firm's entry into the network. Repeated events were

---

[1] Single episode models allow for event transitions from the origin state to the destination state whereas multi-state models allow us to analyze event transitions from the origin state to multiple destination states. In contrast, multi-episode models allow for repeated events or event transitions over time. Finally, multi-state multi-episode models can be applied to analyze both repeated episodes and repeated events. For further details see (Blossfeld et al. 2007).

not considered. Firms were basically considered to have two ways of entering the German laser industry innovation network. The first cooperation event can be either participating in a *Foerderkatalog* project or in a *CORDIS* project. Both types of cooperation event were coded separately by using a dummy variable. All event occurrence dates and durations were recorded in century months. Variables were grouped in the following categories: organizational, relational and contextual.[2]

Organizational variables[3] were included in the dataset to account for heterogeneity across firms. In particular these variables are firm origin *[origin_ev]*[4] and firm size *[firmsize_cat_ev]*.[5] Additionally, a simple dummy variable was created to differentiate between "young" and "mature" firms *[firmage_ev]*.[6] Relational variables were included in the dataset. Both types of publicly funded R&D cooperation projects, *Foerderkatalog* projects as well as *CORDIS* projects, were coded separately *[coop_type_ev]*.[7] Occurrence dates and duration were recorded in century months. Moreover, a set of contextual variables was included in the dataset. For the first set of geographical variables, we split the sample into four geographical regions. We included a set of geographical location variables in our dataset *[region_ev]*[8] indicating whether a firm is located in the northern, southern, eastern or western part of Germany. Finally, we included a set of cluster variables *[clu_ev]* in our dataset indicating whether a firm was located inside or outside of a densely

---

[2] Note that the following variables in the event history dataset can differ from those that were coded and used in the panel dataset. The suffix "_ev" indicates an event oriented variable.

[3] Organizational level variables were coded at the date of a firm's population entry and considered to be time-invariant for the purpose of the non-parametric spell-duration analysis.

[4] Origin dummies are coded on the basis of the following categories: origin_ev1 = new foundation; origin_ev2 = PRO spin-off; origin_ev3 = LSM spin-off; origin_ev4 = other background, such as: spin-offs from other types of organizations, name change and post-merger firm formations.

[5] The following five size categories were used: firmsize_cat_ev1 = "micro firm" = 1–9 employees; firmsize_cat_ev2 = "small firm" = 10–49 employees; firmsize_cat_ev3 = "medium firm" = 50–249 employees; firmsize_cat_ev4 = "larger firms" = more than 250 employees. This categorization is drawn upon the definition proposed by the European Commission (2005). Missing data for the number of employees were extrapolated based on employee data for the same firms but other observation windows.

[6] We used the mean age of the firms (97 months) in the observation period to split the sample. Definition of "young firms": firmage_ev = 0 if firmage_ev $<$ = 97 months (8.1 years); "mature firms": firmage_ev = 1 if firmage_ev $>$ 97months (8.1 years).

[7] The variable coop_type = 1 in the case of a *CORDIS* project; coop_type = 2 in the case of a Foerderkatalog project.

[8] Definition of the four geographical regions: region_ev1 = Baden-Württemberg (BW) Bavaria (BY); region_ev2 = Bremen (HB) Hamburg (HH) Schleswig-Holstein (SH); region_ev3 = Berlin (B) Brandenburg (BB) Mecklenburg-Western Pomerania (MV) Saxony (S) Thuringia (TH) Saxony-Anhalt (SA); region_ev4 = North Rhine-Westphalia (NW) Lower Saxony (NS) Rhineland-Palatinate (RP) Saarland (SR) Hessen (H). The variable for Saarland had to be omitted due to the non-existence of LSMs in this federal state throughout the entire time period.

crowded region. The four geographical clusters were identified based on the descriptive analysis in Sects. 7.1.2 and 7.1.3.[9]

6.1.2 Dataset II: Panel Data Structure

Panel data methods require the data to be in long form, meaning that each individual time pair in the dataset is a separate observation (Cameron and Trivedi 2009, p. 274). The panel dataset for the German laser industry is constructed and organized as follows: The unit of analysis is the firm year meaning that each firm in the sample is observed for each year. Thus, we decided in favor of annual time intervals for the purpose of this study. The panel is unbalanced due to a considerable proportion of firms entering the sample after 1990 (i.e. new foundations, spin-offs etc.) or leaving the sample before 2010 (i.e. mergers, bankruptcies etc.). Unbalanced data usually causes no significant complications as most empirical methods are designed to handle both balanced and unbalanced panel data (Cameron and Trivedi 2009, p. 230).

Over the course of 21 years we have a total of 233 laser source manufacturers (LSMs) and 2,645 firm years. Thus, we have an average of 11.35 observations per firm. The dataset contains time-variant as well as time-invariant variables organized in content-specific groups of explanatory variables.[10]

The first group of variables encompasses all firm-specific variables. A linear firm age variable [firmage] as well as a squared firm age variable [firmage_sq] were generated on the basis of firm entry and firm exit dates. Both age variables in our panel dataset are recorded on an annual basis. Data on yearly turnover for each firm provides the basis for the calculation of a time-invariant average turnover variable for each firm in the sample [avgturnover]. A set of dummy variables was constructed to account for the origin of firms in the sample [origin][11] and a second set of dummies was generated to account for differences in legal status across firms [leg_stat].[12] In addition, the yearly number of employees was used to code a set of firm size dummies [firmsize_cat].[13]

---

[9] These clusters are defined as follows: planning regions: 72, 73, 74, 76 & 77 = clu_ev_bw, located in Baden-Württemberg; planning regions: 86, 90 & 93 = clu_ev_bay, located in Bavaria: planning regions: 54 & 56 = clu_ev_thu, located in Thuringia; planning region 30 = clu_ev_B, located in Berlin.

[10] Note that we had to choose a more detailed categorization for most of the panel data variables due to the theoretical considerations (cf. Chaps. 10, 11 and 12).

[11] Origin dummies are coded on the basis of the following categories: origin1 = new foundation; origin2 = name change; origin3 = post merger firm; origin4 = PRO spin-off; origin5 = LSM spin-off; origin6 = spin-offs from other types of organizations.

[12] Legal status dummies are coded on the basis of the following categories: leg_stat1 = GmbH; leg_stat2 = GmbH & Co; leg_stat3 = GmbH & Co KG; leg_stat4 = OHG; leg_stat5 = AG; leg_stat6 = other.

[13] The following five size categories were used: firmsize_cat1 = "micro firm" = 1–9 employees; firmsize_cat2 = "small firm" = 10–49 employees; firmsize_cat3 = "medium firm" = 50–249 employees; firmsize_cat4 = "large firms" = 250–749 employees; firmsize_cat5 = "very large

The second group of variables encompasses all geographical measures on a firm level, regional level and industry level. All geographical variables are coded on the basis of annually updated address data for three types of laser-related organizations: LSMs, PROs and LSPs. At the firm level a set of geographical dummy variables for each LSM was generated indicating the federal state in which the firm is located *[fed_state]*.[14] Two types of geographical co-location measures *[coloclsm, colocpro]* were included in the dataset which were calculated on the basis of the localized density measures outlined above (cf. Sect. 5.3.1). Additionally we split the PRO sample into two sub-samples.[15] Localized geographical density measures were generated by calculating all distances between LSM and PROs in both sub-samples separately *[colocpro_appl, colocpro_basic]*. In order to get a picture of geographical concentration tendencies at the industry level, HHI indices were calculated for LSMs, LSPs and PROs (cf. Sect. 5.3.2) and included in the dataset *[hhi_lsm; hhi_lsp; hhi_pro]*.

The third group of variables encompasses all cooperation-related variables. Data on publicly funded cooperation projects from both "*Foerderkatalog*" as well as "*CORDIS*" databases were used to generate cooperation counts *[coopcnt_fk; coopcnt_c]*, cumulative cooperation counts *[coopcum_fk; coopcum_c]* and cooperation funding *[coopfund_fk; coopfund_c]* on an annual basis. Based on these measures several combined variables were generated which include both project types *[coopcnt_fkc; coopcum_fkc; coopfund_fkc]*. All cooperation funding variables are measured in thousand euros.

The fourth group of variables encompasses network variables calculated at three levels of analysis[16] (cf. Sect. 5.2). The following network level variables were included in the dataset: overall network density *[nw_density]*, network size *[nw_size]*, clustering coefficients *[nw_clust]* a weighted clustering coefficient *[nw_wclust]*, a network fragmentation measure *[nw_compcnt]* and an average reachability measure *[nw_areach]*. The next set of network variables allows us to quantify the structural configuration of ego network characteristics for each firm in the sample. In particular these ego network variables measure the ego network size *[ego_size]*, the ego network density *[ego_density]* and two ego network-based brokerage indicators *[ego_broker; ego_nbroke]*. The last set of network variables

---

firms" = more than 750 employees. Missing data for the number of employees was extrapolated based on employee data for the same firm but for other firm years.

[14] Definition of federal state dummies: fed_state1 = Baden-Württemberg (BW); fed-state2 = Bavaria (BY); fed_state3 = Berlin (B); fed_state4 = Brandenburg (BB); fed_state5 = Bremen (HB); fed_state6 = Hamburg (HH); fed_state7 = Hessen (H); fed_state8 = Mecklenburg-Western Pomerania (MV); fed_state9 = Lower Saxony (NS); fed_state10 = North Rhine-Westphalia (NW); fed_state11 = Rhineland-Palatinate (RP); fed_state12 = Saarland (SR); fed_state13 = Saxony (S); fed_state14 = Saxony-Anhalt (SA); fed_state15 = Schleswig-Holstein (SH); fed_state16 = Thuringia (TH). The variable for Saarland had to be omitted due the non-existence of laser source manufacturers in this federal state throughout the entire time period.

[15] The full population of 149 PROs is a relatively heterogonous group of organizations. Some predominantly focus on applied research whereas others mainly conduct basic research.

[16] We used UCI-Net 6.2 if not otherwise stated (Borgatti et al. 2002).

measures the strategic network positioning of each firm within the network. The following network centrality measures were included in the dataset: degree centrality *[ctr_degree]*, betweenness centrality *[ctr_between]*, two reach-based measures *[ctr_2step; ctr_ard]* and two power-related measures *[ctr_ev; ctr_bon]*. The last group of variables encompasses innovation indicators measured by patent count variables. Patent application counts *[pacnt]* and patent grant counts *[pgcnt]* were recorded on an annual basis and included in the dataset. Lag variables were generated for both variables with a lag of 1 year *[pacnt1; pgcnt1]*, 2 years *[pacnt2; pgcnt2]* and 3 years *[pacnt3; pgcnt3]*.

6.2   Introducing Selected Econometric Methods

In this section we turn our attention to longitudinal econometric methods. The discussion addresses general issues connected to event-history and panel data estimation methods. Specific issues are addressed in the context of their application later this book.

6.2.1   Event History Analysis Methods

There are basically three classes of event history methods: non-parametric methods, parametric methods and semi-parametric methods (cf. Allison 1984, p. 14). Non-parametric models do not make distributional assumptions. Parametric models assume that the time until an event of interest occurs follows a specific distributional form. Semi-parametric models make no assumption with regard to the distribution of event time. But these models require a specification of a regression model with a specific functional form (cf. Allison 1984, p. 14).

As stated above, we focus on non-parametric event history methods to analyze cooperation events of German laser source manufacturers. Non-parametric estimation methods do not make any assumptions about the distribution of the process under investigation and are well suited for an initial analysis of a specific phenomenon (Blossfeld et al. 2007, p. 58). The most commonly used non-parametric approach is the Life-Table method. However, this approach has some notable limitations (Blossfeld and Rohwer 2002, p. 56). First, the method requires the pre-specification of fixed and discrete time intervals. Second, to ensure the reliability of estimates conditional for each interval, the Life-Table method is usually applied in the case of a relatively large number of episodes. To overcome these restraints an alternative non-parametric approach has been proposed, i.e. the Product-Limit estimator, also known as the Kaplan-Meier method (Kaplan and Meier 1958).

The Kaplan-Meier method is a non-parametric empirical method that estimates the survivor function based upon longitudinal event data (Cleves et al. 2008, p. 93). In general, the survival function gives the probability of surviving past time t, or to put it in another way, the probability of failing after time t (ibid). The Kaplan-Meier method provides some notable advantages. The method is straightforward to use, requires only weak assumptions and allows us to analyze non-repeated events in single-episode event history data. In this study we are interested in the German laser source manufacturers' propensity to cooperate for the first time and enter the innovation network as well as the length of time until the first cooperation occurs on average. Consequently, based on the risk set specified above we define and interpret the survival function as follows: the survival function estimates the firms' probability of having the first cooperation event after time t. In our case the survival function reflects the probability of moving from the origin state ("no cooperation") to the destination state ("first cooperation") at a given point in time. In addition, both the variance and the confidence interval can be calculated by using Greenwood's variance formula of the survival function and the asymptotic variance of the logarithm of the survival function respectively.[17]

In some cases the hazard rate[18] or the cumulative hazard rate function is of interest rather than the survival function itself. We focus on the latter concept as it allows us to measures the overall risk that has been accumulated up to time t (Cleves et al. 2008, p. 8). There is a simple relationship between the survival function and the cumulative hazard rate.[19] A simple interpretation of the cumulative hazard rate is that it records the number of times we would theoretically expect to observe the occurrence of an event (Cleves et al. 2008, pp. 13–15). It is important to note that cumulative hazards must be interpreted in the context of repeated events regardless of whether the event of interest is, due to its very nature, repeatedly observable or not (ibid).[20] The commonly used method to calculate the cumulative hazard rate is the Nelson-Aalen estimator. The reason is that the Nelson-Aalen estimator exhibits better small-sample properties than the Kaplan-Meier estimator (Cleves et al. 2008, p. 108).

Non-parametric estimation methods provide the opportunity to compare survivor functions. The overall population can be divided into two or more subgroups by using an indicator variable to analyze whether the probability of failing after time t significantly differs among these subgroups. The indicator variable defines the membership in a particular subgroup based on firm-specific characteristics

---

[17] For an in-depth description of the calculation methods see Cleves et al. (2008, p. 96).

[18] The hazard rate function h(t) can vary from zero to infinity and is also known as the conditional failure rate. It is defined as "[...] the (limiting) probability that the failure event occurs in a given interval, conditional upon the subject having survived to the beginning of the interval, divided by the width of the interval" (Cleves et al. 2008, p. 7).

[19] The cumulative hazard function H(t) is defined as: $H(t) = -\ln\{S(t)\} = \int_0^t h(u)\, du$, where S(t) represents the survival function and h() gives the hazard function (Cleves et al. 2008, p. 107).

[20] To illustrate this point, cooperation events can occur several times in a firm's lifespan whereas other events such as firm exits occur only once.

(Blossfeld et al. 2007, p. 76). Using these preparatory steps, separate survival functions are calculated for the members of each subgroup. We apply this approach to analyze the extent to which firm-specific characteristics affect the cooperation behavior over time.

The simplest way to check for statistically significant differences in survivor functions is to calculate and compare the confidence intervals for the estimated survivor functions. The survivor functions are said to be significantly different as long as the confidence intervals are not overlapping (Blossfeld et al. 2007, p. 76).

The more comprehensive approach is to calculate a test statistic.[21] For the purpose of this study we make use of the most commonly applied test statistics, i.e. the Log-Rank test, Cox test, Wilcoxon-Breslow test and Tarone-Ware test. These tests are designed to compare globally defined overall survival functions (Cleves et al. 2008, p. 123). Even though these tests provide, in most cases, relatively similar results, it can be useful to calculate and compare alternative test statistics. One reason for this is that some tests (e.g. Wilcoxon-Breslow) emphasize differences in survivor functions at the onset of the observation period whereas other test statistics (e.g. Log-Rank) stress the differences at the end of the observation period (Blossfeld et al. 2007, p. 81). Several alternative test statistics have since been proposed. For instance the Cox test is very similar to the Log-Rank test whereas the Tarone-Ware test, like the Wilcoxon-Breslow test, puts more weight on earlier time slots.[22] Common to all these test statistics is that they are χ^2-distributed with m-1 degrees of freedom.[23] The tests are based on the null hypothesis that the survivor functions do not differ significantly from each other (Blossfeld et al. 2007, p. 81). A significant test result indicates that the null hypothesis must be rejected (ibid), or to put it another way, the rejection of the null hypothesis based on a significant test result supports the alternative hypothesis that the compared survivor functions differ significantly from one another.

6.2.2   Econometric Methods for Panel Count Data

The econometric analysis of firm innovativeness based on patent data requires the use of a particular category of estimation methods, so-called count data methods. Patent data, which is the same as other types of count data,[24] takes discrete non-negative integer values (Wooldridge 2002, p. 645) and is typically highly

---

[21] For a description of the general approach to construct test statistics to compare non-parametric survival functions, see Blossfeld and Rohwer (2002, pp. 79–81).

[22] For an in-depth discussion on other further tests (e.g. Peto-Peto-Prentice test or the Fleming-Harrington two-parameter test), see Cleves et al. (2008, pp. 122–128).

[23] The degree of freedom is determined by the number of pre-specified subgroups. Thus, the variable m takes the value 2 in the case of two subgroups (Blossfeld et al. 2007, p. 81).

[24] For a brief overview of other count variables analyzed in economics and social science see (Wooldridge 2002, p. 645).

skewed making the use of conventional linear models inappropriate (Cincera 1997, p. 266). Moreover, a considerable fraction of patent data observations takes on the value zero so that a natural log transformation of the dependent variable in linear models is not possible (Wooldridge 2002, p. 645). Consequently, for count data it is advisable to model the population regression directly and to choose a functional form that ensures positivity for the vector of explanatory variables and any parameter values possible (ibid). As long as the dependent variable has no upper limit, an exponential function is an appropriate choice to meet these requirements (ibid). In general, count data models can be used to analyze cross-sectional as well as longitudinal data. Due to the aim of the present study, focus is placed on the longitudinal models. In their seminal work on the analysis of firm-level R&D activities, Hausman et al. (1984) propose econometric models which are in many cases still the model of choice for the analysis of longitudinal patent count data.

We will start with a brief discussion on panel data characteristics. Even though panel data methods are more complicated than cross-sectional methods, they provide considerable additional value as they allow data to be analyzed that encompasses both variation across individual units as well as variation over time (Cameron and Trivedi 2009, p. 229). The fundamental advantage of panel data is that it allows great flexibility in modeling differences across individual units (Greene 2003, p. 284). For instance, a set of firms is by no means homogeneous as all firms involved differ from one another in several dimensions. Panel data allows us to cope with the problem of unobserved heterogeneity (Kennedy 2003, p. 302). Both fixed and random effects models are associated with some notable advantages and disadvantages that are discussed later. Panel data alleviates multicollinearity problems by creating more variability through combining variation across individual units and across time (Kennedy 2003, p. 302). In addition, panel data allows the analysis of dynamic adjustments which can be crucial in understanding economic phenomena (ibid). Panel data consists of repeated measurements at different points in time, usually observed in regular time intervals, on a well-defined set of individual units such as a spatially and sectorally delimited set of firms. In other words, panel data are repeated observations of the same cross section of firms (Wooldridge 2002, p. 7).

In general, one can distinguish between short panels (i.e. many individual observations across a few time periods) and long panels (few individual observations across many time periods) or panels where the cross-sectional and the time series dimension are roughly of the same magnitude (Cameron and Trivedi 2009, p. 230; Wooldridge 2002, p. 7). Most panel data methods can handle both balanced as well as unbalanced panels (Cameron and Trivedi 2009, p. 230). In the first case, the full set of individual units is observed over all time periods whereas in the second case a considerable fraction of individual units are observed for fewer time periods. Due to the aim of this study and the structural features of the laser industry panel dataset (cf. Sect. 6.1.2) we focus below on single equation models for analyzing short, unbalanced panels.

Now we turn our attention to the choice of model. In many cases the explanatory variables of our count data model reveal empirical evidence for overdispersion.[25] There are several ways to deal with overdispersion in count data models. Commonly, overdispersion that is induced by unobserved heterogeneity is accounted for by estimating Negative Binomial models (NB model) instead of the intuitive standard Poisson model. The NB model is more general than the Poisson model because it allows for increased dispersion by incorporating an additional parameter α. The variance is a linear function of the mean that can be transformed into a Poisson model (Cameron and Trivedi 1986). The NB model reduces to the Poisson Model as $\alpha \rightarrow 0$ (Winkelmann 2003). It enables us to deal with a predominance of zero and small integer values (Cameron and Trivedi, 1986). Finally, we have to take a brief look at the differences between models that allow for a correlation between the explanatory variables and the time-invariant component of the error term, and models which require the unobserved effects and the explanatory variables to be completely uncorrelated (Wooldridge 2002, p. 668). In the former case we have a fixed effects model and in the latter case a random effects model. Pioneering work on the analysis of unobserved effects in panel count data has been conducted by Hausman et al. (1984) who developed a fixed effects and random effects model under full distributional assumption (Wooldridge 2002, p. 668). Panel data shows two types of variation. Variation from one observation to another observation for an individual unit is called "within-variation" whereas the variation from one individual unit to another individual unit is called "between-variation". The fixed effects model ignores variation across individuals and uses only within-variation for all individual units over all observation windows (Kennedy 2003, p. 307). It is important to note that the coefficient of the regressor in fixed effects models will be incorrectly estimated or not identified with little or no within-variation (Cameron and Trivedi 2009, p. 238). Fixed effects models have some important advantages. The fixed effects estimator is unbiased because it includes dummy variables for the different intercepts and is more robust against selection bias problems compared to the random effects estimator (Kennedy 2003, p. 304). However, fixed effects models also have two considerable drawbacks. Firstly, all time-invariant explanatory variables are thrown out because the estimation procedure fails to estimate a slope coefficient for variables that do not vary within an individual unit (Kennedy 2003, p. 304). Secondly, using only within-variation leads to less efficient estimates and the model loses explanatory power (Cameron and Trivedi 2009, p. 259). The random effects model compensates for this disadvantage.

The random effects estimator takes advantage of within-variation as well as between-variation in panel data (Cameron and Trivedi 2009, p. 256) by using cross-sectional variation in panel data and by running OLS estimation on the average values for each individual unit in order to calculate a (matrix) weighted average of both between-estimators and within-estimators (Kennedy 2003, p. 307). The random effects model has several advantages compared to the fixed effects

---

[25] The procedure to check for overdispersion was proposed by Cameron and Trivedi (1990).

model. On the one hand random effects estimators make better use of the information values of patent data and generate efficient estimates with higher explanatory power. In addition, a random effects estimator can generate coefficient estimates of both time-variant and time-invariant explanatory variables (Kennedy 2003, p. 307). However, these advantages are not without a downside. The major drawback of the random effects model is that correlations between the error term and explanatory variables generate biased estimates (Kennedy 2003, p. 306). In other words, the random effects estimator generates inconsistent results when the model assumptions are violated.

In summary, the main difference between the estimation techniques is that fixed effects models allow for correlation between the unobserved individual effect and the included explanatory variables whereas random effects models require the unobserved individual effect and the explanatory variables to be uncorrelated (Greene 2003, p. 293).

The question remains whether fixed effects models or random effects models should be applied. Hausman (1978) has proposed a specification test to select the appropriate model. The basic idea of the Standard Hausman specification test is to test the null hypothesis that the unobserved effect is uncorrelated with the explanatory variables (Greene 2003, p. 301). In the case that the null hypothesis cannot be rejected, both the fixed effects estimates and the random effects estimates are consistent and the model of choice is the random effects model due to its higher explanatory power. Under the alternative, random effects and fixed effects estimators differ and it can be argued that the latter model is the appropriate choice (Cameron and Trivedi 2009, p. 260).

Nonetheless, choosing the model based on the Standard Hausman specification test is controversial for two reasons. First, there is an interdisciplinary controversy of whether consistency should be preferred over efficiency. Microeconometric literature advocates the use of fixed effects models whereas most other branches of applied statistics tend to give preference to random effects models due to their higher explanatory power (Cameron and Trivedi 2009, p. 230). Second, the Standard Hausman specification test itself has serious shortcomings because it requires the random effects estimator to be efficient (Cameron and Trivedi 2009, p. 261). In the case of unbalanced panels, a robust version of the specification test can alleviate the latter point (Cameron and Trivedi 2009, pp. 261–262).

References

Allison PD (1984) Event history analysis – regression for longitudinal event data. Sage, London

Blossfeld H-P, Rohwer G (2002) Techniques of event history analysis – new approaches to causal analysis. Lawrence Erlbaum, London

Blossfeld H-P, Golsch K, Rohwer G (2007) Event history analysis with Stata. Lawrence Erlbaum, London

Borgatti SP, Everett MG, Freeman LC (2002) Ucinet for windows: software for social network analysis. Analytic Technologies, Harvard

Box GEP (1979) Robustness in the strategy of scientific model building. In: Launer RL, Wilkinson
     GN (eds) Robustness in statistics. Academic, New York
Cameron CA, Trivedi PK (1986) Econometric models based on count data: comparisons and
     applications of some estimators and tests. J Appl Econ 1:29–53
Cameron CA, Trivedi PK (1990) Regression based tests for overdispersion in the Poisson model.
     J Econ 46(3):347–364
Cameron CA, Trivedi PK (2009) Microeconometrics using Stata. Stata Press, College Station
Cincera M (1997) Patents, R&D, and technological spillovers at the firm level: some evidence
     from econometric count models for panel data. J Appl Econ 12(3):265–280
Cleves MA, Gould WW, Gutierrez RG, Marchenko YU (2008) An introduction to survival
     analysis using Stata, 2nd edn. Stata Press, College Station
European Commission (2005) The new SME definition – user guide and model declaration.
     Enterprise and Industry Publications, Brussels
Greene WH (2003) Econometric analysis, 5th edn. Prentice Hall, Upper Saddle River
Hausman JA (1978) Specification tests in econometrics. Econometrica 46(6):1251–1271
Hausman JA, Hall BH, Griliches Z (1984) Econometric models for count data with an application
     to the patents – R&D relationship. Econometrica 52(4):909–938
Kaplan EL, Meier P (1958) Nonparametric estimation from incomplete observations. J Am Stat
     Assoc 53:457–481
Kennedy P (2003) A guide to econometrics. Blackwell, Oxford
Winkelmann R (2003) Econometric analysis of count data, 4th edn. Springer,
     Heidelberg/New York
Wooldridge JM (2002) Econometric analysis of cross sectional and panel data. MIT Press,
     Cambridge, MA

Part III
Descriptive Analysis

Chapter 7
Industry Dynamics and Geographical Concentration

Use a picture. It's worth a thousand words.

(Arthur Brisbane 1911)

Abstract A natural starting point for the descriptive part of this book is to look at the industry as a whole. The results of our initial industry level analysis provide the basis for exploring cooperation activities in the following sections. This chapter is divided into two sections. Section 7.1 focuses on industry dynamics and geographical concentration patterns in the German laser industry. The initial descriptive exploration provides a comparison of industry dynamics, geographical concentration indices and spatial distribution patterns for three types of laser-related organizations – laser source manufacturers (LSMs), laser system providers (LSPs) and laser-related public research organizations (PROs). Section 7.2 focuses on LSMs that constitute the core of the industry due to their central position along the industry value chain. Our analysis reveals some interesting insights by uncovering entry and exit dynamics of LSMs on an annual basis and illustrating the size distribution of firms at the regional and national level. Finally, we take a closer look at the public research landscape in the German laser industry by exploring the structural composition of all PROs in the sample.

7.1 Exploring the German Laser Industry from Various Angles

This section will begin by focusing on industry dynamics[1] and geographical concentration patterns in the German laser industry.

---

[1] For an in-depth discussion on industry evolution in the German laser industry see Buenstorf (2007). We had to identify all organizational entities under investigation at firm or business unit level to meet the requirements of our study. The consequence is that the following descriptive findings can differ from the industry evolution patterns reported by Buenstorf (2007).

© Springer International Publishing Switzerland 2015
M. Kudic, *Innovation Networks in the German Laser Industry*,
Economic Complexity and Evolution, DOI 10.1007/978-3-319-07935-6_7

7.1.1   Industry Dynamics – An Overview of the Major Trends

Figure 7.1 shows the total number of LSPs (black line), PROs (gray line) and LSMs (dotted gray line) in Germany in the past 20 years. It is immediately apparent that the yearly number of LSPs exceeds the number of LSMs and PROs throughout the entire period under observation. The early period between 1990 and 1992 is characterized by a strong growth tendency for LSPs followed by a stagnating period between 1992 and 1996. The following decade is characterized by an almost stable growth trend with minor fluctuations followed by a peak in 2005. The last 5 years are characterized by a slight decrease in numbers. With the exception of some minor differences, the overall LSM trend mirrors the long-term LSP trend for the most part. However unlike the LSPs, the number of LSMs stagnates between 1992 and 1996 and is at a significantly lower level throughout. The number of LSMs decreases slightly after 2005, however in the last 5 year period, there are some notable differences between the LSM curve and the LSP curve. This is highlighted by a short but accentuated increase in LSMs followed by a relatively high number of firm exits in 2008.

The PRO line on the graph shows the total number of laser-related universities and public research organizations per annum. The pronounced increase in PROs between 1990 and 1991 is mainly the result of the integration of former GDR research facilities into the FRG's sectoral laser industry innovation system. In general, there is a less marked increase in PROs than in LSMs and LSPs. After 1991, the number of PROs remains remarkably stable during the entire period under observation.

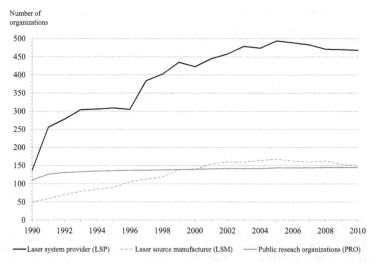

Fig. 7.1 Industry dynamics – overall trends between 1990 and 2010 (Source: Author's own calculations and illustration)

To obtain a more detailed picture of the industry we will now turn our attention to geographical aspects. More precisely, we will explore geographical concentration indices at the industry level and spatial distribution patterns at the state level.

7.1.2 Geographical Concentration

Figure 7.2 illustrates the geographical Herfindahl-Hirschman Indices (HHI-indices) which are calculated on an annual basis and broken down by type of organization (LSP, PRO, LSM).[2]

The graph also includes an average trend line seen here as a dotted black line. It represents the average concentration for all organizations in our sample: LSPs, PROs and LSMs. We can observe a general decrease in concentration which amounts to an increasing geographical dispersion of laser-related organizations in Germany over time. The HHI indicates an overall industry concentration of 0.062 index points at the beginning of the observation period in 1990. Average concentration decreases until 2003 after which the trend remains stable at around 0.04 index points.

However, a closer look at the geographical concentration tendencies, broken down by organizational type, reveals some interesting insights. LSPs (black line) have the highest geographical concentration at the beginning of the observation period of about 0.13 index points. This is followed by a comparably sharp decrease in concentration over time.

In contrast, the geographical dispersion tendency is less pronounced for LSMs (gray dashed line). The LSM concentration level starts at about 0.08 index points in 1990 and, after decreasing sharply in the first 2 years, they level off at around 0.06 index points in 1996. After some minor fluctuations between 1996 and 2001, the LSM trend stabilizes at about 0.05 index points and remains relatively stable until the end of the observation period.

Finally, a look at the geographical concentration patterns of PROs (gray line) reveals a different picture. In contrast to LSMs and LSPs, PROs display an increasing geographical concentration over time. Between 1990 and 1991 there is a short but pronounced increase in the geographical concentration of PROs. The curve remains relatively stable over the course of the next 19 years showing little fluctuation and remaining at between 0.036 and 0.038 index points.

7.1.3 Spatial Distribution Patterns

Next, we refine our initial findings by changing our analytical perspective and illustrating the location of laser-related organizations within a geographical space. Figure 7.3 shows the spatial distribution of LSPs, LSMs and PROs based on laser

---

[2] For a detailed description of the calculation procedure, see Sect. 5.3.2.

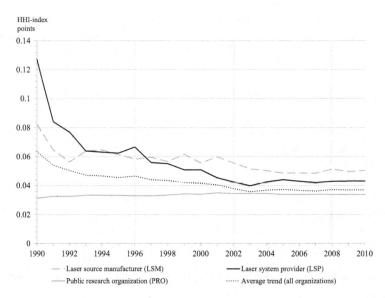

Fig. 7.2 Geographical concentration indices in the German laser industry (Source: Author's own calculations and illustration)

industry maps at four distinct points in time.[3] These maps are divided into 97 *"Raumordnungsregionen"* (i.e. planning regions) in order to provide a fine-grained picture of the organizations' positioning within the geographical space.[4] LSMs and PROs are illustrated by differently shaped and sized elements on the maps whereas the number of LSPs is reflected in the shading of the regions.[5]

The data reveals that the German laser industry included 138 LSPs, 50 LSMs and 110 PROs in 1990. It should be noted that this year saw a relatively high number of LSMs and PROs in three planning regions: Munich (planning region 93: with LSMs = 10, PROs = 7), Berlin (planning region 30: with LSMs = 6, PROs = 8), and Stuttgart (planning region 72: with LSMs = 3, PROs = 7). In addition, Fig. 7.3a indicates that PROs are quite equally dispersed over the geographical space. This confirms our previous findings. However, a look at the spatial distribution of LSPs in this first year provides a somewhat different picture (cf. Fig. 7.3a). The largest number of LSPs was located in Munich with a total of 44 firms (planning region 93). With eleven firms in Starkenburg (planning region 52), 7 firms in Dusseldorf (planning region 42) and 6 firms in the Rhine-Main region (planning region 51), LSPs were concentrated in quite different regions than LSMs and PROs at that point in time.

---

[3] We chose the years 1990, 1996, 2002 and 2008 based on the findings in Sects. 7.1.1 and 7.1.2 since these yearly snap-shots reflect some remarkable turning points for the organizations under observation.

[4] Appendix 2 provides a complete list of the 97 planning regions as applied in this analysis.

[5] The ESRI ArcMap 10.0 software package was applied to visualize the spatial distribution patterns in the German laser industry. We would like to thank Mr. Michael Barkholz for his support.

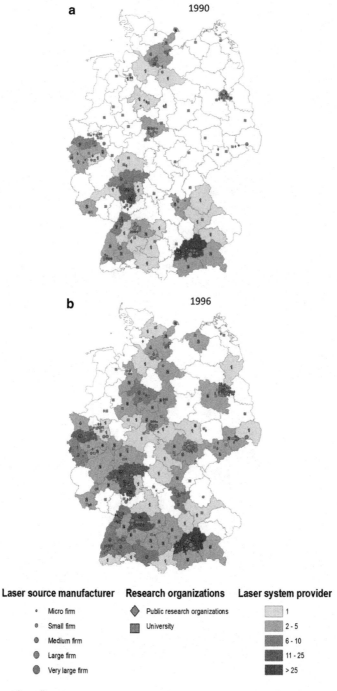

Fig. 7.3 (continued)

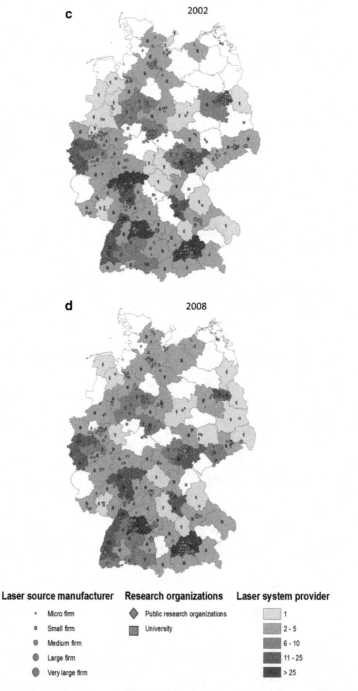

Fig. 7.3 Spatial distribution of LSPs, PROs and LSMs (Source: Author's own calculations and illustration)

In 1996, the total number of organizations increased among all types of organizations to a total of 306 LSPs, 137 PROs and 106 LSMs. The spatial distribution of PROs remained nearly unchanged over the entire observation period aside from a few notable exceptions (cf. Fig. 7.3, a–d). However, a comparison of the geographical locations of PROs in 1990 and 1996 reveals some interesting patterns. Data indicates that about 50 % of the total increase in PROs between 1990 and 1996 took place in only three planning regions – the Upper Elbe Valley (planning region 58: from 3 to 7 PROs), East Thuringia (planning region 56: from 1 to 4 PROs) and Berlin (planning region 30: from 8 to 13 PROs). All of these regions are located in the eastern part of Germany (cf. Fig. 7.3b). In 1996, the number of PROs in Munich remained just as high as in the years before. A closer look at the geographical distribution of LSMs between 1990 and 1996 shows that these firms entered the scene in 15 additional planning regions. In other words, in 1996, we can find at least one LSM in 39 out of every 97 planning regions. Once again Munich has the highest number of LSMs in a given year with a total of 15 firms. The sharp increase in LSMs in Berlin (from 6 to 12 LSMs) and East Thuringia (from 0 to 10 firms) by 1996 is quite remarkable. Finally, this period is marked by the emergence of LSPs throughout the entire landscape with the highest increases in LSPs in the western and the southern parts of Germany (cf. Fig. 7.3b). This is reflected in the doubling of LSPs in the Rhine-Main area (planning region 51), Dusseldorf (planning region 42), and Starkenburg (planning region 52). The number of LSPs in Stuttgart (planning region 72) rose considerably (from 2 LSPs in 1990 to 13 LSPs in 1996). Not surprisingly, the Munich region shows the highest presence of system providers (a total of 65 LSPs) at that time.

By 2002, the total number of organizations had again increased throughout all three categories. Data for this year shows there were 458 LSPs, 160 LSMs and 142 PROs. A comparison of 1996 and 2002 reveals some remarkable patterns. We will start by looking at LSMs (cf. Fig. 7.3c). Compared to 1996, the number of firms in the dominant southern regions increased on average by about 30 % – Stuttgart (planning region 72: by 33 %), Southern Upper Rhine (planning region 77: by 25 %), and Munich (planning region 93: by 33 %). In contrast, the eastern regions present a rather heterogeneous picture between 1996 and 2002. The total number of LSMs increased at quite a different rate. For instance, Berlin (planning region 30) shows a pronounced increase of 83 % to a level of 22 LSMs in 2002. By contrast, data for East Thuringia (planning region 56) indicates a moderate increase of 20 % to a level of 12 LSMs in 2002. Neither the number nor the positioning of the PROs in the geographical space changed substantially compared to the situation in 1996. A closer look at the PROs reveals that research facilities in the regions of the Upper Elbe Valley (planning region 58: with 7 PROs), East Thuringia (planning region 56: with 4 PROs), Berlin (planning region 30: with 15 PROs) and Munich (planning region 93: with 7 PROs) clearly dominated the scene. Figure 7.3c illustrates the increasing dispersion of LSPs throughout the planning regions. Unlike in the previous years, the number of LSPs in the western regions of Germany increased at a significantly lower rate. The number of LSPs in Dusseldorf (planning region 42) increased slightly to a level of 18 firms in 2002 whereas Starkenburg (planning region 52) lost about 4.5 % of its LSPs compared to previous years. The same is true for some southern regions like Munich where the number of LSPs remained constant at 65 firms. Surprisingly, the industrial region of Central Franconia (planning region 86) exhibited a remarkable growth tendency in the last 6 years with LSPs increasing from 8 to 17 firms. Finally, two regions in the eastern part of Germany made significant gains in terms of LSP presence in 2002. The number of laser source providers in East Thuringia (planning region 56) and Berlin (planning region 30) nearly doubled over the course of 6 years.

We can identify a total of 472 LSPs, 163 LSMs and 145 PROs for 2008. A comparison between 2002 and 2008 reveals no great surprises in terms of geographical concentration for either PROs or for LSMs. In 2008 the number of PROs in the regions of the Upper Elbe Valley (planning region 58), East Thuringia (planning region 56), Berlin (planning region 30) and Munich (planning region 93) was at the same level as 6 years previously. In Central Franconia (planning region 86) one new public research facility entered the scene. 2008 saw an increase of at least one or more LSMs in 52 % of the planning regions. The number of LSMs in the dominant regions did not change considerably over the course of 6 years. During this period the dispersion of LSPs increased slightly. Figure 7.3d illustrates the increasing emergence of LSPs in regions around Berlin (planning region 30), East Thuringia (planning region 56), Munich (planning region 93), Stuttgart (planning region 72) and Central Franconia (planning region 86).

In summary, the comparably sharp increase in PROs between 1990 and 1996 can be explained to a large extent by the fact that former GDR research facilities were being integrated into the German laser industry innovation system after the reunification. The spatial distribution of PROs in subsequent time periods remained nearly unchanged. The pronounced increase in LSMs in Thuringia during the early 1990s was largely driven by the reorganization and integration of former state-owned companies such as VEB Carl Zeiss Jenoptik into the German sectoral innovation system. In other words, spin-offs are strongly influenced by the dominate actors in the region. In 1996 about 40 % of all laser source manufactures were located in only five of the 97 planning regions. Berlin had an especially high number of LSM entries during that time. The following decade was characterized by industry growth and geographical dispersion tendencies. After a short but pronounced increase in LSPs in West Germany this trend slowed down in 2002.

In summary, our analysis shows that the laser organizations were concentrated quite early on in the regions of Munich, Thuringia, Berlin, and in and around Stuttgart. These geographical areas still constitute the centers of the German Laser Industry.

7.2 A Closer Look at the Core of the Industry

In this section we focus on LSMs and PROs for the following reasons. As outlined before, LSMs constitute the core of the German laser industry due to their central position along the industry value chain. Thus this section explores the entry and exit dynamics as well as size distribution patterns for the entire population of LSMs between 1990 and 2010. Breaking up the data into entries and exits reveals some details that would otherwise go undetected. Because focus is on innovation networks, the technological dimension of the industry has to be taken into particular consideration. Consequently, we provide some descriptive statistics and give an overview of all laser-related public research organizations active in the period between 1990 and 2010.

7.2.1   Exploration of LSM Entry and Exit Dynamics

The upper half of Fig. 7.4 illustrates the number of actively operating LSMs on an annual basis (cf. Buenstorf 2007; Kudic et al. 2011). Starting with a total of 50 active LSMs in 1990 we observed a total of 183 entries and 83 exits resulting in a total number of 233 firms throughout the entire observation period. The lower half of Fig. 7.4 illustrates firm entries (darkly shaded bars) and firm exits (lightly shaded bars).

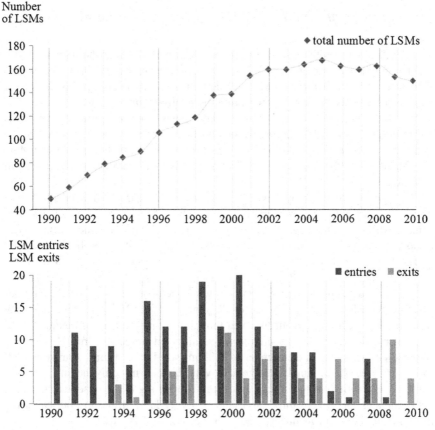

Fig. 7.4 Industry dynamics – LSM entries and exits (Source: Author's own calculations and illustration)

The overall trend indicates a 3.4-fold increase in firms over the course of just 15 years peaking in 2005 at 168 firms. This is followed by an overall decrease to 150 LSMs in 2010. Our data indicates the highest number of firm entries in the years 1995, 1999 and 2001 with firm entries peaking in 2001. In contrast, both 1990 and 2010 are characterized by no firm entries at all. However, a relatively high number of firm exits in 2001 resulted in the steepest net increase in LSMs having occurred in 1999. No LSMs left the industry from 1991 to 1993, nor did any leave in 1996 or 1999. The total number of firm exits peaked in 2000 with eleven LSMs leaving the industry. However, due to twelve firm entries, the overall industry growth trend remained unbroken resulting in a marginal increase for this year. Three years later there is another case of a high fluctuation of entries and exits that are not reflected in the overall industry growth trend. More precisely, both the number of LSM entries as well as the number of LSM exits amounts to a total of nine firms in 2003. Unlike in previous years, firm exits significantly exceeded firm entries in 2006 and 2007. After an increase in the total number of firms in 2008 we can observe a slightly increased number of exits at the end of the observation period.

7.2.2   Size Distribution of LSMs at the National Level

Figure 7.5 illustrates the size distribution of German LSMs at the national level. The bar graph at the top of Fig. 7.5 shows the absolute number of LSMs divided into five distinct size categories. The line graph at the bottom of Fig. 7.5 shows the changes in size distribution by presenting the relative terms for each size category. Firm size categories are based on the number of employees in a firm. Smaller firms are represented by lighter colored bars and lines while darker shades symbolize larger firms. To enhance visibility, micro firms are represented by black-hashed bars and the dotted black line. As before, the period under observation lasted from 1990 to 2010.

We start our analysis by looking at the absolute figures displayed in the bar graph. To start with, the comparably high number of micro firms and small-sized firms is striking. At the beginning, more than half of all firms are micro firms. After a short increase in the number of micro firms in the early 1990s the absolute number of LSMs remained roughly constant for nearly a decade at around 45–55 firms before starting to decline after 2005. We can also observe an increase in the number of small-sized firms over time. Starting with twelve firms in 1990 the number increases five-fold over the course of 20 years to 59 LSMs. The trend is nearly the same for medium-sized firms, even though the total number of medium-sized firms is roughly one half the number of small-sized firms. Accordingly, we can observe a 4.5-fold rise in the number of medium-sized firms, from six firms in 1990 to 27 firms in 2010. Finally, large and very large firms only begin to play a significant quantitative role after 1994. At the beginning of the observation period data indicates there were three large and two very large firms. In both cases the absolute number of firms quadrupled over the course of 20 years.

The relative values enable us to get a clearer picture of the firm size distribution within the industry. The line graph provides the relative terms for all five firm size

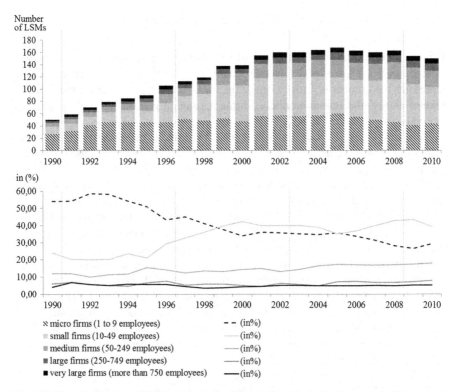

Fig. 7.5 Size distribution of LSMs at the national level (Source: Author's own calculations and illustration)

categories. The black dotted line represents the micro firms in the sample. It becomes obvious that micro firms dominated the industry at the onset of the observation period, not only in absolute terms but also in relative terms. However, the decreasing trend indicates a diminishing relevance over time of micro firms in comparison to larger firms. A closer look at small firms reveals a completely different picture. The line graph clearly indicates a rise in the significance of small-sized firms compared to other firms in the sample. The same is true for medium-sized firms in terms of relative figures. In other words, both small and medium-sized firms gain in importance over time. Finally, the proportion of large firms and very large firms remains remarkably stable over the entire observation period.

In summary, our descriptive analysis reveals that micro firms, in particular, lost ground in the German laser industry over time. One possible explanation is that micro firms outgrow their infancy and, in time, turn into small firms. Small firms show the highest average growth rates, followed by medium, large and very large firms. At the end of the observation period small firms dominate the scene. It should also be stated that due to the moderate but continuous growth of medium-sized LSMs (by about 20 % up until 2010) there is a clear increase in the presence of both small and medium-sized LSMs in the German laser industry during the observation period. Next we explore the size distribution of LSMs at the regional level.

7.2.3   Size Distribution of LSMs at the Regional Level

Figure 7.6 shows the size distribution of LSMs in a geographical space.[6] The level of analysis is the state level ("Bundesländer"). The vertical axis in the bar graph represents the number of firms whereas the horizontal axis represents time. For the sake of clarity, the annual number of firms per region is grouped into 3-year time intervals.

To start with, we look at the federal states of Bavaria and Baden-Wurttemberg, both located in the southern part of Germany. Bavaria shows the highest absolute number of LSMs throughout the entire period of observation. Looking more closely at firm size distribution reveals that the high number of small and medium-sized firms is responsible for the above-average presence of LSMs in Bavaria.

In contrast, the majority of large and very large firms are located in Baden-Wurttemberg. Baden-Wurttemberg also shows a relatively stable trend in firm growth throughout the entire period of observation. The federal state of Thuringia, located in the eastern part of Germany, reveals a very similar picture to that of Bavaria. In both cases we see a pronounced growth phase in the early 90s followed by a shakeout at the end of the observation window. Even though the total number of firms in Thuringia is lower than in both southern states, we can again observe a relatively high number of very large firms. Moreover, in all three federal states – Bavaria, Baden-Wuerttemberg and Thuringia – micro firms lose ground over time whereas small firms and medium-sized firms are on the rise. In summary, firm size distribution in these three states follows very similar patterns with only minor exceptions.

The situation looks somewhat different in Lower Saxony since no very large firms are located in this state. Nonetheless, the relatively high number of micro firms, small firms and exceptionally the presence of some large firms highlights the importance of Lower Saxony as a location for LSMs in Germany. The situation in Berlin is characterized by a comparatively high number of both micro and small firms. The most plausible explanation for this seems to be the comparably high number of PRO spin-offs in Berlin. Medium, large and very large firms are completely missing here.[7] This is important to note since otherwise there is a danger of overemphasizing Berlin in terms of LSM presence. At first glance, the situation in North Rhine-Westphalia looks quite similar to that of Berlin. However, the main difference is the existence of a solid stock of medium-sized firms through-out the entire observation period. Finally, we look at the federal states of Hesse and Hamburg. The comparatively low presence of LSMs conceals the fact that a small number of highly relevant actors are located in these regions. As we will see later the same is true for the federal state of Rhineland Palatinate. The remaining federal states had a very low number of LSMs throughout the 21 year period. Saarland shows no LSM presence at all.

---

[6] A similar analysis was previously conducted by Kudic et al. (2011).

[7] A contemporary study that focuses on the laser industry in Berlin-Brandenburg confirms this finding. It found that 94 % of all firms studied have less than 50 employees (TSB 2010, p. 9).

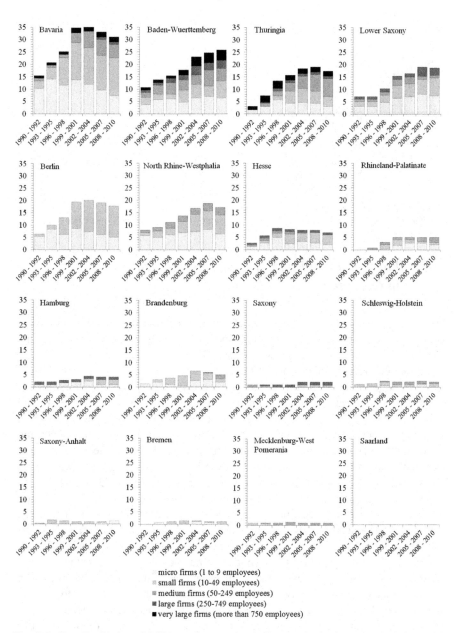

micro firms (1 to 9 employees)
small firms (10-49 employees)
medium firms (50-249 employees)
large firms (250-749 employees)
very large firms (more than 750 employees)

Fig. 7.6 Size distribution of LSMs at the state level (Source: Author's own calculations and illustration)

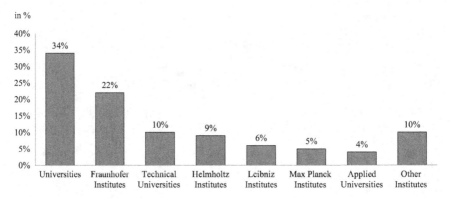

Fig. 7.7 Composition of laser-related PROs in Germany (Source: Author's own calculations and illustration)

7.2.4 Laser-Related Public Research Organizations in Germany

Figure 7.7 illustrates the composition of public laser-related research facilities in Germany. Average values are reported since only minor changes occurred to the composition of the research landscape during the observation period.

Public research organizations (PROs) in Germany that are actively operating in the field of laser research can be grouped into eight categories. Universities are divided into three categories – technical universities, universities (in general), and universities of applied science – and make up about half of all laser-related PROs in Germany. Laser-related research activities were identified at the chair level and thereafter aggregated at the overall university level. In total, the proportion of technical universities, universities and universities of applied science was 10 %, 34 %, and 4 % respectively. Data reveals hardly any fluctuation in terms of population entries or exits among these organizations over the entire observation period. Next we turn to non-university research facilities. The German research landscape is characterized by four large non-university research societies. To start with, Max Planck Society is a publicly funded, non-governmental and non-profit organization. Its 80-plus institutes conduct fundamental research in the areas of natural sciences, life sciences, social sciences and the humanities.[8] The proportion of Max Planck Institutes active in the field of laser research amounts to 5 % on average. The Helmholtz Association is a community of 18 scientific-technical and bio-medical research centers which conduct research in the fields of energy, earth and environment, health, key technologies, structure of matter, aeronautics, space and transport.[9] Our data shows that about 9 % of the laser-related research facilities

---

[8] Information from: http://www.mpg.de (Accessed: February 2012).

[9] Information from: http://www.helmholtz.de (Accessed: February 2012).

identified belong to the Helmholtz Association. The Leibniz Association comprises 86 scientifically and organizationally independent research institutions that conduct research in the areas of natural science, engineering, environmental science, economics, social science, infrastructure research and the humanities.[10] Our data shows a sharp increase in Leibniz Institutes at the beginning of the observation period. Between 1990 and 1993 the number of institutes active in the field of laser research nearly quadrupled. This can be explained to a large extent by the integration of former GDR research facilities into the FRG science landscape. Leibniz Institutes make up, on average, about 6 % of all PROs in the sample.

The last group of institutes is organized under the umbrella of the Fraunhofer Society. The Fraunhofer Society is Germany's largest application-oriented research organization which is made up of around 60 institutes. These institutes primarily conduct applied research in the fields of health, security, communication, energy and the environment.[11]

Our data reveals two interesting facts. Firstly, at the beginning of the observation period we can again witness a steep increase in population entries. Between 1990 and 1991 we registered a rise by over 50 %; thereafter there was hardly any change in terms of population entries or exits. The explanation for this is similar to that of the Leibniz Institutes. Secondly, Fraunhofer Institutes make up the largest percentage of non-university research organizations in our sample at about 22 %. Finally, about 10 % of the overall population, a notable percentage of laser-related PROs, do not belong to one of the four large German research societies.

References

Buenstorf G (2007) Evolution on the shoulders of giants: entrepreneurship and firm survival in the German laser industry. Rev Ind Organ 30(3):179–202

Kudic M, Guhr K, Bullmer I, Guenther J (2011) Kooperationsintensität und Kooperationsförderung in der deutschen Laserindustrie. Wirtschaft im Wandel 17(3):121–129

TSB (2010) Laser technology report – Berlin Brandenburg. TSB Innovationsagentur GmbH, Berlin

---

[10] Information from: http://www.wgl.de (Accessed: February 2012).

[11] Information from: http://www.fraunhofer.de (Accessed: February 2012).

Chapter 8
Evolution of the Industry's Innovation Network

> *In the long history of humankind (and animal kind, too),*
> *those who learned to collaborate and improvise most*
> *effectively have prevailed.*
>
> (Charles Darwin)

Abstract At the very heart of this book is the analysis of R&D cooperation and networking activities of firms in science-driven industries. As outlined before, we have used two official databases to gather data on nationally and supra-nationally funded R&D cooperation projects (cf. Sect. 4.2.3). These two data sources provided the basis for the construction of the German laser industry innovation network. Network analysis methods (cf. Sect. 5.2) provide us with a broad range of instruments to explore and analyze structural characteristics of networks (Wasserman and Faust 1994; Degenne and Forse 1999; Carrington et al. 2005; Borgatti et al. 2013). These methods can be used to analyze both network snap-shots at a particular point in time, and evolving network patterns over time. This chapter is divided into three sections. Section 8.1 gives an overview of the organizations involved in publicly funded R&D cooperation projects from various angles. Based on these findings, we explore the proportion of LSMs and PROs participating in two types of publicly funded research projects – "CORDIS" and "Foerderkatalog". Then, we take an initial look at the large-scale topology of the German laser industry innovation network. Next we focus on the evolutionary change patterns of the German laser industry innovation network. In Sect. 8.2 we start our longitudinal exploration by analyzing a set of basic node-related and tie-related network measures over time. In Sect. 8.3 we provide an in-depth analysis of the network topology by testing for the existence of three distinct large-scale network properties. First, we analyze the overall degree distribution and check for the emergence of scale-free properties (Barabasi and Albert 1999). Then we test whether the German laser industry's innovation network exhibits small-world properties by applying the method proposed by Watts and Strogatz (1998). Finally, we use different but complementary methodological approaches to check for the existence of a core-periphery structure (Borgatti and Everett 1999). We finish off the descriptive analysis by visualizing the evolution of the German laser industry innovation network over time.

© Springer International Publishing Switzerland 2015   153
M. Kudic, *Innovation Networks in the German Laser Industry*,
Economic Complexity and Evolution, DOI 10.1007/978-3-319-07935-6_8

8.1   Laser-Related Publicly Funded R&D Cooperation Projects

The aim of this section is threefold. First we provide some basic descriptive statistics on publicly funded R&D cooperation projects broken down by cooperation type. Then we explore the involvement of LSMs and PROs in the cooperation projects over time. Finally, we take a look at all cooperation activities between German laser source manufacturers and laser-related public research organizations between 1990 and 2010. In other words, we illustrate the transition from a dyadic perspective to a network perspective by exploring the large-scale topology of the German laser industry innovation network over the entire observation period.[1]

8.1.1   Summary Statistics on Publicly Funded R&D Cooperation

Table 8.1 shows some descriptive statistics on publicly funded R&D cooperation projects based on both *Foerderkatalog* and *CORDIS* data for the period between 1990 and 2010.

The Foerderkatalog data encompasses, in total, information on approximately 110,000 completed or ongoing subsidized research projects. We were able to identify 416 laser-related R&D cooperation projects for the entire population of 233 German laser source manufacturers. A total of 2,656 organizations were involved in these projects. Data exploration revealed an overall involvement of 643 LSMs and 570 laser-related PROs. In other words, we found at the project level a significant degree of interconnectedness among organizations in our sample. Data on the remaining 1,443 organizations was fully recorded but due to the focus of this study and related network boundary specifications, they were not included. At the project level our data reveals a minimum of 2 and a maximum of 33 partners. An average of 6.39 organizations was involved in each project with a standard deviation of 3.96.

The overall number of project files in the *CORDIS* database is considerably smaller and consists of 31,000 files.[2] We identified a total of 154 R&D cooperation projects for the entire LSM population. We found that a total of 189 LSMs and 132 PROs were involved in these projects. As before, other types of organizations were fully registered but not included as they are not the subject of this analysis. *CORDIS* projects are considerably larger than *Foerderkatalog* projects. The

---

[1] We use the standard routines implemented in UCI-Net 6.2 (Borgatti et al. 2002) to calculate network measures and we employ the software package NetDraw (Borgatti 2002) for the visualization of the German laser industry innovation network.

[2] This figure refers to our database extract provided by the *CORDIS* Service Team, European Commission (latest update: end of 2010).

Table 8.1 Publicly funded R&D cooperation projects – broken down by cooperation type

Descriptives	Foerderkatalog projects	CORDIS projects
Overall number of project files	110,000	31,500
Total number of laser-related projects	416	154
Total number of organizations	2,656	1,607
Other types of organizations	1,443	1,286
Total number of LSMs	643	189
Total number of PROs	570	132
Max. no. of organizations at the project level	33	53
Min. no. of organizations at the project level	2	2
Avg. project size (no. of partners)	6.385	10.435
Std. dev. project size (no. of partners)	3.955	8.019

Source: Author's own calculations

minimum and maximum number of project partners involved at the project level was two and 53 respectively. The average project size, based on the number of partners, came to 10.44 with a standard deviation of 8.02.

8.1.2 R&D Cooperation Involvement of LSMs and PROs

Figure 8.1 shows LSM (black line) and PRO (dotted line) participation in publicly-funded cooperation projects between 1990 and 2010 as expressed in terms of percentages at the national level. The line graph on the left shows the proportion of LSMs and PROs participating in *CORDIS* projects whereas the line graph on the right illustrates their involvement in *Foerderkatalog* projects. The line graph below these illustrates LSM and PRO participation in either *CORDIS* or *Foerderkatalog* projects. Basic descriptive statistics are reported below each of the three line charts.

In general, we can observe an increasing percentage of organizations participating in publicly funded research projects. *CORDIS* project data indicates a maximum percentage of LSM and PRO involvement at 16.23 % and 17.24 % respectively. On average, cooperation project participation in *CORDIS* projects is slightly higher for LSMs (at 8.94 %) compared to PROs (at 8.85 %). In both types of organizations we see only minor deviations from the upwards-sloping long-term trend. In contrast, the involvement of LSMs and PROs in *Foerderkatalog* projects is significantly higher than in *CORDIS* projects. The exploration of our data reveals a maximum participation in *Foerderkatalog* projects of 44.16 % for LSMs and 54.48 % for PROs. The average participation of LSMs and PROs in *Foerderkatalog* projects is 34.11 % and 39.72 % respectively. In addition, we can observe higher fluctuations for PROs (standard deviation = 10.82 %) compared to LSMs (standard deviation = 6.12 %) for the period in question.

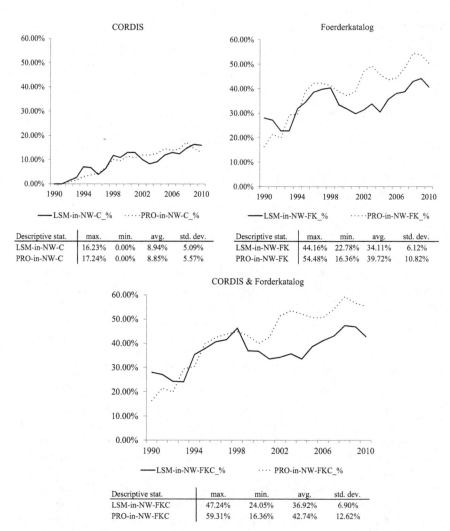

Fig. 8.1 Participation OF LSMs and PROs in publicly funded cooperation projects (Source: Author's own calculations)

The overall participation in both types of publicly funded cooperation projects is displayed in the line graph below. Data on LSMs indicates a minimum participation of 24.05 % in 1990, a maximum participation of 47.24 % in 2008 and an average participation of 36.92 %. In contrast, PROs show a significantly lower rate of involvement in cooperation projects (16.36 %) at the onset. This initially low involvement in cooperation projects quickly changes direction after a rather short period of time. The overall participation of PROs in either *CORDIS* or *Foerderkatalog* projects increases about 2.5 times between 1990 and 1998. This trend continues with nearly the same intensity and some minor fluctuations until the end of the observation period. As a consequence, the average percentage of PROs participating in cooperation projects is 42.74 % and the maximum percentage of cooperation reaches nearly 60 % in 2008.

8.1.3   Large-Scale Network Topology of the Innovation Network

In general, the visualization of network is no trivial matter. It allows the researcher to obtain an initial and initiative understanding of the structural configuration of the system (Borgatti et al. 2013, p. 124). Figure 8.2 illustrates all cooperation activities between German laser source manufacturers and laser-related public research organizations between 1990 and 2010.

According to Borgatti et al. (2013, p. 101) there are three basic approaches to network layout: (a) an attribute-based scatter plot, (b) a multidimensional scaling (MDS) layout, and (c) graph theory-based layout algorithms.

We visualize the network by using a simple random layout (cf. Fig. 8.2a) and by applying a spring-embedded layout (Fig. 8.2b) which was originally proposed by Eades (1984) and Fruchterman and Reingold (1991) and it is still one of the most commonly used graph theoretical layout algorithms. The basic idea behind the algorithm is simple. "Its effect is to distribute nodes in a two-dimensional plane with some separation, while attempting to keep connected nodes reasonably close together" (Golbeck and Mutton 2005, p. 173). We employ the geodesic distance criterion, which is defined as the shortest path connecting any pair of nodes in the network (Wasserman and Faust 1994), to compute the layout. We used NetDraw 2.0 to visualize the network (Borgatti 2002).

These two simple initial explorations already contain some important information. For instance, the density of the network structure indicates a pronounced cooperation propensity among firms and other organizations in the industry. The size of the node is determined by the network actors' degree of connectedness (i.e. the number of direct linkages). Figure 8.2 indicates that some network actors seem to attract nodes at a higher rate than others. Both types of actors, LSMs[3] as well as PROs,[4] seem to be spread out over the entire network and occupy positions in densely as well as sparsely connected areas of the networks.

However, an in-depth exploration and analysis of the network properties requires a decomposition of the network. As a result, we apply a time-discrete

---

[3] Each ID in Fig. 8.3 with the syntax: "LSMxxx" represents one of the 233 laser manufacturing firms. Note that the sequential ID number can be larger than the total number of firms in our sample.

[4] Public research organizations are symbolized by the following abbreviations: University = "RxxxU", University of Applied Sciences = "RxxxA", Technical University = "RxxxT", Fraunhofer Institute = "RxxxF", Max Planck Institute "RxxxM", Helmholtz Institute "RxxxH", Leibniz Institute "RxxxL" and other laser-related PROs = "RxxxD".

a

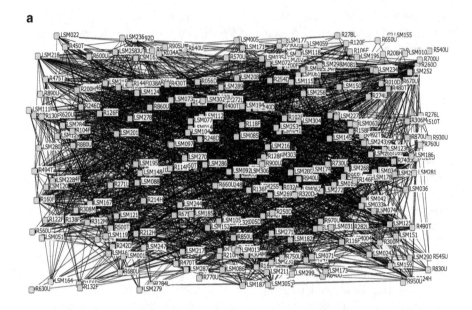

b

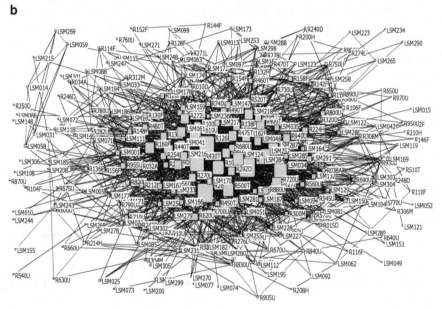

Fig. 8.2 The German laser industry innovation network. (**a**) Random layout. (**b**) Spring-embedded; degree-based node size (Source: Author's own calculations and illustration)

approach and analyze structural changes to both node-related and tie-related network characteristics broken down by year.

8.2 Longitudinal Exploration of Basic Network Characteristics

In this section we apply an exploratory social network analysis approach (De Nooy et al. 2005). The primary objective of this method is to reveal structural network particularities and make them measurable. Emphasis is not on refuting established structural hypotheses but rather on measuring, exploring and visualizing network properties. In other words, "[...] instead of testing pre-specified structural hypotheses, we explore social networks for meaningful patterns" (De Nooy et al. 2005, p. 5). Exploratory social network analysis is conducted in four steps: network definition, network manipulation, identifying network features and visualization (De Nooy et al. 2005, pp. 5–6).

8.2.1 Basic Network Change Patterns: Measures at the Node Level

In order to explore basic network measures at the node level over time, we chose a time-discrete approach and separated the network into annual slices of time. All network measures are calculated on a yearly basis by using both *Foerderkatalog* and *CORDIS* data. Note that this exploration differs significantly from the analysis reported before (cf. Sects. 8.1.1 and 8.1.2).

Now we are not focusing on the organizations' participation in different types of R&D multi-partner collaborations at the project level but on the involvement of both types of network actors in the overall German laser industry innovation network. Figure 8.3 illustrates the network boundaries and the size of the network.

The number of all actively operating LSMs and PROs determines the outer boundary of the innovation network. In other words, these are all organizations which are at risk of cooperating, irrespective of whether they are part of the network or not. All organizations with at least one dyadic R&D linkage to another LSM or PRO in the sample are considered to be an integral part of the innovation network. Thus, the outer circle (dotted line) in Fig. 8.3 illustrates the network's outer boundaries whereas the inner circle (solid line) reflects the actual size of the network over time. In 1990, only 20.1 % of all LSMs and PROs in the industry were actively involved in the innovation network. This comparably small participation rate nearly doubles over the course of just 5 years. In 1995, we register a network participation rate of 38.94 %. Despite some minor fluctuations, the participation rate continues to grow over the next 10 years. The percentage of LSMs and

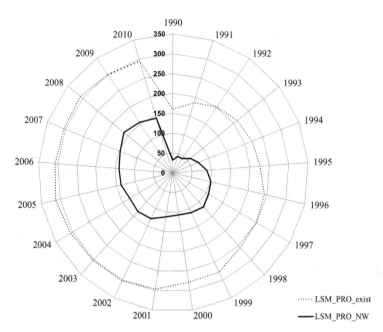

Fig. 8.3 Network boundaries and network size (Source: Author's own calculations and illustration)

PROs actively involved in the German laser industry innovation network ranges from 37.35 % in 2001 to 45.53 % in 1998. After 2005 we again record a noticeable increase in network entries with a maximum network participation rate of 52.92 % in 2008. Thereafter both trend lines begin to decrease.

In addition to network size, the connectedness of a network is arguably one of the most salient network features if one wants to get an in-depth understanding of the structural network configuration itself (Wasserman and Faust 1994, p. 109) and to understand the evolutionary network change processes over time (Amburgey et al. 2008, p. 178). A network is called "connected" as long as there is at least one path that connects all pairs of actors in a network (Newman 2010, p. 142). In contrast, a "disconnected" network consists of at least two components where a component is defined as a subgroup of network actors that are connected with one another but have no connection to other connected network subgroups (Newman 2010, p. 142). Figure 8.4 illustrates the fragmentation of the German laser industry innovation network broken down by cooperation type. The ordinate records the number of network components and the abscissa captures the time dimension. On the left we see the *CORDIS* network (dotted line), on the right is the *Foerderkatalog* network (gray line), and at the bottom is the overall network consisting of both cooperation types (black line). On average, the *CORDIS* network is characterized by a higher fragmentation (average component count = 4) and exhibits less pronounced fluctuation tendencies (standard deviation = 2.1) compared to the

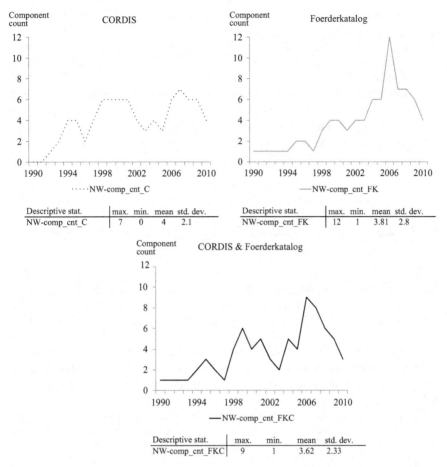

Fig. 8.4 Network fragmentation – annual component counts (Source: Author's own calculations and illustrations)

Foerderkatalog   network   (average   component   count = 3.81;   standard deviation = 2.8).

Comparing the two networks at two separate time intervals gives us a more detailed picture. Between 1990 and 2000 the connectedness of the *Foerderkatalog* network is clearly more pronounced than that of the *CORDIS* network. This tendency, however, changes after 2000. The fragmentation of the *Foerderkatalog* increases considerably and reaches a maximum of 12 unconnected network components in 2006. The component structure of the overall network reveals a slightly different picture. Just like with the two separate networks we can see an increasing tendency towards fragmentation for the overall network over time. This trend, however, is accompanied by some pronounced fluctuations. The overall network consists of 3.62 components on average with a standard deviation of 2.33. Between

1990 and 1993 and in the year 1997 the network is fully connected and consists of one single giant component. Only 2 years later, in 1999, we can observe a total of six components in the overall network. The fragmentation reaches a maximum of nine unconnected components in 2006 and decreases considerably in subsequent time periods.

In summary, we gain some interesting insights by exploring the size and component structure of the innovation network. Nonetheless, several questions remain unanswered. For instance, the node structure and tie structure within and between the network components remains entirely unconsidered. These issues will be addressed later. First, however, we focus on the exploration of some basic tie-related network characteristics.

8.2.2   Basic Network Change Patterns: Measures at the Tie Level

The overall network density measure provides an initial indication of a network's structural configuration. It simply indicates to what extent the network actors are connected to each other. Figure 8.5 shows the density measure for the German laser industry innovation network over time.

The overall density of an unvalued network is defined as the total number of ties divided by the total number of possible ties. If all nodes of a graph are adjacent, then it is equal to 1 and the graph is said to be complete (Wasserman and Faust 1994, p. 102).

The German laser industry innovation network had a maximum overall network density of 0.441 in 1990. The density decreased continuously until 1998. After a

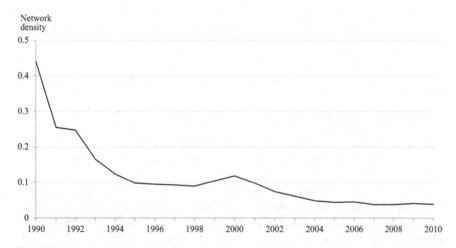

Fig. 8.5 Network density – overall network density (Source: Author's own calculations and illustration)

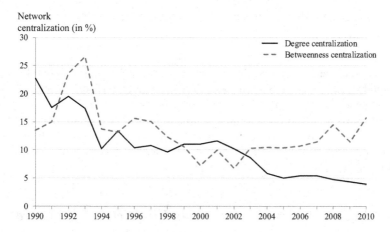

Fig. 8.6 Network centralization – degree and betweenness centralization indices (Source: Author's own calculations and illustration)

short-lived density peak in 2000 ($NW_d = 0.118$), the overall network density began decreasing again, reaching a minimum network density of 0.038 in 2010.

Now we turn our attention to global centrality measures originally proposed by Freeman (1979). Figure 8.6 displays two network centralization indices – degree and betweenness centralization – for the German laser industry network between 1990 and 2010.

The degree centralization index indicates an alignment of the network actors' degree centralities over time. The index has a maximum value of 22.74 % in 1990 and decreases with some marginal fluctuations. In 2010, the index reaches a minimum value of 3.96 % indicating that network actors show only minor disparities in terms of their degree centralities. Nonetheless, it should be noted that the degree centralities are by no means equally distributed.

The betweenness centralization index provides quite a different picture. Most remarkably, the index shows a much higher volatility compared to the degree centralization index. During the initial years we can observe a pronounced increase in the index from 13.52 % in 1990 to 26.56 % in 1993. The following years are characterized by an alignment of the network actors' betweenness centralities over time. In 2002 the inequalities among network actors in terms of their brokerage activities reach a minimum with an index value of 6.73 %. In subsequent years the index increases again until the network finally reaches a betweenness centralization of 15.76 % in 2010.

8.3   Exploring the Emergence of Large-Scale Network Properties

This section addresses large-scale network properties. Real world networks differ from random networks in many respects. Accordingly, we check for the existence and emergence of three types of network properties – scale-free distribution, small-world phenomenon, and core-periphery structure – to demonstrate that the structural configuration of the German laser industry network exhibits fairly different patterns than randomly generated reference networks. Finally, we visualize the network topology at four distinct points in time.

8.3.1   Degree Distribution and Scale-Free Network Structure

In random networks the placement of links is purely random which means that the resulting system is characterized by nodes that have approximately the same number of links (Barabasi and Bonabeau 2003, p. 52). In contrast, real-world networks typically show very different large-scale patterns. In a seminal paper on large-scale network properties Barabasi and Albert (1999, p. 510) suggest that "[...] large networks self-organize into a scale-free state." This, however, implies that some actors attract ties at a higher rate than others. The reasons for this can be manifold. For instance, some actors have simply more to offer than others or show a higher capability in establishing or sustaining interorganizational partnerships. In this context, sociologists have highlighted the importance of reputation, status (Podolny 1994) and interorganizational endorsement effects (Stuart et al. 1999). However, these actors are usually called "hubs" (Newman 2010, p. 245) and have a much higher degree than the majority of other network actors.

The exploration of a network's degree distribution provides a simple but powerful diagnostic indicator of whether tie formation in a network is equiprobable (simply random) for all pairs of nodes or systematically biased (Powell et al. 2005, p. 1151). In other words, the existence of these network hubs should be reflected in the overall degree-distribution of the network. "Unlike the tail of a random bell curve whose distribution thins out exponentially as it decays, a distribution generated by a popularity bias has a "fat" tail for the relatively greater number of nodes that are highly connected" (Powell et al. 2005, p. 1151).

Figure 8.7 illustrates the degree distribution of the German laser industry innovation network (above) and a randomly generated Erdös-Renyi network (below). In order to analyze the large-scale properties of the German laser industry innovation network we have generated a random network which is comparable in terms of network size and network density. This procedure was repeated several times to obtain reliable average degree values.

The abscissa represents the degree k and the ordinate measures the fraction of nodes in the network p(k) for each degree value. The right-skewed distribution indicates that the German laser industry innovation network consists of a few

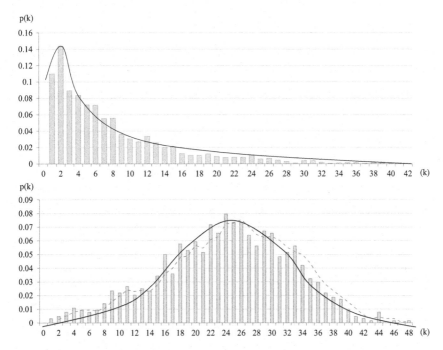

Fig. 8.7 Degree distribution – German laser industry innovation network vs. random network
(Source: Author's own calculations and illustrations)

extremely well-connected actors (with a degree of up to 46) whereas the majority of
network actors are rather sparsely connected (with a nodal degree of 1 or 2).

We follow the procedure proposed by Newman (2010, pp. 247–260) to detect
power-law behavior in networks.[5] The logarithm of the degree distribution p(k) is a
linear function of the degree k with a negative sloping gradient and a constant
y-intercept which can be written as a logarithmic equation by simply taking the
exponential of both sides (Newman 2010, p. 247). This leads to a function p(k) that
is defined by the degree k with a negatively defined constant exponent α, which is
known as the "exponent of the power law", and a constant multiplier C (Newman
2010, p. 248). A simple histogram or scatter graph of the degree distribution plotted
on a log-log scale provides the easiest way to detect power law behavior in real
world networks. A true power-law distribution monotonically decreases over its
entire range and appears in the log-log plot as a negatively sloping straight line
(Newman 2010, p. 249). Figure 8.8 provides the log-log scatter plots of the degree
distributions for the German laser industry network and for a comparable Erdös-
Renyi random network.[6]

---

[5] These types of networks are called scale-free networks (Barabasi and Bonabeau 2003, p. 52).

[6] To provide a solid benchmark for the real world network, we proceeded as follows: First we
calculated the network size and density measures of each real world network on a yearly basis.

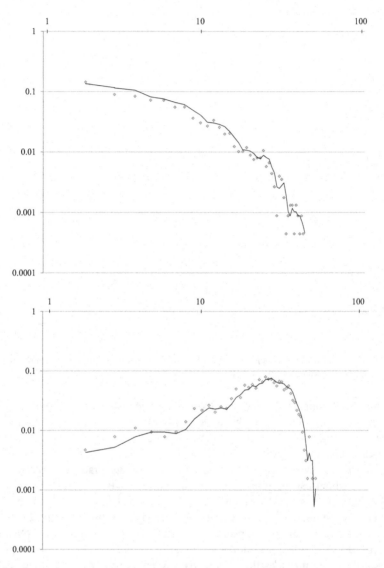

Fig. 8.8 Power law and sale free patterns – German laser industry innovation network versus random networks (Source: Author's own calculations and illustrations)

Then we used the Erdös-Renyi procedure implemented in UCI-Net 6.2 (Borgatti et al. 2002) to generate random networks on an annual basis. Each annual random network corresponded exactly to its real-world equivalent in terms of network size and network density. Finally the annual degree distributions for both the real world network and the random network were accumulated and the results were plotted on a log-log scale.

We aggregated all log degrees (abscissa) and the log node fraction in the network (ordinate) over all time periods. Even though the German laser industry innovation network shows no perfect power law behavior, we can clearly detect the tendency towards the emergence of a straight line in the log-log plot. In other words, the degree distribution of our real world network reveals systematically different structural patterns compared to a purely random network. This indicates a pronounced tendency towards the emergence of scale-free properties. Our analysis reveals .quite similar structural patterns as were reported by Powell and his colleagues (2005) for the degree distribution of the interfirm network (one-mode network: DBFs – DBFs) and the interorganizational network (two-mode network: DBFs – universities) for the US biotech industry.

8.3.2 Small-Word Properties

Now we turn our attention to small-world network properties. Even though the underlying idea of small-world networks can be traced back to a series of network experiments conducted by Stanley Milgram and his team in the late 1960s, it took nearly 30 years before scholars were able to quantify the concept (Watts and Strogatz 1998).

Milgram (1967) showed in his letter-passing experiment that people in the United States are separated, more or less, by six degrees of separation (i.e. letters that have been sent even reach far-off targets after roughly six distinct steps on average). He concluded that a small-world network is characterized by a short path length despite a high level of clustering (Uzzi et al. 2007, p. 78). Small-world properties have some far-reaching implications for innovation networks. As we will discuss in more detail later (cf. Chap. 11), it is plausible to assume that macro-level network properties affect firm innovativeness. However, in this section, we apply the method proposed by Watts and Strogatz (1998) to check for the existence of small-world properties in the German laser industry innovation network. According to this methodological approach, two conventional network measures can be used[7]: the overall clustering coefficient and average path length clustering (Uzzi et al. 2007, p. 78).

We proceeded as follows to check for the existence of small-world properties in the German laser industry network. First we generated a total of 21 Erdös-Renyi random networks for the period under observation, one network for each year.[8] In order to ensure comparability between real world and random networks, both the size and the density parameters were adapted to the actual proportions of the real networks. In general, random networks are characterized by a short average path length and a low clustering tendency as neighboring nodes have the same

---

[7] For details on the calculation and interpretation of both measures, see Sect. 5.2.3.

[8] To gain a more robust random benchmark this procedure has to be repeated several times. However, this may be dispensed with for the purpose of this analysis.

probability of being connected as non-neighboring nodes (Uzzi et al. 2007, p. 79). Then, based on the procedure proposed by Watts and Strogatz (1998), we calculated "clustering" and "reach" measures for both the annually constructed German laser industry networks and for the annually constructed Erdös-Renyi networks. Finally, we calculated the "clustering coefficient ratio", the "path length ratio" (Watts and Strogatz 1998) and the "small-world Q" (Uzzi and Spiro 2005) to compare network properties.

The "clustering coefficient ratio" is defined as the real world network clustering coefficient divided by the random network clustering coefficient. The "path length ratio" is defined as the real world average path length divided by the random network average path length. The small world Q is defined as the "clustering coefficient ratio" divided by the "path length ratio" (Watts and Strogatz 1998; Uzzi and Spiro 2005; Uzzi et al. 2007).[9]

Standard procedures implemented in UCI-Net 6.2 (Borgatti et al. 2002) were applied to calculate both the overall clustering coefficient and the weighted overall clustering coefficient. In accordance with Schilling and Phelps (2007, pp. 1117–1118) we chose the latter measure here since the weighted clustering coefficient provides exactly the same measure as the transitivity index of each transitive triple (Borgatti et al. 2002).

According to Watts and Strogatz (1998) and with reference to Uzzi et al. (2007, p. 79) small-world networks have to fulfill at least one of the following two conditions: (I) a "clustering coefficient ratio" that is many times greater than 1.0 and a "path length ratio" that is approximately 1.0 or (II) a "small-world Q" that is much greater than 1.0.

The threshold values are not exactly specified as they can differ slightly for different types of real world networks. In our case we chose a threshold value of 2.5 for both the "clustering coefficient ratio" and the "small-world Q" and a band of accepted "path length ratio" values ranging from 0.7 to 1.3. Areas between minimum and maximum thresholds are shaded in light gray.

Figure 8.9 shows the "clustering coefficient ratio" (cf. Fig. 8.9, top), the "path length ratio" (cf. Fig. 8.9, center) and the "small-world Q" (cf. Fig. 8.9, bottom) for the German laser industry network between 1990 and 2010. The illustrations show that the conditions specified above are fulfilled with very few exceptions (e.g. year 1990).

In summary, our data clearly shows an increasing tendency towards small-world properties over time.

Concerns were expressed that, unlike unipartite networks, bipartite[10] networks significantly exaggerate the network's true level of clustering and understate the true path length (Uzzi and Spiro 2005, p. 453). Based on the pioneering work of Watts and Strogatz (1998) a new interpretation of small-world indicators for

---

[9] For further details, see Sect. 5.2.3.

[10] Bipartite networks are based on the assumption that all members of a team form a fully connected clique (Uzzi and Spiro 2005, p. 453). We explicitly checked for this issue, as our network data is compiled on the basis of multi-partner R&D cooperation projects.

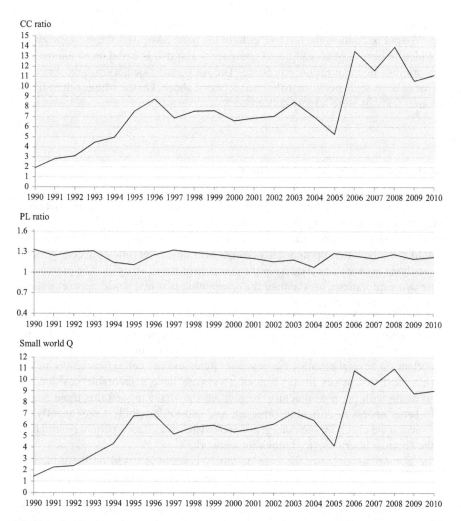

Fig. 8.9 Small-world properties in the German laser industry innovation network (Source: Author's own calculation and illustration)

bipartite networks was proposed by Newman et al. (2001). They showed that the "path length ratio" in bipartite networks are interpreted in the same way as unipartite networks (Uzzi and Spiro 2005, p. 454). In contrast, according to Newman et al. (2001) and Uzzi and Spiro (2005), the "clustering coefficient ratio" has to be interpreted in a different way. A clustering coefficient ratio of about 1.0 indicates within-team clustering whereas an exceeding clustering coefficient ratio indicates an increase in between-team clustering (Uzzi and Spiro 2005, pp. 454–455).

In our case, both the comparably low path length ratio throughout the observation period, ranging from 1.05 to 1.3, and the increasing tendency towards comparably high clustering coefficient ratios over time, confirms our initial findings. We put our data to the test to check for the issue addressed above. Appendix 3 provides the results of an additional consistency test that is based on an alternative network data decomposition procedure. The additional calculations reveal nearly the same large-scale network patterns as reported above. On the whole, our initially reported findings were largely confirmed.

8.3.3 Core-Periphery Structure

As we have shown in Sect. 7.1.1, the overall German laser industry network consists of an average of 3.6 components with a standard deviation of 2.33. However, the question remains as to what the size proportions of these components look like and how these proportions change over time. The following core-periphery analyses goes way beyond a simple component-based analysis. The previously presented explorations substantiate the assumption that real world networks show quite unique structural patterns.

Several authors have suggested that interorganizational networks typically display core-periphery structures (Rank et al. 2006; Amburgey et al. 2008; Muniz et al. 2010). The identification of core-periphery structures in real world networks is important for several reasons. For instance, Rank and her colleagues (2006, p. 76) have argued that actors in the core of a network have a favorable position for negotiating with peripheral actors. In addition they have argued that these actors have better access to critical information and knowledge (ibid). Consequently, in this section we check for the emergence and existence of a core-periphery structure in the German laser industry innovation network.

In its most basic sense, the core-periphery concept is based on the notion of "[...] a dense, cohesive core and a sparse, unconnected periphery" (Borgatti and Everett 1999, p. 375). In addition, the core of the network occupies a dominant position in contrast to the subordinated network periphery (Muniz et al. 2010, p. 113). Several formalizations of the concept have so far been proposed. We argue that using single indicators runs the risk of providing a somewhat biased picture of the actual network structure. Thus, in order to identify a core-periphery structure in longitudinal network data, we propose the simultaneous use of four distinct indicators, each of which addresses different network characteristics.

According to Doreian and Woodard (1994, p. 269) a core of a network is simply a more cohesive and richly connected area of the network, relative to the overall structure of the entire network. Technically spoken, the specification of a network core is nothing else but the specification of a cohesive subgraph by using concepts such as n-cliques, k-plexes, k-cores and related concepts (ibid).

To start with, we focus on a concept that basically draws an actor-based k-core analysis. Amburgey et al. (2008) have applied this concept to conduct a k-core decomposition at the overall network level in order to analyze the emergence of a core-periphery structure in the biotech industry. The basic idea behind the concept is straightforward. "A k-core is a subgraph in which each node is adjacent to at least a minimum number, k, of the other modes in the subgraph" (Wasserman and Faust 1994, p. 266). The repeated calculation of k-core values in well-specified time intervals enables network actors to be categorized and grouped according to their nodal degree. For instance, a subgroup consisting of network actors with a k-core of $k = 6$ indicates that all of these actors have at least six direct linkages to other network actors. Amburgey et al. (2008) have argued that the exploration of the coreness strata (i.e. coreness layers for $k = 1...n$) allows us to check over time for the existence and emergence of a core-periphery structure. This approach provides a very valuable initial look at the network's core-periphery structure.

However, when focusing on "connectedness" as one of the most important features of networks (Wasserman and Faust 1994, p. 109) the measure creates a distorted picture for the following reasons. Firstly, and most importantly, the k-core concept is not a component-based concept. It allows us to identify cohesive sub-graphs in a network based on the actors' nodal degree. This, however, implies that high degree nodes can be found in both peripheral components as well as in the main component. In other words, nodes with the same k-core value can be spread over the whole network regardless of whether they belong to the main component or a peripheral component. Secondly, the k-core concept concentrates exclusively on the tie dimension. This means that the size distribution of the core component versus peripheral components remains ignored. In other words, the proportion of nodes that fills the main component is not captured by the concept.

Consequently we argue that additional measures are needed to substantiate and complement a coreness analysis. The next two measures are as simple as the previous one but they provide a quite different view of the same phenomenon. Newman (2010, p. 235) shows that the majority of real world networks are not fully connected and the main component usually fills more than 90 % of the whole network. Our data confirms this finding and indicates that peripheral components are not only considerably smaller but also quite heterogeneous in terms of size and structure. In other words, we can distinguish between at least two elementary types of components in real world networks – the main component and peripheral component(s). Based on these considerations, two simple ratios – M-P tie ratio & M-P node ratio – can be calculated which allow us to quantify the proportion of ties or nodes that fill the main component versus peripheral components. The values for both ratios range between 0 and 1. These two ratios do not claim to provide comprehensive core-periphery indicators. Instead they give a valuable initial idea of size and density proportions between the fully connected main component and the scattered periphery of a network.

The last of our four core-periphery indicators was originally proposed by Borgatti and Everett (1999). They introduced two different concepts – discrete model and continuous model – that can be used to conduct a coreness analysis based on directed or undirected as well as valued or non-valued graphs. The underlying idea of the core-periphery identification procedures is based on a comparison of a hypothetically optimal core-periphery structure in an artificially generated network with a network structure that has actually been observed in a real-world network. Borgatti and Everett (1999, pp. 377–378) argued that an optimal core-periphery structure is characterized by a few core nodes that are adjacent to other core nodes, core nodes that are adjacent to some periphery nodes, and a notable proportion of periphery nodes that are not connected with other periphery nodes. As real-world networks are very unlikely to fit this theoretically optimal pattern, Borgatti and Everett (1999, pp. 377–378) have proposed an algorithm that measures how well the real-world network structure approximates the optimal core-periphery structure. The discrete model categorizes all network nodes into two classes – core nodes and peripheral nodes – whereas the procedure implemented in the continuous model simultaneously matches a core-periphery model to the overall network and estimates the coreness parameters of each actor in the network (Borgatti and Everett 1999; Borgatti et al. 2002). The parameter ρ is a measure of the network coreness. The measure ranges from 0 to 1 whereas large values indicate a high fit between an optimal core-periphery structure and an empirically observed network.

Figure 8.10 illustrates the calculation results for all four core-periphery indicators for the German laser industry innovation network between 1990 and 2010. As before, we chose a time-discrete approach and calculated all four indicators on an annual basis.

Figure 8.10a displays the k-core decomposition results. In contrast to Amburgey et al. (2008) we did not plot and interpret each k-core strata layer separately. Instead we grouped all network actors into three groups based on their k-core values: high ($k \geq 8$), medium ($8 > k > 4$) and low ($4 \geq k > 0$). We argue that a high spread between the first and the last group indicates the existence of a core-periphery structure at a given point in time. Our k-core decomposition analysis indicates that between 1994 and 1997, and between 2002 and 2010 there was a pronounced tendency towards having a core-periphery structure.

Figure 8.10b and c shows the M-P tie ratio and the M-P node ratio indicating the proportion of ties as well as nodes that fill the main component. Both ratios considerably decrease between 1994 and 1997, and between 2004 and 2008. In addition, a closer look at the M-P node ratio points to the fact that between 1998 and 2002 a notable proportion of nodes are located in peripheral components.

Figure 8.10d illustrates the results of a core-periphery analysis according to the approach proposed by Borgatti and Everett (1999). For the purpose of this study we have applied the continuous core-periphery model for undirected graphs. The estimation procedures implemented in UCI-Net 6.2 (Borgatti et al. 2002) were used to calculate coreness values on an annual basis. Figure 8.10d reports the gini-based core-peripheriness measure. Large values indicate a tendency towards a core-periphery structure. Results reveal that the German laser industry innovation network approximated a hypothetically optimal core-periphery structure quite

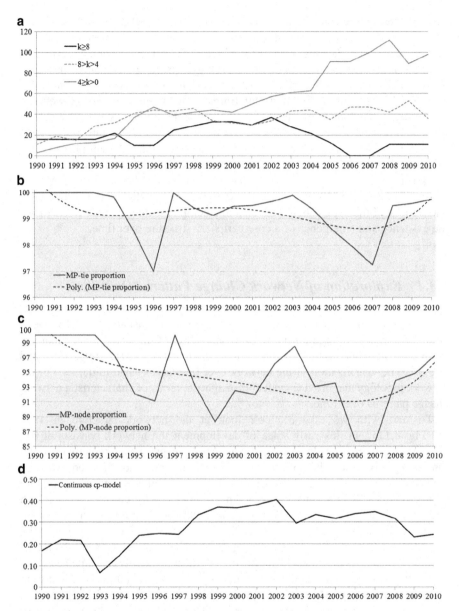

Fig. 8.10 Core-periphery structure in the German laser industry – a comparison of four indicators (Source: Author's own calculations and illustration)

well between 1994 and 2008. Surprisingly, the model indicates that the highest core-peripheriness occurred between 1999 and 2002.

In summary, our analysis gives us good reasons to assume that the German laser industry innovation network exhibited a comparably pronounced core-periphery structure during three time periods – (I) 1994–1997, (II) 1999–2002 and (III) 2004–2008. In all three time periods at least two out of four indicators substantiate this finding. In addition, the long-term trends indicate the tendency toward an increasing coreness of the German laser industry network over time. At first glance, the k-core analysis fails to indicate the pronounced core-peripheriness between 1999 and 2002 indicated by the continuous model of Borgatti and Everett (1999). However, a closer look at the dotted gray line in Fig. 8.10a (medium group, with k-core values: $8 > k > 4$) reveals a structural transition between 1998 and 1999. Obviously, there seem to be some hidden structural processes that strengthen the periphery at that time. It is interesting to note that the comparably simple M-P node ratio points to the second time period and reveals patterns that would have maybe remained unseen if only degree-based indicators were used. To conclude, both node-related and tie-related indicators should be used in a complementary manner to check for the existence and emergence of a core-periphery structure over time.

8.3.4   Exploration of Network Change Patterns Over Time

The last step in an exploratory network analysis is visualization (De Nooy et al. 2005, pp. 5–6). Figure 8.11 gives us snap-shots of the German laser industry innovation network at four distinct points in time (i.e. 1991, 1995, 1999 and 2007).[11] The visualization of the network over time gives us an initial idea of the network topology and provides valuable insights in terms of characteristic network change patterns over time.

To start with we take a closer look at the network structure in 1991 (cf. Fig. 8.11a). In this early stage of development the network consists of one component. Thus, the network is fully connected. Nevertheless, the network structure is by no means homogeneous. We can generally identify one densely connected area in the network whereas the majority of the network actors are relatively sparsely connected. This finding is in line with the comparably high degree centralization index of about 17.5 % in 1991 as reported earlier.

Only a few years later the picture changes considerably (cf. Fig. 8.11b). In 1995, the network consists of three distinct components. The main component is by far the largest. The size proportions among the peripheral components are quite heterogeneous. We have a dyadic component on the one hand, and a multi-node component that consists of five LSMs and PROs on the other. As we will see later (cf. Chap. 9) this has some important implications for the theoretical conceptualization of evolutionary network change processes. In addition, the network plot reveals the existence of several densely connected areas – hot spots – within the main component of the network. It turns out that initially central actors, such as LSM287,

---

[11] We used NetDraw 2.0 to visualize the network (Borgatti 2002).

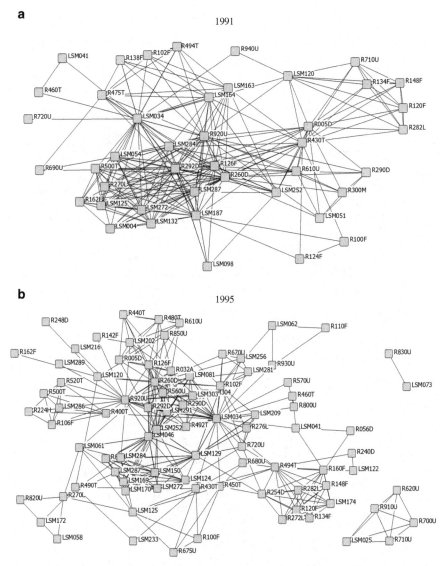

Fig. 8.11 The evolution of the German laser industry innovation network, 1991 & 1995 (Source: Author's own calculations and illustrations)

LSM272, R126F, R920U, R260D, were able to further develop their position in terms of their nodal degrees.

Two patterns in 1999 are striking (cf. Fig. 8.12a). Firstly, the number of components has grown considerably and the main component continues to dominate in terms of size. Now the periphery consists of two dyadic and three triadic components. Furthermore it is interesting to note that the previously identified

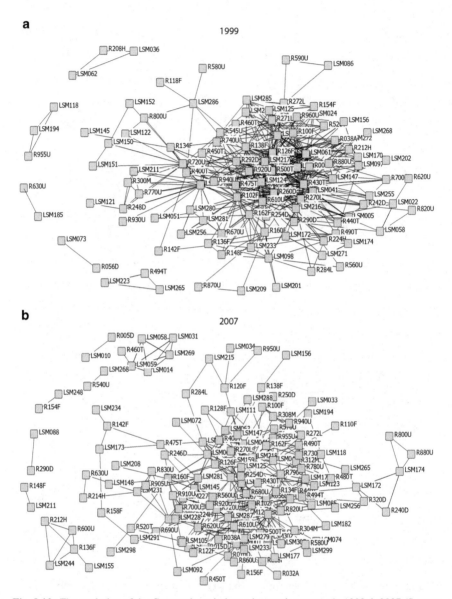

Fig. 8.12 The evolution of the German laser industry innovation network, 1999 & 2007 (Source: Author's own calculations and illustrations)

peripheral multi-node component has meanwhile been integrated into the main component. In other words, between 1995 and 1999 at least one of the five LSMs or PROs in the multi-node component was able to establish a bridging tie to an actor within the main component. Secondly, we can observe an increasing concentration tendency in the main component. Some nodes are quite loosely linked to the main component whereas others are embedded in the core of the main component and show an above-average nodal degree. This observation clearly supports the results of our core-periphery analysis (cf. Sect. 8.3.3). Not surprisingly, the initially identified high-degree actors are still positioned at the core of the network. But what is perhaps more interesting are the high-degree nodes in the network that entered the scene later. For instance, the firms LSM124 and LSM061 entered the industry in 1994 and 1995, respectively. This indicates that some nodes seem to reach the core of the network much faster than others.

Finally, the last network plot (cf. Fig. 8.12b) illustrates the network topology in 2007. The network structure is clearly more fragmented than it was in 1999. In addition, we can observe a large number of peripheral components. These components are quite heterogeneous in terms of size. More precisely, we see five dyadic components and two multi-node components consisting of five and six nodes respectively. Moreover, it is remarkable that neither of the peripheral network organizations identified in 1999 are still in the network periphery in 2007.

References

Amburgey TL, Al-Laham A, Tzabbar D, Aharonson BS (2008) The structural evolution of multiplex organizational networks: research and commerce in biotechnology. In: Baum JA, Rowley TJ (eds) Advances in strategic management – network strategy, vol 25. Emerald Publishing, Bingley, pp 171–212

Barabasi A-L, Albert R (1999) Emergence of scaling in random networks. Science 286 (15):509–512

Barabasi A-L, Bonabeau E (2003) Scale-free networks. Sci Am 288(5):50–59

Borgatti SP (2002) NetDraw: graph visualization software. Analytic Technologies, Harvard

Borgatti SP, Everett MG (1999) Models of core/periphery structures. Soc Networks 21 (4):375–395

Borgatti SP, Everett MG, Freeman LC (2002) Ucinet for windows: software for social network analysis. Analytic Technologies, Harvard

Borgatti SP, Everett MG, Johnson JC (2013) Analyzing social networks. Sage, London

Carrington PJ, Scott J, Wasserman S (2005) Models and methods in social network analysis. Cambridge University Press, Cambridge

De Nooy W, Mrvar A, Batagelj V (2005) Exploratory social network analysis wit PAJEK. Cambridge University Press, Cambridge

Degenne A, Forse M (1999) Introducing social networks. Sage, London

Doreian P, Woodard KL (1994) Defining and locating cores and boundaries of social networks. Soc Networks 16(1994):267–293

Eades P (1984) A heuristic for graph drawing. Congr Numer 42:149–160

Freeman LC (1979) Centrality in social networks: I. conceptual clarification. Soc Networks 1 (3):215–239

Fruchterman T, Reingold E (1991) Graph drawing by force-directed placement. Softw Pract Exp 21(11):1129–1164

Golbeck J, Mutton P (2005) Spring-embedded graphs for semantic visualization. In: Geroimenko V, Chen C (eds) Visualizing the semantic web. Springer, Heidelberg/New York, pp 172–182

Milgram S (1967) The small-world problem. Psychol Today 1(1):60–67

Muniz AS, Raya AM, Carvajal CR (2010) Core periphery valued models in input–output field: a scope from network theory. Pap Reg Sci 90(1):111–121

Newman ME (2010) Networks – an introduction. Oxford University Press, New York

Newman ME, Strogatz S, Watts D (2001) Random graphs with arbitrary degree distributions and their applications. Phys Rev E 64:1–17

Podolny JM (1994) Market uncertainty and the social character of economic exchange. Adm Sci Q 39(3):458–483

Powell WW, White DR, Koput KW, Owen-Smith J (2005) Network dynamics and field evolution: the growth of the interorganizational collaboration in the life sciences. Am J Sociol 110 (4):1132–1205

Rank C, Rank O, Wald A (2006) Integrated versus core-periphery structures in regional biotechnology networks. Eur Manag J 24(1):73–85

Schilling MA, Phelps CC (2007) Interfirm collaboration networks: the impact of large-scale network structure on firm innovation. Manag Sci 53(7):1113–1126

Stuart TE, Hoang H, Hybles RC (1999) Interorganizational endorsements and the performance of entrepreneurial ventures. Adm Sci Q 44(2):315–349

Uzzi B, Spiro J (2005) Collaboration and creativity: the small world problem. Am J Sociol 111 (2):447–504

Uzzi B, Amaral LA, Reed-Tsochas F (2007) Small-world networks and management science research: a review. Eur Manag Rev 4(2):77–91

Wasserman S, Faust K (1994) Social network analysis: methods and applications. Cambridge University Press, Cambridge

Watts DJ, Strogatz SH (1998) Collective dynamics of 'small-world' networks. Nature 393 (6684):440–442

Part IV
Econometric Analysis

Chapter 9
Causes and Consequences of Network Evolution

Scholars are slowly shifting from positing simple systems to using more complex frameworks, theories, and models to understand the diversity of puzzles and problems facing humans interacting in contemporary societies

(Source: Ostrom 2009)

Abstract In this chapter we analyze a firm's propensity and timing to cooperate and enter the industry's innovation network. The conceptual framework considers three groups of determinants – organizational, relational and contextual. Selected factors within these groups are assumed to cause network change processes at the micro-level – tie formations and tie terminations – and shape the structural network configuration at the overall network level. The elements of the framework are substantiated by drawing upon evolutionary ideas and concepts from organization science, sociology and evolutionary economics. The following chapter is organized as follows: We start with a brief introduction in Sect. 9.1. Section 9.2 provides a literature review and introduces the theoretical cornerstones needed for an in-depth discussion on evolutionary network change. Based on these ideas we derive our conceptual framework in Sect. 9.3 and formulate a set of testable hypothesis in Sect. 9.4. Section 9.5 addresses some methodological issues and provides an overview of data and variables used. In Sect. 9.6 we introduce our empirical approach and present estimation results from our non-parametric event history model. Section 9.7 concludes with a summary and discussion on the implications of our key findings.

9.1 On the Evolutionary Nature of Innovation Networks

In this investigation we seek to understand the relationship between network change determinants, network change processes at the micro-level and structural consequences at the overall network level.[1] We employ an event history dataset on

---

[1] This chapter draws upon a joint research project conducted by Andreas Pyka, Chair for the Economics of Innovation, University of Hohenheim, and Jutta Guenther and Muhamed Kudic

© Springer International Publishing Switzerland 2015

M. Kudic, *Innovation Networks in the German Laser Industry*,
Economic Complexity and Evolution, DOI 10.1007/978-3-319-07935-6_9

publicly-funded R&D cooperation projects in the German laser industry to analyze one specific facet of the entire network evolution process, i.e. a firm's propensity and timing to cooperate and enter the industry's innovation network.[2]

Innovation networks have been the subject of a broad range of theoretical and empirical studies over the past decades.[3] Both organizational scholars and economists agree that the evolutionary change of complex networks still represents a widely unexplored area of research (Parkhe et al. 2006, p. 562; Brenner et al. 2011, p. 5). Quite recently scholars from various scientific disciplines such as physics (Albert and Barabasi 2000; Jeong et al. 2003), biology (Nowak et al. 2010), sociology (Doreian and Stokman 2005; Snijders 2004; Powell et al. 2005), organization and management science (Walker et al. 1997; Gulati and Gargiulo 1999; Koka et al. 2006; Zaheer and Soda 2009), economic geography (Glueckler 2007) and economics (Jackson and Watts 2002; Cowan et al. 2006; Jun and Sethi 2009) have started to intensify their research efforts in this area in order to understand the determinants and mechanisms affecting the structural evolution of networks. Despite this progress, we still face more questions than answers and empirical evidence remains scarce.

There are many reasons for this. Firstly, network evolution is a complex phenomenon encompassing causes and consequences of network change among multiple levels of analysis. In the most basic sense, all types of networks consist of nodes and connections among these nodes (Wasserman and Faust 1994). The concept of network evolution "[...] captures the idea of understanding change via some *understood* process [...]" whereas these underlying processes can be defined as a "[...] *series of events that create, sustain and dissolve* [...]" the network structure over time (Doreian and Stokman 2005, pp. 3–5). Thus, network change processes at the micro-level – i.e. tie formations or tie terminations – as well as changes with regard to network nodes – i.e. node entries or node exits – affect the structural configuration of overall networks over time. These processes of creative destruction are clearly Schumpeterian in nature and provide the basis for explaining the evolution of networks (Boschma and Frenken 2010, p. 129).

However, due to both the conceptual ambiguities caused by the complex nature of networks and the extensive data requirements needed to analyze the evolution of these entities, research in this field is still in its inception. Secondly, micro-level network change processes are determined by several factors which can be grouped

---

from the Department for Structural Economics at the Halle Institute for Economic Research. An early draft was presented at the 14th ISS Conference in Brisbane, Australia (Kudic et al. 2012). We are grateful to Wilfried Ehrenfeld for his helpful suggestions. This chapter has greatly benefited from the comments made by audience members at the Buchenbach Workshop on evolutionary economics in 2009 and is strongly influenced by the ideas and concepts discussed at the summerschool on organizational ecology in 2007 taught by Terry Amburgey, Rotman School of Management, Toronto, Canada. I take full responsibility for any errors in this chapter.

[2] We used STATA 10.1 (Stata 2007), a standard software package for statistical data analysis.

[3] For a comprehensive overview of the research conducted in this field see Pittaway et al. (2004) or Ozman (2009).

into three categories: organizational, relational and contextual. Previous research has predominantly concentrated on network formation processes affected by individual factors within one of these three groups. Surprisingly little research has been conducted on network formation processes affected by both endogenous and exogenous factors. Finally, even though tie terminations are as important as tie formations in understanding network evolution, there is a strong bias in the literature towards the presence of relationships versus their absence (Kenis and Oerlmans 2008, p. 299). This arises, on the one hand, from data availability issues as the majority of empirical studies in this field are based on alliance network databases in which tie terminations are systematically underrepresented.[4] On the other hand we can observe a construct validity problem in most studies as often no distinction is made conceptually between tie failures and intended tie terminations (Kenis and Oerlmans 2008, p. 299).

Against the backdrop of these issues, the aim of this analysis can be summarized as follows. On the one hand, an in-depth analysis of network change determinants requires a comprehensive understanding of network evolution in general. Thus, we propose a conceptual framework that consists of three building blocks: determinants, micro-level network change processes and structural consequences. Starting from an evolutionary economic perspective (Hanusch and Pyka 2007b) we consider innovation networks as an integral part of an innovation system that can be both spatially and sectorally delimited (Cooke 2001; Malerba 2002). We apply an interdisciplinary approach to substantiate the building blocks of our framework by drawing upon concepts from evolutionary economics, sociology and organizational science. On the other hand, we derive and test a set of hypotheses that addresses some selected facets of evolutionary network change processes.

The analytical part is inspired by two empirically observable large-scale network properties of the German laser industry's innovation network. Firstly, the German laser industry innovation network shows a fat-tailed degree distribution indicating that some nodes attract ties at a higher rate than others once they have entered the network (cf. Sect. 8.3.1). The same properties have been observed in other real-world networks such as in the US biotech innovation industry (Powell et al. 2005). Secondly, a substantial number of potential network entrants do not cooperate at all (cf. Sect. 8.2.1).[5]

In this analysis we are especially interested in analyzing network entry processes. More precisely, we ask the following research question: what are the endogenous or exogenous determinants affecting a firm's propensity and timing

---

[4] For an overview of the most frequently used alliance databases and their limitations, see Schilling (2009).

[5] The descriptive analysis reveals a minimum network participation rate of 20.1 % in 1990 and a maximum network participation rate of 52.92 % in 2008 for LSMs and PROs under observation.

to cooperate for the first time and enter the industry's innovation network? To answer this question we employ a single-episode event history dataset (cf. Sect. 6.2.1). This dataset allows for an exact time tracking of all node entries and exits as well as all tie formations and tie terminations.

9.2   State of the Art and Theoretical Background

This section starts with a brief review of the literature on the dynamics of alliances and networks. Then we turn our attention to some evolutionary concepts that provide the theoretical basis for our conceptual framework.

9.2.1   Literature on the Dynamics of Alliances and Networks

The literature on the dynamics of alliances and networks is quite heterogeneous. Several scholars have provided schemes to systematize the work that is been done in this field.[6] In this chapter we draw upon a general systematization scheme originally proposed by Van De Ven and Poole (1995) which has been applied and adapted to categorize dynamically oriented conceptualizations in the field of alliance research (De Rond and Bouchiki 2004) and network research (Parkhe et al. 2006) into three[7] groups: life-cycle model, teleological approaches and evolutionary approaches.

The use of life-cycle analogies is not new to economics and has been employed to capture product exploitation stages (Levitt 1965) as well as change patterns of industries (Klepper 1997) or clusters (Menzel and Fornahl 2009) over time. Life-cycle conceptualizations of alliance and network change are based on the notion of "[...] linear, irreversible and predictable progressions of events or states over time" (Parkhe et al. 2006, p. 562). The basic idea that underlies most of these models is that one can identify ideal development stages like initialization, growth, maturity and decline. Thus, some authors often refer to these models as phase models (Schwerk 2000; Sydow 2003). Change is imminent in life-cycle models which indicate that the developing entity has an underlying logic within itself that regu-

---

[6] For instance, Sydow (2003) has proposed a separation of dynamic network approaches in five model categories: life-cycle models, non-linear process models, intervention oriented process models, evolutionary models and co-evolutionary models. For other systematization schemes see for example Schwerk (2000) or Tiberius (2008).

[7] In contrast to De Rond and Bouchiki (2004) our review does not consider the dialectic approach. This is in line with the systematization applied by Parkhe et al. (2006). Hence, we end up with three instead of four categories.

lates the process of change (Van De Ven and Poole 1995, p. 515). The change process itself is regarded as a linear sequence of events where all development stages are traversed only once without disruptions or feedback loops along the way. These events are cumulative in nature which means that each development stage in both alliance and network life-cycle models can be seen as a precursor to successive stages (Van De Ven and Poole 1995, p. 515; De Rond and Bouchiki 2004, p. 57).

Literature often contains examples of life-cycle or phase models that address alliance and network change. For instance, Dwyer and his colleagues (1987) have proposed a model of buyer-seller linkages in which relationships evolve in general phases: awareness, exploration, expansion, commitment and dissolution. Murray and Mahon (1993) have proposed a somewhat similar phase model for strategic alliances that contains five distinct stages: courtship, negotiation, start-up, mainte-nance, and ending. Other authors have proposed phase models that encompass four stages. For instance, Forrest and Martin (1992) suggest an alliance process model based on their findings from an interview-based survey of senior executives in 70 North American biotech firms. Their model consists of four distinct stages: matching, negotiation, agreement and implementation. The last category comprises three-stage life-cycle models that are predominantly growth-oriented. For instance, Larson (1992) has proposed an entrepreneurial dyad formation model whose stages consist of: preconditions to exchange, conditions to build, integration and control. In contrast to this dyadic conceptualization Lorenzoni and Ornati (1988) introduce one of the first growth-oriented network formation models by arguing that firms that are expanding pass through three cooperation stages: unilateral relationships, recip-rocal relationships and network constellations. Critics of life-cycle models have argued that the phase specification and the length of stages in these models may vary arbitrarily (Sydow 2003, p. 332). In addition, the notion of a linear change process that does not consider disruptions or feedback loops is – to put it mildly – questionable at least.

According to the teleological school of thought, change in organizational entities is explained by relying on a philosophical doctrine according to which the purpose or goal is the ultimate cause of change (Van De Ven and Poole 1995, p. 515). From this point of view development is regarded as a "[. . .] repetitive sequence of goal formulation, implementation, evaluation and modification of goals [. . .]" whereas all of these sequences are affected by the experiences and intentions of an adaptive entity (Van De Ven and Poole 1995, p. 516). This means that organizational entities are able to learn at each stage of the repetitive sequences and reformulate their goals. In response to the limitations of the previously discussed lifecycle concep-tualizations, scholars have applied this teleological perspective in order to gain more open-ended and iterative process models of alliance and network change in which the final goal guides the underlying change process (De Rond and Bouchiki 2004, p. 57). Teleological alliance and network change models do not explicitly refer to life cycle analogies. In summary, this view emphasizes "[. . .] purposeful cooperation by entities toward desired end states" (Parkhe et al. 2006, p. 562). As these models allow for learning and adaptation processes in all development stages, some authors refer to these models as non-linear process models (Schwerk 2000; Sydow 2003).

Non-linear process models operating on a dyadic level are the most prominent applications of teleological ideas in an alliance and network context (Ring and Van De Ven 1994; Doz 1996; Kumar and Nti 1998; Arino and De La Torre 1998). The advantages of these models over life-cycle models are obvious. Non-linear process models provide a basis for analyzing dynamics but also the instability of dyadic alliances by considering endogenous factors like social embeddedness, trust, learning and knowledge transfer processes. In addition these models integrate the idea of feedback loops which affect the alliance development process. They take formation and catalyst processes of alliances into consideration and place a greater importance on unplanned terminations (Schwerk 2000, p. 230). This means there is no fixed assumption with regard to phase transition patterns (ibid). One prominent example of a non-linear process model was proposed by Ring and Van De Ven (1994). This model seeks to explain how and why interorganizational relationships emerge, evolve and dissolve over time. It considers three basic processes (negotiation, commitment and realization) and refers to the idea that formal and informal aspects need to be balanced in every process. Another influential non-linear process model is the conceptualization by Doz (1996). This model includes several internal and external dimensions – environment, task, process skills and goals – which are assumed to affect the processes of alliance change over time. The change process itself is characterized by sequences of interactive learning processes, reevaluation and readjustment. It explains both the successful development of alliances over time as well as the alliance failure as a result of little or divergent learning or frustrated expectations among partners (De Rond and Bouchiki 2004, p. 57).

Next, research delved further into network process models (Sydow 2003, p. 336). This approach has been strongly influenced by the contributions of the IMP research group (Hakansson and Johanson 1988; Hakansson and Snehota 1995; Halinen et al. 1999) and focuses predominantly on business relation networks. In these models, network change is driven by market access and internationalization goals. For instance, Halinen and colleagues (1999) have proposed a dynamic network model that includes radical and incremental change processes at the dyadic and network level. The framework integrates the ideas of mechanisms, nature and forces of change and contains two interdependent circles of radical and incremental change which are affected by external drivers of change and stability. In summary, the strength of teleological alliance and network change models lies in the rejection of simplistic, uniform and predictable sequences of change towards more realistic non-linear process models which recognize that unplanned events, unexpected results, as well as conflicting interpretations and interests can and do affect the change process over time (De Rond and Bouchiki 2004, p. 58).

Evolutionary conceptualizations of alliance and network change draw our attention to "[...] change and development in terms of recurrent, cumulative, and problematic sequences of variation, selection and retention." (Parkhe et al. 2006, p. 562). Evolutionary approaches seek to understand the forces that cause network change over time (Doreian and Stokman 2005, p. 5) which means that focus is placed on the underlying determinants and mechanisms of network change processes. In other words, understanding "[...] the 'rules' governing the sequence of change through time [...]" (Doreian and Stokman 2005, p. 5) provides an in-depth understanding of the network change process itself. These conceptualizations encompass the determinants that trigger the change processes at the micro-level, the mechanisms that generate change, and the structural consequences over multiple aggregation levels. Evolutionary conceptualizations of network change can be grouped into three partially overlapping categories: network emergence, network evolution and co-evolutionary approaches.

The first category – so-called network emergence or network growth approaches – focuses predominantly on determinants and mechanisms affecting alliance formations and associated network change patterns at the overall network level (Walker et al. 1997; Gulati 1995; Gulati and Gargiulo 1999; Hagedoorn 2006; Kenis and Knoke 2002). These growth oriented models consider both endogenous as well as exogenous factors of alliance and network change and recognize the importance of previous network structures in current cooperation decisions (Gulati and Gargiulo 1999). However, these studies clearly place little emphasis on tie termination processes and the associated structural consequences for the overall network configuration.

In response to these limitations, network evolution explicitly encompasses both network formation processes as well as network fragmentation processes by simultaneously considering the determinants and mechanisms behind these processes (Venkatraman and Lee 2004; Powell et al. 2005; Amburgey and Al-Laham 2005; Doreian and Stokman 2005; Glueckler 2007). The main point of network evolution models is to understand why and how networks emerge, solidify and dissolve over time. For instance, Powell and his colleagues (2005) have analyzed the underlying mechanisms such as "cumulative advantage", "homophily", "following the trend" and "multiconnectivity" in order to explain the structural evolution of complex networks in the US biotech industry. Organizational scholars have analyzed the impact of tie formations and tie terminations on the component structure and connectivity of networks (Amburgey and Al-Laham 2005). Economic geographers have argued that evolutionary processes of retention and variation in network structure are affected by a spatial dimension (Glueckler 2007). Co-evolutionary approaches concentrate on simultaneous change processes between networks and other subjects of change such as industries (Ter Wal and Boschma 2011), technologies (Rosenkopf and Tushman 1998) or even other types of networks between the same actors (Amburgey et al. 2008). The analytical focus is on understanding the interdependencies between simultaneously evolving network change patterns.

9.2.2   An Evolutionary View on Interorganizational Change

Despite the differences among evolutionary schools of thought, one can identify some cornerstones that create the common ground for evolutionary thinking in economics and related disciplines (Witt 2008b; Aldrich and Ruef 2006; Amburgey and Singh 2005; Dopfer 2005; Stokman and Doreian 2005).

Firstly, the preceding discussion reveals that evolutionary theories generally focus on dynamic change over time rather than on analyzing static or comparatively static snap-shots of economic activity. Closely related to the first point is the fact that evolutionary theories agree on the notion of path dependencies and irreversibilities, in other words, that past and present events affect the current decisions and behavior of economic actors (Arthur 1989; David 1985). Thirdly, the idea that change occurs simultaneously across multiple levels of analysis is common to most evolutionary approaches. For instance, organizational ecology scholars have analyzed intraorganizational evolution, organizational evolution, population evolution and institutional evolution (Amburgey and Rao 1996). Economists have proposed a differentiation between three levels of analysis: "micro", "meso" and "macro" (Dopfer et al. 2004). Thus, the majority of evolutionary theories are in line with the notion that change occurs simultaneously and interdependently across multiple levels (Amburgey and Singh 2005, p. 327). Finally, evolutionary theories explicitly include the underling mechanisms – the drivers or rules – that guide the change process. Most evolutionary scholars would agree that evolution includes an understanding of the forces that initiate or drive change (Doreian and Stokman 2005) and the mechanisms of modification or replacement of existing entities (Amburgey and Singh 2005). For instance, Glueckler (2007) proposes applying general evolutionary principles such as selection, retention and variation on relationships in networks. Below we concentrate on the neo-Schumpeterian school of thought (cf. Sect. 2.3).

Neo-Schumpeterian economics has its intellectual roots in evolutionary economics, industry life-cycle theory, complexity theory and systems theory and incorporates the ideas of path dependencies, irreversibilities, bounded rationality and collective innovation processes among heterogeneous actors (Hanusch and Pyka 2007a).

Research in this field is centered on the role of knowledge and innovation for the development and economic prosperity of firms and societies. Witt (2008a, p. 555) identifies the following topics as being at the core of the neo-Schumpeterian research agenda: innovation, R&D, firm routines, industrial dynamics, competition, growth and the institutional basis for innovation. Hanusch and Pyka (2007a, pp. 276–277) argue that the focus on novelty and uncertainty is what primarily sets neo-Schumpeterian economics apart. They highlight the following constitutive normative principles of neo-Schumpeterian economics: qualitative change affects all levels of economy; an idea of punctuated equilibria encompassing smooth as well as radical change; and change processes characterized by non-linearities and feedback effects responsible for pattern formation and spontaneous structuring.

The neo-Schumpeterian, or knowledge-based approach, regards innovation as a collective process of interacting heterogeneous economic actors (Pyka 2002). These actors can be characterized as bounded rational agents with incomplete knowledge bases and capabilities (Pyka 2002). The importance of formal as well as informal networks for the creation of novelty was recognized quite early on as "[. . .] networks were shown to be essential both in the acquisition and in the processing of information inputs" (Freeman 1991, p. 501). Networks allow firms to share knowledge, learn from each other and innovate (Pyka 2002; Hanusch and Pyka 2007a). In addition, networks are not static; they change over time. New relationships are established and existing relationships may be adjusted or even dissolved depending on the needs, capabilities and cooperation strategies of the actors involved. Due to the very nature of these underlying processes, networks are regarded as evolving organizational entities. Most recently we can observe the emergence of interesting intersections with related disciplines like economic geography (Boschma and Martin 2010) which provide a fertile ground for a greater consideration of the spatial dimension in evolutionary change processes.

In summary, the neo-Schumpeterian approach provides a powerful framework for analyzing knowledge transfer and interorganizational learning processes among heterogeneous economic actors in sectoral and spatial delimited systems in their efforts to innovate. It also takes into consideration the evolutionary change of complex collaborative systems driven by endogenous as well as exogenous determinants and mechanisms of micro-level change processes.

9.3   Linking Micro-Level Processes and Macro-Level Change

Drawing upon our previous considerations, in this section we introduce and discuss five general principles of network evolution models proposed by Stokman and Doreian (2005) in light of innovation networks, and incorporate the notion of network evolution according to Glueckler (2007) and Doreian and Stokman (2005). Based on these theoretical underpinnings we derive a conceptual framework that aims to provide an in-depth understanding of evolutionary change in innovation networks.

9.3.1   General Principles of Network Evolution Models

Stokman and Doreian (2005, pp. 244–251) recommend five general principles for constructing network evolution models which guide the following discussion.

Firstly, the instrumental character of networks provides the starting point for modeling network evolution. This means that the motives or goals of the actors involved have to be taken into consideration right from the very beginning. Innovation research has identified a broad range of reasons for why firms participate in innovation networks (Parkhe 1993; Pyka 2002) whereas the exchange of knowledge and initialization of mutual learning processes can be regarded as the most salient for successfully generating novelty.

Secondly, in order to gain an in-depth understanding of the actors' actions and the structural consequences of those actions it is appropriate to assume that a network actor possesses only partial or limited local information. This means that network actors possess global knowledge in the rarest cases. Instead, Stokman and Doreian (2005, p. 245) argue that network actors should be seen and modeled as adaptive entities that learn through experience and imitation. This principle is consistent with the neo-Schumpeterian notion of bounded rational agents with incomplete knowledge bases and capabilities (Pyka 2002).

The third principle highlights the importance of the relational dimension of cooperation. This means that the parallel tracking of goals by network actors affects the emergence of ties in a sense that both entities have to agree upon common goals and parallelize decisions. From an innovation network perspective, this principle highlights the importance of integrating concepts that operate primarily on the dyadic level, such as mutual trust or tensions between partners.

The fourth basic principle refers to the complexity of evolutionary processes in networks. Consequently, Stokman and Doreian (2005, p. 247) recommend designing network evolution models that are as simple as possible.

The fifth principle refers to the falsifiability of network evolution models. The authors suggest that network evolution models should have sufficient empirical reference and conclude that "statistical models are strongly preferred, as they enable the estimation of essential parameters and test the goodness of fit of the model" (Stokman and Doreian 2005, p. 249).

9.3.2 Building Blocks of the Network Evolution Framework

Network evolution is neither random nor determined (Glueckler 2007, p. 620). This means that mechanisms have to be considered that create cumulative causation and lead to path-dependency and mechanisms that produce contingency in the sense that the agent's strategies and actions may deviate from existing development paths that result in path destruction (ibid). In line with Doreian and Stokman (2005, p. 5) we regard the designations "network dynamics" or "network development" as more general terms to describe networks change over time. In contrast, network evolution "[...] has a stricter meaning that captures the idea of understanding change via some *understood* process [...]" whereas these underlying processes can be defined as a "[...] *series of events that create, sustain and dissolve* [...]" the network structure over time (Doreian and Stokman 2005, pp. 3–5). In addition, we have to note "[...] that the unit of analysis is always dyadic tie formation, whereas the object of knowledge is network structure" (Glueckler 2007, p. 622).

Based on the ideas outlined above, we specify three elementary building-blocks in our conceptual framework (cf. Fig. 9.1): (**I**) determinants of network change (**II**) micro-level network change processes and (**III**) structural consequences over multiple levels.

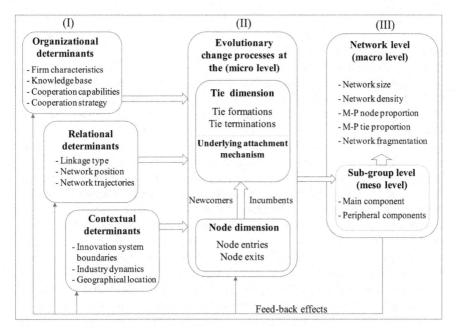

Fig. 9.1 Conceptual framework – causes and consequences of evolutionary network change processes (Source: Author's own illustration)

9.3.2.1 Determinants of Evolutionary Micro-level Network Change Processes

Due to their very nature, determinants that affect evolutionary micro-level network change processes can be categorized as organizational, relational and contextual.

To start with, we turn our attention to contextual determinants (cf. Fig. 9.1, left). Firms and organizations in interorganizational networks are considered to be an integral part of a spatial-sectoral innovation system (Cooke 2001; Malerba 2002). Innovation systems have several characterizing features.[8] Firstly, they consist of heterogeneous economic actors that are dispersed throughout geographical space within the system boundaries.[9] Secondly, populations of actors in the system can change over time which means that, for instance, firms or other types of organizations can, over time, enter the system (i.e. new company founding, spin-offs etc.)

---

[8] For the purpose of this study we focus on some selected features of innovation systems. Note that the innovation system approach is much richer than described here (cf. Sect. 2.3.3).

[9] Network actors are simultaneously embedded in multiple proximity dimensions (Boschma 2005) each of which is likely to affect a firm's cooperation behavior (Boschma and Frenken 2010). For the sake of simplicity, we include only the geographical dimension in the framework.

and exit the system (i.e. closures, failures, bankruptcies etc.). Thirdly, the system's elements do not exist in isolation; they are interconnected by various types of formal or informal linkages.

This leads to the relational determinants in our framework. Dyads consist of at least one directed or undirected tie connecting two nodes in a well-defined population and, at the same time, constitute the most basic building block of a network (Wasserman and Faust 1994). Triadic components are more complex network building blocks (ibid). Below, we refer to all components with more than two nodes as multi-node components. For the purpose of this analysis we specify innovation networks as formal, knowledge-related and publicly funded R&D partnerships among a well-defined population of firms and public research organizations.[10] The existence of a tie among two nodes in an innovation network implies a certain degree of partner fit, mutual trust, cooperation capabilities and commitment to common goals between both parties. The sum of these dyadic network ties spans the overall innovation network within the system boundaries. Firms and organizations occupy qualitatively different positions within the overall network structure. These network positions are the result of cooperation decisions taking place in the shadow of the past (Gulati and Gargiulo 1999). Soda and Zaheer (2004) argue that networks have a "memory" in the sense that past and present networks affect current actions. Doreian (2008) refers to this issue by introducing the concept of "network trajectories" in the context of the evolutionary change process of networks.

Finally, we move on to organizational determinants in our framework. As we will establish in more detail later, firm characteristics such as size, age, origin, knowledge stock and cooperation capabilities etc. are likely to affect knowledge-related cooperation behavior in innovation networks.

9.3.2.2 Micro-Level Network Change Processes at the Core of the Model

We continue the debate by moving on to micro-level network change processes at the core of the model (cf. Fig. 9.1, center). In a similar vein, Hite (2008) highlights in her model the importance of micro-level network change processes in the context of network evolution. Glueckler (2007, p. 623) argues that "[. . .] a complete theory of network evolution [. . .] has to theorize both the emergence and disappearance of ties *and* nodes".

We will start by turning our attention to the node dimension. In the most basic sense we can differentiate between system actors who participate and those who do not participate in a particular network. The first group includes all actively

---

[10] Informal partnerships and other structural collaborative forms such as short-term contracts, licensing and franchise agreements, consultancy contracts, consortia, non-funded long-term partnerships or joint ventures were deliberately excluded from the framework.

cooperating network actors, whereas the second group provides a pool of potentially available network actors. We follow the suggestion made by Guimera
et al. (2005) and differentiate between two groups of potential network actors:
"incumbents" and "newcomers". Both groups are subject to change due to dynamics at the industry level. Entries and exits affecting actors within the first group
(i.e. active network actors) have direct consequences for the structural configuration
of the network, whereas the same events affecting actors in the second group
(i.e. potential network actors) have an indirect impact by enlarging or reducing
the pool of cooperation partners that are potentially available. To control for this
node-related dimension of change in the German laser industry innovation network,
one needs to have an exact picture of all laser source manufacturers and laser-
related public research organizations over time. In this analysis we choose yearly
time period to capture the industry's configuration.

Now we will take a closer look at the tie dimension by considering two types of
events – tie formations and tie terminations – to explain the structural change of the
network. In line with Hite (2008) we refer to these events below as micro-level
network change processes. Moreover, tie formation and tie termination processes can
be coupled or uncoupled. A good example of coupled micro-level network change
processes are joint R&D projects with a fixed timeframe. In contrast, strategic long-
term partnerships have no predefined end date and provide a concrete example of
uncoupled micro-level network change processes. For reasons of simplicity, we
focus on coupled events. This approach has two considerable advantages. Firstly,
we have an exact time tracking of all tie termination events which are, from a
structural point of view, as important as tie formation events. Secondly, we considerably reduce complexity as tie termination processes do not follow their own
underlying logic. We argue, in line with Nelson and Winter (2002) and with
reference to Glueckler (2007), that micro-level network change processes can be
explained by the general evolutionary mechanisms of variation, selection and retention.[11] At the same time, the formation and termination of partnerships are affected
by the previously discussed determinants and follow the logic of underlying network
change mechanisms. The preferential attachment concept provides one of the most
frequently discussed tie formation mechanisms in network studies. The underlying
logic is quite simple: highly connected nodes are more likely to connect to new nodes
than sparsely connected nodes (Barabasi and Albert 1999; Albert and Barabasi
2002). The mechanism generates quite a unique structural pattern at the overall
network level which is characterized by a power law degree distribution (cf. Sect.
8.3.1). Several other mechanisms and underlying logic of network formation processes have been discussed in the literature. These include "homophily" according to
which actors with similarities are more likely to connect to one another (McPherson
et al. 2001), "heterophily" according to which heterogeneous actors attract one
another (Amburgey et al. 2009), "herding behavior" where actors follow the crowd

---

[11] For an in-depth discussion, see Glueckler (2007, pp. 623–630).

(Kirman 1993; Powell et al. 2005) and "transitive closure" where two nodes, which are both connected to a third partner, attract one another (Snijders et al. 2010).

9.3.2.3   Structural Consequences of Micro-Level Network Change Processes

Only a few previous studies have analyzed the structural consequences of micro-level network change processes (Elfring and Hulsink 2007; Baum et al. 2003; Amburgey and Al-Laham 2005). We draw upon evolutionary ideas and network change models proposed by Amburgey et al. (2008), Guimera et al. (2005) and Glueckler (2007) to substantiate this part of the puzzle in our framework.

We start by looking at the model proposed by Amburgey et al. (2008). The authors provide a conclusive theoretical explanation for structural consequences of tie formations and tie terminations by introducing four distinct structural processes: (a) the creation of a bridge between components, (b) the creation of a new component, (c) the creation of a pendant to an existing component and (d) the creation of an additional intra-component tie (Amburgey et al. 2008, pp. 184–186). The framework provides us with very valuable insights. Nonetheless, we argue that these considerations have to be extended and refined in several ways.

Firstly, we argue that tie formations and tie terminations, as well as subsequent structural consequences, depend on the actor's strategic orientation. Strategies and actions of network actors can result in the destruction of existing network paths (Glueckler 2007, p. 620) and they determine, at the same time, the scope of future cooperation options and possibilities. Therefore, we propose and integrate three basic types of knowledge-related cooperation strategies into our framework: progressive, moderate and conservative. Progressive strategies are characterized by a firm's objective to considerably improve its knowledge base by accessing multiple knowledge sources simultaneously or by establishing and controlling global knowledge streams that connect entire groups of actors in the networks. The underlying objective of moderate strategies is to gradually improve the knowledge base through linkages to a few selected individual partners or through the establishment and control of local knowledge streams. Conservative strategies aim to secure a firm's knowledge base by protecting the existing knowledge stock or by securing and sustaining existing local or global knowledge channels.

Secondly, the framework of Amburgey et al. (2008, pp. 184–186) primarily focuses on the tie dimension and neglects the importance of different types of actors for the structural evolution of networks. As outlined above, not all innovation system actors are involved in a particular type of innovation network. Instead, a considerable number of system actors are not embedded at all, whereas others cooperate repeatedly with the same partners. To account for this fact we follow the suggestion of Guimera et al. (2005, p. 698) and split the population into "newcomers" and "incumbents". This gives us four distinct partnership constellations: "newcomer-newcomer" (NN), "incumbent-newcomer" (IN), "incumbent-incumbent" (II) and "repeated incumbent-incumbent" (RI).

Partner types	Cooperation options		Knowledge-related cooperation strategies			Structural consequences				
			Progressive	Moderate	Conservative	Size	Density	Fragmentation	M-P node proportion	M-P tie proportion
N	NN	N1			No cooperation	╱	╱	╱	╱	╱
		N2		New dyadic component		⇧	╱	⇧	⇩	╱
		N3	New multi-node component			⬆	╱	⬆	⬇	╱
	NI	N4			No extension	╱	╱	╱	╱	╱
		N5		P component extension		⇧	╱	╱	⇩	╱
		N6	M component extension			⇧	╱	╱	⇧	╱
I	II	I1			M-comp. fragmentation	╱	⇩	⇧	⇩	╱
		I2		P component consolidation		╱	⇧	╱	╱	⇩
		I3	M component consolidation			╱	⇧	╱	╱	⇧
		I4			P-P component merger (dyad)	╱	⇧	⇩	⇩	╱
		I5		P-P component merger		╱	⇧	⬇	⬇	╱
		I6	M-P component merger			╱	⇧	⬇	⬆	╱
	RI	I7		Network solidification		╱	⇧	╱	╱	╱
		I8			Network fragmentation	╱	⬆	╱	╱	╱

Legend:

N = Newcomer	NN = Newcomer-Newcomer	M = Main component	⬆⬇	Pronounced structural effect
I = Incumbent	NI = Newcomer-Incumbent	P = Peripheral component	⇧⇩	Moderate structural effect
	II = Incumbent-Incumbent			
	RI = Repeated Incumbent -Incumbent		╱	No structural effect

Fig. 9.2 Partner constellations, cooperation strategies and structural consequences (Source: Author's own illustration)

Thirdly, under real-world conditions we can frequently observe the formation and termination of both dyadic ties connecting two actors but also of large-scale multi-partner projects that encompass a large number of actors. Consequently, we differentiate between dyadic and multi-node components in our framework.

Finally, in the majority of real world networks, the main component usually fills more than 90 % of the entire network (Newman 2010, p. 235).[12] This substantiates the assumption that essential elements of industry-specific technological knowledge are tied to the main component. In contrast, peripheral components are likely to entail only small, rather specific fragments of the industry's technological knowledge. Thus, we argue that there is a qualitative difference between whether network change processes affect the core or the periphery of the network.

Figure 9.2 (left) summarizes our previous considerations and illustrates the anticipated structural consequences at the overall network level (Fig. 9.2, right).[13] To address the structural consequences at the network level we now take a closer

---

[12] For the German laser industry network we found that the main component fills 94.51 % of the network on average (cf. Sect. 8.3.3).

[13] In line with Amburgey et al. (2008) we use three simple indicators to discuss structural network change: network size, network density and overall network fragmentation. To account for processes affecting the core-periphery structure of the network, we introduce two additional ratios to measure the proportion of nodes and ties in peripheral components in relation to the size and density of the main component (cf. Sect. 8.3.3).

look at newcomers who have basically two possible partner constellations (NN and NI) and six cooperation options (N1–N6). We start our discussion on structural consequences by focusing on the moderate knowledge-related cooperation strategy of newcomers.

Actors aiming to gradually improve their knowledge base through selected individual collaborations basically have two options: either they can cooperate with another potential newcomer, which would lead to the creation of a new dyadic component (N2), or they can connect with an incumbent who is embedded in a peripheral component (N5). The structural consequences are consistent with the structural processes (b) and (c) identified by Amburgey et al. (2008). However, we have to consider two additional knowledge-related cooperation strategies. Conservatively oriented actors who predominantly aim to protect their existing knowledge stock are likely to isolate themselves from other newcomers or incumbents. Thus, neither is a new component created (N1) nor an existing component extended (N4). In both cases, the structural configuration of the network is not affected. Even though these two cooperation strategies have no direct structural consequences they are important in understanding what prevents potential network entrants from cooperating for the first time. In contrast, progressively oriented actors seek to improve their knowledge stock considerably by accessing multiple diverse knowledge bases simultaneously. The initialization of multi-partner projects among newcomers (N3) leads, from a structural standpoint, to the creation of a multi-node component. In contrast, the establishment of a linkage to an incumbent in the main component of the network offers a broad variety of direct and indirect knowledge-accessing opportunities (N6) and is reflected in the extension of the main component.

The structural consequences at the network level for the cooperation options (N2) and (N3) are quite similar but less pronounced in the former. The creation of new ties affects the number and size distribution of components (Amburgey et al. 2008, p. 186). This leads to increasing network fragmentation and a decreasing proportion of nodes in the main component in relation to the number of nodes in peripheral components. A look at the cooperation options (N5) and (N6) reveals that the number of components remains constant but the network size is affected. This is in line with structural implications anticipated by Amburgey et al. (2008, p. 186). However a closer look at the proportion of nodes in the main and peripheral components reveals two opposing structural effects for the cooperation options (N5) and (N6). Moderate cooperation strategies produce a situation in which the main component shrinks in relation to the network's periphery. On the other hand progressive strategies lead to a relative growth in the main component versus the network periphery.

Now we turn our attention to incumbents who, like the newcomers, basically have two possible partner constellations (II and RI). In this context, Amburgey et al. (2008, p. 186) differentiate between two structural processes: the creation of a bridge between two components and the creation of intra-component ties. This

distinction provides valuable insight into the structural consequences of cooperation events between previously unconnected or indirectly connected network actors
(I1–I6).

However, in order to refine the picture we have to separate consolidation
processes from solidification and fragmentation tendencies in the network. Thus,
we explicitly consider the structural consequences of repeated ties between already
connected incumbents (I7–I8). Moreover, we account for path dependencies in our
framework. By referring to Glueckler (2007, p. 620) we argue that the initial
cooperation strategy of a network entrant affects its later cooperation path. In
other words, the initial cooperation event is hereditary in a sense that it does restrict
cooperation opportunities, yet at the same time it opens up new cooperation options.
Below, we refer to this very specific type of network path dependency as "cooperation imprinting".

Figure 9.2 illustrates six potentially achievable cooperation options (I1–I6)
among previously unconnected incumbents (II). Newcomers who have pursued a
moderate network entry strategy start the next cooperation round out of a dyadic
component located in the periphery of the network. In contrast, the situation looks
quite different for newcomers who have a progressive strategic orientation at the
onset. These actors started their cooperation path by creating a new multi-node
component and linking themselves to the main component. In both cases the initial
conditions for the next cooperation round are considerably better than for network
entrants with a moderate strategy.

The previous considerations imply that incumbents, who are located in the
network periphery and are still pursuing a moderate cooperation strategy, are likely
to look for cooperation opportunities in their direct neighborhood. This case
addresses the creation of alternative knowledge channels in peripheral components
(I2). In contrast, there are peripheral incumbents who change their strategic orientation towards a more progressively oriented cooperation behavior. These actors
actively search for novel knowledge stocks and tend to establish or control knowledge streams to other groups of network actors. This case is reflected, from a
structural standpoint, in the emergence of brokerage ties among peripheral incumbents (I5). In summary, we can observe the consolidation of a connected peripheral
subgraph on the one hand, and the amalgamation of two previously unconnected,
peripheral sub-graphs on the other. Both structural processes are in line with the
model proposed by Amburgey et al. (2008). However, it is important to note that the
cooperation options (I2) and (I5) in our framework exclusively address structural
consequences that occur in the periphery of the network due to the network
entrants' cooperation imprinting.

Now we look at incumbents who entered the network by pursuing a progressive
cooperation strategy (using N3 or N6). Network entrants who linked themselves to
the main component (using N6) face quite a comfortable situation in the next
cooperation round. On the one hand, they can expand their position in the main
component by establishing direct links to new partners in the main component
(I3) or they can wait for new specific knowledge-accessing opportunities to pop up
in the network periphery in order to establish bridging ties (I6). However, main

component actors can also pursue a conservative strategy in order to protect and secure the existing knowledge stock. In other words, a main component actor can decide to withdraw from the main component by leaving the main component either alone or together with a handful of strategic partners. The structural consequences are far-reaching, especially in the latter case (I1). The overall network density decreases, the fragmentation of the network increases and the component shrinks in relation to the periphery.

Actors with a progressive cooperation imprinting who entered the network through the creation of a new multi-node network component (using N3) start the second cooperation round from a peripheral position. However, multi-partner projects provide a better starting point than dyadic components because they are much more visible and prestigious. Incumbents with a progressive strategy can establish a bridging tie to an actor in the main component (I6).[14] This strategy provides access to essential elements of an industry-specific technological knowledge pool tied to the main component and leads to an amalgamation of a peripheral component with the main component. Incumbents pursuing a moderate cooperation strategy will try to gain access to the much more specific knowledge pool by bridging the gap to another peripheral multi-node component (I5) or, in the case of a conservative cooperation strategy, to another dyadic component (I4).

A comparison of options I2 and I3 reveals some interesting structural implications. In both cases the network density is affected. This is in line with structural implications anticipated by Amburgey et al. (2008, p. 186). At the same time the ratio of main-component ties to peripheral-component ties reveals an opposing structural effect. The amalgamation of two previously unconnected network components affects the density and fragmentation of the network (Amburgey et al. 2008, p. 186). Furthermore, the differentiation between main and peripheral components (I5 and I6) once again shows an opposing structural effect.

Finally, we take a look at repeated incumbent-incumbent partnerships. Repeated partnerships can occur sequentially (at different points in time) or in parallel (at the same point in time). Not only the former but also the latter case is quite important but frequently neglected in network evolution studies. We refer to these ties as redundant network ties. These ties secure access to external knowledge sources on the one hand, while providing the opportunity to exchange qualitatively different stocks of knowledge among the same partners. In addition, redundant ties have far-reaching implications for the overall network structure. We argue that redundant ties can affect the stability of the network in several ways. Basically we can distinguish between two cases. The previously outlined ideas substantiate the argument that a network in which progressive and moderate cooperation strategies dominate is likely to show a solidification tendency over time (I7). In contrast, a

---

[14] Note that there is a qualitative difference when comparing the cooperation option (I6) of an incumbent who is embedded in the main component with an incumbent who is embedded in a peripheral component. The former case reflects a strategically important gate-keeping position. This position allows an actor to control who gets access to essential elements of the industry's technological knowledge pool tied to the main component.

network in which moderate and conservative cooperation strategies dominate is likely to show fragmentation tendencies over time (I8).

9.4   Hypotheses Development for Network Entry Processes

Based on our previously introduced framework we now derive a set of hypotheses that address only a few selected facets of the entire evolutionary network change process described above. In order to answer the research question raised initially, we exclusively concentrate on network entry processes. As a consequence, the analytical part is confined to a firm's initial cooperation event. Secondly, each group of determinants in our framework contains a broad variety of factors that are likely to affect a firm's cooperation behavior. The hypotheses outlined below are centered on only a small selection of factors that are assumed to play a key role in explaining network entry processes of German laser source manufacturers.

Initially we take a closer look at firm-specific determinants. The resource-based view (Wernerfelt 1984; Barney 1991; Peteraf 1993) suggests that a firm's ability to achieve and maintain a profitable market position and outperform competitors depends, to a large extent, on its ability to exploit both internal resources (Barney 1991) and external resources (Gulati 2007) and to generate a competitive advantage.[15] In this context, it has been argued that small firms face some substantial disadvantages compared to larger firms in the form of limited reputational, human capital and financial resources (Lu and Beamish 2006). Small firms can overcome their resource constraints and counteract their comparably high risk of failure – also known as "liability of smallness" (Barron et al. 1994) – by forming alliances with external partners (Baum et al. 2000). Proponents of the knowledge-based view have argued that alliances allow firms to gain access to external knowledge stocks (Grant and Baden-Fuller 2004) and learn from cooperation partners (Hamel 1991) in order to gain competitive advantages (Dierickx and Cool 1989; Coff 2003) and resist the increasing pressure of global competition. Both resource-based as well as knowledge-based arguments provide solid theoretical arguments to substantiate high cooperation propensities of small firms in science-based industries.

However, given the need and willingness of these firms to cooperate, there are several factors that are likely to hamper their ability to cooperate for the first time or which delay network entry. Firstly, in the pre-cooperation phase it can be quite difficult to assess a potential partner's intentions (Dacin et al. 1997, p. 7). This enhances the level of uncertainty, especially in international alliances (ibid). Secondly, potential network entrants have to make a considerable effort and spend both time and limited resources on identifying potential cooperation partners (Dacin et al. 1997, p. 4). From a New Institutional Economic standpoint we would argue that a firm faces considerable screening costs to overcome information

---

[15] For an in-depth discussion on the resource-based view, see Sect. 2.4.2.

asymmetries and lower the adverse selection risk (Ackerlof 1970; Spence 1976, 2002). These search costs, however, are likely to cause a disproportional burden on small firms due to their comparably low resource endowment in the pre-cooperation phase. Once potential partners are identified, other obstacles are likely to delay network entry. Small firms lack alliance management capabilities (Schilke and Goerzen 2010) and standardized cooperation interfaces (Goerzen 2005). Finally, Lu and Beamish (2006) point to the fact that SMEs are usually owned and managed by the founders and decision-making is much more centralized compared to larger firms. This, however, is likely to delay the responsiveness of decision makers at lower hierarchy levels and may hamper the firm's ability to react rapidly to newly emerging cooperation opportunities. The arguments outlined above substantiate our first hypothesis:

H1 Small firms take longer than large firms to enter an innovation network for the first time.

With regard to relational determinants the question arises as to how the type of cooperation impacts the time it takes a firm to initialize its first cooperation event. During the past decades substantial efforts were undertaken by both the EU and by the German government to support key industries. The funding of R&D cooperation projects is regarded as a key policy instrument. The main difference between these two types of cooperation is that EU-framework projects explicitly aim to encourage scientific and technological cooperation between member states whereas national funding initiatives predominantly aim to address domestic applicants. There are some clear benefits associated with international R&D project environments. According to Gunasekaran (1997, p. 639) these include access to new and different technologies, enhanced scope of potentially accessible technological knowledge stocks, better access to qualified employees and a broad range of training opportunities for technical personnel. Nonetheless, there are also some difficulties that go hand in hand with international R&D projects. The pre-formation phase is characterized by higher search costs to identify potential partners. In the post-formation phase, international alliances require greater investment in communication and transportation to support interaction among the partners involved (Lavie and Miller 2008, p. 625). Project governance costs tend to be higher due to a higher level of uncertainty (ibid). It is also well recognized that cross-national cultural differences may affect interaction between firms and organizations in multiple ways (Hofstede 2001). Firms entering cross-national cooperation projects face the challenge of adjusting to both a foreign country and to an alien corporate culture (Barkema et al. 1996, p. 154; Lavie and Miller 2008, p. 626). Differences in national culture are reflected in differing managerial ideologies of decision makers and have the potential to significantly affect strategic decisions in both the pre and post alliance formation phase (Dacin et al. 1997, p. 6). As a consequence, it has been argued that cross-national cultural differences are likely to affect a firm's attitude towards cooperation and thus the predisposition to enter international R&D consortia (Nakamura et al. 1997, p. 155). These considerations underpin our second hypothesis:

H2 A firm will enter a national innovation network sooner than an international network (mode of entry).

Finally, we take a closer look at the contextual dimension. Based on a proximity framework originally proposed by Boschma (2005), Boschma and Frenken (2010) have argued that network change is likely to be affected by other dimensions of proximity such as cognitive, organizational, institutional or geographical proximity. Like other science-driven industries (Owen-Smith et al. 2002), the German laser industry shows a pronounced tendency to cluster geographically (Kudic et al. 2011). Consequently, we focus on the relationship between geographical proximity and a firm's cooperation timing. More precisely, we distinguish between inside-cluster and outside-cluster firms and analyze the extent to which cluster membership affects cooperation timing. Firstly, it is important to note that cluster membership does not require or imply network membership. Firms can be located in a densely crowded region (agglomeration) without having formal partnerships with other firms or organizations in their immediate geographical surroundings. Theoretically, there are three potential ways in which cluster membership can affect a firm's propensity to cooperate and its timing to do so. A firm's cluster membership may have an accelerating impact, a decelerating impact or no impact at all on its propensity to cooperate for the first time and its timing to do so.

We follow the traditional line of argument which assumes a positive relationship between a firm's location in a geographically crowded region and its initial cooperation activities. In this context, it has been argued that the local environment generates positive externalities in terms of knowledge spillovers (Feldman 1999; Audretsch and Feldman 1996). Social interactions between employees and decision makers within a regional agglomeration are an important source of information. As a result, firms located in densely crowded industrial regions become aware of local cooperation opportunities sooner than others. It is therefore plausible that regional environments can speed up a firm's successful search for potential partners and shorten the time needed to enter the network. However, geographical proximity may also be accompanied by negative effects. Boschma (2005, p. 70) argues that highly specialized regions can become too inward-looking and this sensitizes them to the problem of spatial lock-in effects because of their lack of openness to the outside world. While there is a great deal of empirical evidence for the importance of spatial proximity over functioning spillover channels, other dimensions of proximity such as cognitive proximity (Boschma and Frenken 2010), might outperform spatial proximity in certain cases. In line with Feldman (1999) and with Audretsch and Feldman (1996) we formulate our last hypothesis:

H3 The time it takes to first enter an innovation network is shorter for firms located in densely crowded regions (agglomeration areas) than for firms located in remote regions.

9.5   Data and Variable Specification

This analytical section employs the previously introduced event history dataset (cf. Sect. 6.1.1) which is based on three main data sources: industry data, organizational data and cooperation data.

Industry data came from a proprietary dataset containing detailed information on firm entries and exits for the entire population of German laser source manufacturers between 1969 and 2005 (Buenstorf 2007). This initial industry dataset has been modified in several ways to meet the requirements of this analysis (cf. Sect. 6.1.1). We ended up with an industry dataset encompassing 233 laser source manufacturers for the entire observation period from 1990 to 2010. To analyze the transition from the origin state ("no-cooperation") to the destination state ("first cooperation") we had to account for all firms with "incomplete" cooperation histories to avoid left truncation and left censoring problems (Blossfeld and Rohwer 2002, pp. 39–41). In cases where the number of censored observation units is small, it is acceptable to simply exclude them (Allison 1984, p. 11). Starting with a full population of 233 LSMs in our sample, we identified 39 firms which were founded before 1990 and excluded them from the dataset. Thus, a total of 194 firms were potentially at risk of conducting the first cooperation event. Out of this population we ended up with a total of 112 cooperating firms whose first cooperation event unambiguously fell between 1990 and 2010.

Organizational level data was basically taken from the same raw data sources that were used at the industry level (cf. Sect. 4.2.1). Moreover, we used annually compiled count data on different types of laser related organizations – laser source manufacturers (LSMs), laser-related public research organizations (PROs) and laser system providers (LSPs) – which was supplied by the LASSSIE project consortium (Albrecht et al. 2011). Data was available at the planning region level. This allowed us to identify planning regions with an above-average number of LSMs, PROs and LSPs and to group these planning regions into clusters.

Network data came from two electronically available archive data sources: the *Foerderkatalog* database provided by the German Federal Ministry of Education and Research and the *CORDIS* databases provided by the European Community Research and Development Information Service (cf. Sect. 4.2.3).

We are not the first to use these archive data sources to construct knowledge-related innovation networks (cf. Broekel and Graf 2011, p. 6; Fornahl et al. 2011; Scherngell and Barber 2009, 2011; Cassi et al. 2008). There are solid arguments that advocate for the use of these archive data sources for analyzing the evolution of innovation networks. Organizations that participate in R&D cooperation projects subsidized by the German federal government have to agree on a number of regulations that facilitate mutual knowledge exchange and provide incentives to innovate (Broekel and Graf 2011, p. 6). In a similar vein, the European Commission has funded thousands of collaborative R&D projects in order to support transnational cooperation activities, increase mobility, strengthen the scientific and technological bases of industries and foster international competitiveness (Scherngell and Barber 2009, p. 534). Moreover, both data sources provide exact information on the timing of tie formation as well as tie termination processes. They were used to construct a single-episode event history dataset for the German laser industry (cf. Sect. 6.1.1).

The variables in this dataset were grouped into the following three categories: organizational, relational and contextual. An organizational variable was created to account for differences in firm size *[firmsize_cat_ev]*. The following size categories were used: firmsize_cat_ev1 = "micro firm" = 1–9 employees; firmsize_cat_ev2 = "small firm" = 10–49 employees; firmsize_cat_ev3 = "medium firm" = 50–249 employees; firmsize_cat_ev4 = "large firm" = more than 250 employees. A simple relational variable was included in the dataset to account for the type of cooperation. Thus, nationally funded and supra-nationally funded R&D cooperation projects were coded separately *[coop_type_ev]*. The variable was coded coop_type = 1 in the case of a *CORDIS* project and coop_type = 2 in the case of a Foerderkatalog project. Cooperation dates and duration were recorded in century months. Finally, we included a set of cluster variables *[clu_ev]* in our dataset indicating whether a firm was located inside or outside of a densely crowded region. The four geographical clusters were identified and defined as follows: planning regions: 72, 73, 74, 76 and 77 = clu_ev_bw, located in Baden-Württemberg; planning regions: 86, 90 and 93 = clu_ev_bay, located in Bavaria; planning regions: 54 and 56 = clu_ev_thu, located in Thuringia; region 30 = clu_ev_b, located in Berlin.

9.6   Empirical Model and Estimation Results

Non-parametric event history methods were used to test our hypotheses (cf. Sect. 6.2.1). We applied the product-limit estimator (Kaplan and Meier 1958).

9.6.1   Empirical Estimation Approach

The Kaplan and Meier (1958) estimation method has several advantages. Most importantly, it is straightforward to use, requires only weak assumptions and allows non-repeated events in single-episode event history data to be analyzed (Cleves et al. 2008, p. 93). In general, the survival function represents the probability of surviving past time t, or in other words, the probability of failing after time t (ibid).

The event of interest is the first cooperation for all LSMs which are at risk in the time period from 1990 to 2010. The unit of analysis is the firm. The time axis is defined on the basis of century months. All firm foundation dates as well as all start and end dates of cooperation events are given in century months. The dataset allows us to analyze the transition from the origin state ("no-cooperation") to the destination state ("first cooperation"). Repeated events were not taken into account. Thus, the survival function has to be interpreted as follows: the survival function estimates the firm's probability of having the first cooperation event after time t.

Non-parametric estimation methods provide the possibility of comparing survivor functions (cf. Sect. 6.2.1). The overall population can be divided into two or more subgroups by using an indicator variable to analyze whether the probability of failing after time t significantly differs among these subgroups. The indicator variable defines membership in a particular subgroup (Blossfeld et al. 2007, p. 76). We applied this approach to analyze the extent to which organizational, relational and contextual determinants affect cooperation behavior over time.

For the purpose of this analysis we make use of four commonly applied test statistics: i.e. the Log-Rank test, Cox test, Wilcoxon-Breslow test and Tarone-Ware test. These tests are designed to compare globally defined overall survival functions (Cleves et al. 2008, p. 123). The tests are based on the null hypothesis that the survivor functions do not differ significantly from one another (Blossfeld et al. 2007, p. 81). A significant test result indicates that the null hypothesis must be rejected (ibid). Or to put it another way, rejecting of the null hypothesis based on a significant test result supports the alternative hypothesis that the compared functions differ significantly from one another.

9.6.2   Estimation Results

A natural starting point for the presentation of our exploratory findings is to look at the overall survivor function. Figure 9.3 displays a plot of the survivor function.

The vertical axis contains values between zero and one whereas the horizontal axis represents time measured in century months. The interpretation is straightforward. The survivor function represents the firm's propensity and timing to move from the origin state ("no cooperation") to the destination state ("first cooperation"). To illustrate this, after 50 century months (i.e. 4 years and 2 months) about 66 % of all firms in our sample have entered the network, while about 34 % of all firms were still unable to initiate their first cooperation event. Only 50 century months later (i.e. 8 years and 4 months) about 84 % had achieved their first cooperation event and after 150 century months (i.e. 12 years and 6 months) 99.6 % of all firms had moved from the origin state to the destination state.

To test our hypotheses we have used several indicator variables to split the sample, compare survivor functions and analyze the extent to which the probability of entering the network is affected by organizational, relational or contextual factors.

We start the presentation and discussion of our findings by looking at firm size. A comparison of survivor functions for micro, small, medium and large firms reveals some unexpected but quite interesting findings (cf. Fig. 9.4). What we

Fig. 9.3 Timing and
propensity to cooperate and
enter the network (Source:
Author's own calculations
and illustration)

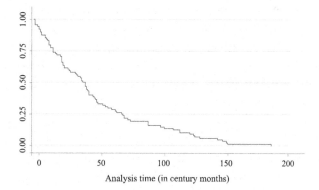

Fig. 9.4 The Kaplan-Meier
approach – comparison of
survivor functions based on
firm size (Source: Author's
own calculations and
illustration)

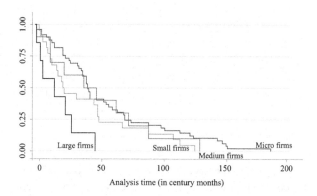

observe is that micro firms enter the network significantly later than small and large firms. The sequence in which micro, small and large firms enter the network remains unchanged and stable throughout the entire observation period. The test statistics reported in Table 9.1 indicate that the null hypothesis must be rejected, in other words, the compared survivor functions differ significantly from one another.

These results seem to confirm, at least at first glance, Hypothesis H1 which states that smaller firms have higher resource constraints and cooperate later than larger firms. However, the group of medium-sized firms complicates the story. At some point in time (e.g. after 50 months) medium-sized firms enter the network significantly later than both large firms and micro and small-sized firms.

In a nutshell, we found only partial support for Hypothesis H1. The findings for micro, small and large firms are in line with our expectations. Moreover, the results clearly indicate that there must be another underlying process affecting a firm's timing in entering the network. An in-depth analysis of additional organizational level determinants is needed to understand what factors cause the delayed entry of medium-sized firms.

Next, we look at the relational dimension. Our initial assumption was that the type of cooperation used by a firm to enter the network is likely to affect how long it

Table 9.1 Test statistics – comparison of Kaplan Meier survivor functions based on firm size

	Log-rank test for equality of survivor functions (by size)			Wilcoxon (Breslow) test for equality of survivor functions (by size)		
	Events observed	Events expected		Events observed	Events expected	Sum of ranks
1	49	57.97	1	49	57.97	-496
2	22	17.53	2	22	17.53	233
3	10	9.97	3	10	9.97	-26
4	7	2.53	4	7	2.53	289
Total	88	88.00	Total	88	88.00	0
	chi2(3) =	11.05		chi2(3) =	9.97	
	Pr>chi2 =	0.0114		Pr>chi2 =	0.0188	

	Cox regression-based test for equality of survival curves (by size)			Tarone-Ware test for equality of survivor functions (by size)		
	Events observed	Events expected	Sum of ranks	Events observed	Events expected	Sum of ranks
1	49	57.97	0.8674	49	57.97	-62.730142
2	22	17.53	1.3254	22	17.53	30.036592
3	10	9.97	1.0490	10	9.97	-2.7858996
4	7	2.53	3.0565	7	2.53	35.479449
Total	88	88.00	1.0000	88	88.00	0
	chi2(3) =	8.26		chi2(3) =	10.42	
	Pr>chi2 =	0.0409		Pr>chi2 =	0.0153	

(Source: Author's own calculations)

would take for the first cooperation event to occur. Figure 9.5 compares the survivor function based on cooperation type. Surprisingly, a comparison of nationally and supra-nationally funded R&D cooperation projects shows no significant differences (cf. Table 9.2). All four test statistics indicate that the null hypothesis must be confirmed, meaning that there is no significant difference between the compared survivor functions. In other words, it makes no difference whether a firm favors nationally funded (i.e. *Foerderkatalog*) or supra-nationally funded (i.e. CORDIS) R&D cooperation projects.

This result implies that the problem of "double layered acculturation" inherent to international cooperation projects (Barkema et al. 1996, p. 154) seems to play a subordinate role in this context. As a consequence we have to reject Hypothesis H2. One potential explanation for this result is that the previously existing interpersonal network between decision makers relativizes culturally contingent cooperation barriers.

Finally, we address here only one of several other contextual determinants by taking a closer look at the geographical proximity dimension. To analyze the extent to which cluster membership affects a firm's timing for entering the network we identified several planning regions with an above-average number of LSMs, PROs and LSPs and grouped them into four clusters: cluster_Th, cluster_Bay, cluster_B, cluster_Bw.

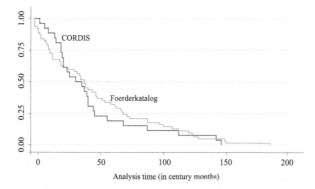

Fig. 9.5 Duration analysis based on cooperation type (Source: Author's own calculations and illustration)

Table 9.2 Test statistics – comparison of Kaplan Meier survivor functions based on cooperation type

Log-rank test for equality of survivor functions (by coop.-type)

	Events observed	Events expected
0	26	23.79
1	62	64.21
Total	88	88.00

chi2(1) = 0.29
Pr>chi2 = 0.5874

Wilcoxon (Breslow) test for equality of survivor functions

	Events observed	Events expected	Sum of ranks
0	26	23.79	-5
1	62	64.21	5
Total	88	88.00	0

chi2(1) = 0.00
Pr>chi2 = 0.9820

Cox regression-based test for equality of survival curves (by coop.-type)

	Events observed	Events expected	Sum of ranks
0	26	23.79	1.0971
1	62	64.21	0.9662
Total	88	88.00	1.0000

chi2(1) = 0.28
Pr>chi2 = 0.5948

Tarone-Ware test for equality of survivor functions (by coop.-type)

	Events observed	Events expected	Sum of ranks
0	26	23.79	7.4968264
1	62	64.21	-7.4968264
Total	88	88.00	0

chi2(1) = 0.07
Pr>chi2 = 0.7910

(Source: Author's own calculations)

Figure 9.6 illustrates our empirical results. Perhaps the most interesting finding is that cluster membership can have quite different effects on a firm's timing in entering the network. Our results show that firms located in the Thuringia Cluster (clu_Th) cooperate significantly earlier than firms that are located elsewhere. Exactly the opposite is true for firms located in the Bavarian Cluster (clu_Bay). In both cases test statistics (cf. Table 9.3) indicate that the compared survivor functions for inside-cluster and outside-cluster firms differ significantly.

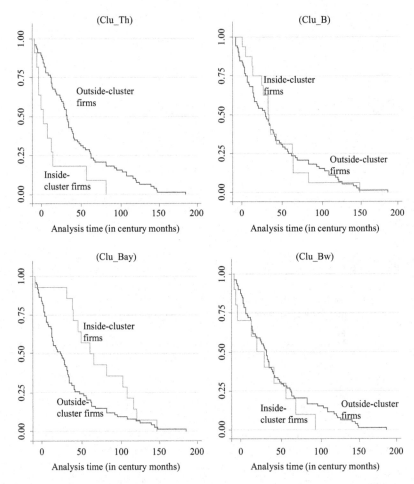

Fig. 9.6 Duration analysis based on cluster membership (Source: Author's own calculations and illustrations)

However, this is only half of the story. Our results for the Berlin Cluster (clu_B) and the Bavarian Cluster (clu_Bw) reveal quite a different picture (cf. Fig. 9.6, bottom). In both clusters we found no empirical evidence for significantly different survivor functions when comparing inside-cluster and outside-cluster firms (cf. Table 9.3). In summary, clusters can, but do not necessarily, affect a firm's timing in cooperating and entering the network. Thus, we found empirical support for each of the three cases proposed by Hypothesis H3. Our findings show that cluster membership is not generally associated with a higher propensity to cooperate. Instead, we need to take a closer look at the clusters themselves in order to disentangle the effects of cluster membership on the timing and propensity to cooperate.

Table 9.3 Test statistics – comparison of Kaplan Meier survivor functions based on cluster membership

Cluster Th

Log-rank test,
equality of survivor functions

	Events observed	Events expected
0	77	82.93
1	11	5.07
Total	88	88.00

chi2(1) =   7.6746
Pr>chi2 =   0.0056

Wilcoxon (Breslow) test,
equality of survivor functions

	Events observed	Events expected	Sum of ranks
0	77	82.93	412
1	11	5.07	-412
Total	88	88.00	0

chi2(1) =   9.85
Pr>chi2 =   0.0017

Cox regression-based test,
equality of survival functions

	Events observed	Events expected	Sum of ranks
0	77	82.93	0.9507
1	11	5.07	2.2829
Total	88	88.00	1.0000

chi2(1) =   5.77
Pr>chi2 =   0.0163

Tarone-Ware test,
equality of survivor functions

	Events observed	Events expected	Sum of ranks
0	77	82.93	-48.520825
1	11	5.07	-48.520825
Total	88	88.00	0

chi2(1) =   9.04
Pr>chi2 =   0.0026

Cluster B

Log-rank test,
equality of survivor functions

	Events observed	Events expected
0	72	71.18
1	16	16.82
Total	88	88.00

chi2(1) =   0.05
Pr>chi2 =   0.8205

Wilcoxon (Breslow) test,
equality of survivor functions

	Events observed	Events expected	Sum of ranks
0	72	71.18	136
1	16	16.82	-136
Total	88	88.00	0

chi2(1) =   0.50
Pr>chi2 =   0.4803

Cox regression-based test,
equality of survival functions

	Events observed	Events expected	Sum of ranks
0	72	71.18	1.0121
1	16	16.82	0.9505
Total	88	88.00	1.0000

chi2(1) =   0.05
Pr>chi2 =   0.8211

Tarone-Ware test,
equality of survivor functions

	Events observed	Events expected	Sum of ranks
0	72	71.18	11.394461
1	16	16.82	-11.394461
Total	88	88.00	0

chi2(1) =   0.21
Pr>chi2 =   0.6461

Cluster Bay

Log-rank test,
equality of survivor functions

	Events observed	Events expected
0	74	65.62
1	14	22.38
Total	88	88.00

chi2(1) =   4.46
Pr>chi2 =   0.0346

Wilcoxon (Breslow) test,
equality of survivor functions

	Events observed	Events expected	Sum of ranks
0	74	65.62	542
1	14	22.38	-542
Total	88	88.00	0

chi2(1) =   7.95
Pr>chi2 =   0.0048

(continued)

Table 9.3 (continued)

Cox regression-based test,
equality of survivor functions

	Events observed	Events expected	Sum of ranks
0	74	65.62	1.1670
1	14	22.38	0.6359
Total	88	88.00	1.0000

chi2(1) =   4.80
Pr>chi2 =   0.0284

Tarone-Ware test,
equality of survivor functions

	Events observed	Events expected	Sum of ranks
0	74	65.62	69.092798
1	14	22.38	-69.092798
Total	88	88.00	0

chi2(1) =   7.20
Pr>chi2 =   0.0073

Cluster BW

Log-rank test,
equality of survivor functions

	Events observed	Events expected
0	78	80.40
1	10	7.60
Total	88	88.00

chi2(1) =   0.86
Pr>chi2 =   0.3540

Wilcoxon (Breslow) test,
equality of survivor functions

	Events observed	Events expected	Sum of ranks
0	78	80.40	107
1	10	7.60	-107
Total	88	88.00	0

chi2(1) =   0.54
Pr>chi2 =   0.4622

Cox regression-based test,
equality of survivor functions

	Events observed	Events expected	Sum of ranks
0	78	80.40	0.9735
1	10	7.60	1.3283
Total	88	88.00	1.0000

chi2(1) =   0.77
Pr>chi2 =   0.3791

Tarone-Ware test,
equality of survivor functions

	Events observed	Events expected	Sum of ranks
0	78	80.40	-14.540845
1	10	7.60	14.540845
Total	88	88.00	0

chi2(1) =   0.61
Pr>chi2 =   0.4357

(Source: Author's own calculations)

9.7   Discussion and Implications

The first empirical part was motivated by a desire to deepen our understanding of how interorganizational innovation networks evolve. This quite demanding task was approached from two directions. On the one hand we proposed a conceptual framework that consists of three elementary building blocks – (**I**) "determinants", (**II**) "micro-level network change processes" and (**III**) "structural consequences" – to provide the theoretical basis for an in-depth analysis of evolutionary network change. On the other hand we conducted a non-parametric event history analysis to provide some empirical evidence on the propensity of LSMs to cooperate for the first time and enter the German laser industry innovation network.

The results of our analysis have interesting implications for both policy makers and practitioners. Firstly, our findings show that micro firms enter the network significantly later than small-sized and large firms but fail to explain the late entry of medium-sized firms. The underlying logic of this finding is straightforward. Even though SMEs depend more on access to external knowledge sources through interorganizational R&D linkages in order to keep pace with larger competitors, there are several factors hampering their ability to initiate R&D linkages for the first time. This finding supports the view of many European countries and regions that have instigated innovation policy programs for SMEs in order to strengthen R&D cooperation and innovation networks (e.g. Muldur et al. 2006; OECD 2008). This enables the joint research potential of SMEs to become effective more quickly. In further research it would be interesting to disentangle the extent to which factors such as search costs, a lack of alliance management capabilities or simply the absence of standardized cooperation interfaces explain the delayed entry of SMEs.

Our second result is surprising. The findings show that the choice of cooperation type (national or international) has no significant impact on a firm's timing in entering the network. Differences between nationally oriented and internationally oriented R&D cooperation projects seem to only play a subordinate role in the German laser industry. This can be taken as an indication of the high degree of internationalization of this technology; it is a cross-sectional technology with many applications in a truly interdisciplinary scientific field. Both factors clearly contribute to creating strongly internationalized networks. A second potential explanation is that previously existing interpersonal networks between decision makers relativize culturally contingent cooperation barriers.

The findings of the final empirical analysis indicate that cluster membership can have quite different effects on a firm's timing in entering the network. Traditionally, it has been argued that a geographically crowded region provides several benefits for firms. It appears that firms in some regions (e.g. Thuringia) tend to cooperate earlier and to have a significantly higher propensity to cooperate than those in other regions (e.g. Bavaria). A plausible explanation for this finding can be found in the spatial lock-in argument (cf. Boschma 2005; Boschma and Frenken 2010). In terms of policy making, this finding means that clustering processes are important but no remedy in and of themselves. Very specialized industries, like the laser industry, depend heavily on cooperation partners located anywhere in Germany and beyond. This corresponds to the findings on national versus international networks mentioned above.

This analysis provides us with interesting insights into firm-specific cooperation patterns and network entry processes. Nonetheless the analysis only reflects a very first step towards a better understanding of network change and a lot remains to be done.

Like any empirical study, this analysis also has some appreciable limitations (cf. Sect. 13.2) and we still face some theoretical and empirical challenges in obtaining a deeper understanding of causes and consequences of evolutionary network change processes. These challenges constitute the next steps in our research agenda (cf. Sect. 14.2).

References

Ackerlof GA (1970) The market for "lemons". Quality uncertainty and the market mechanism. Q J Econ 84(3):488–500

Albert R, Barabasi A-L (2000) Topology of evolving networks: local events and universality. Phys Rev Lett 85(24):5234–5237

Albert R, Barabasi A-L (2002) Statistical mechanics of complex networks. Rev Mod Phys 74 (1):47–97

Albrecht H, Buenstorf G, Fritsch M (2011) System? What system? The (co-) evolution of laser research and laser innovation in Germany since 1960. Working paper, pp 1–38

Aldrich HE, Ruef M (2006) Organizations evolving, 2nd edn. Sage, London

Allison PD (1984) Event history analysis – regression for longitudinal event data. Sage, London

Amburgey T, Al-Laham A (2005) Islands in the net. Conference paper: 22nd EGOS colloquium, Bergen, pp 1–42

Amburgey TL, Rao H (1996) Organizational ecology: past, present, and future directions. Acad Manag J 39(5):1265–1286

Amburgey TL, Singh JV (2005) Organizational evolution. In: Baum JA (ed) The Blackwell companion to organizations. Blackwell, Malden, pp 327–343

Amburgey TL, Al-Laham A, Tzabbar D, Aharonson BS (2008) The structural evolution of multiplex organizational networks: research and commerce in biotechnology. In: Baum JA, Rowley TJ (eds) Advances in strategic management – network strategy, vol 25. Emerald Publishing, Bingley, pp 171–212

Amburgey T, Aharonson BS, Tzabbar D (2009) Heterophily in inter-organizational network ties. Conference paper: 25th EGOS colloquium, Barcelona, pp 1–40

Arino A, De La Torre J (1998) Learning from failure: towards an evolutionary model of collaborative ventures. Organ Sci 9(3):306–325

Arthur BW (1989) Competing technologies, increasing returns, and lock-in by historical events. Econ J 99(394):116–131

Audretsch DB, Feldman MP (1996) R&D spillovers and the geography of innovation and production. Am Econ Rev 86(3):630–640

Barabasi A-L, Albert R (1999) Emergence of scaling in random networks. Science 286 (15):509–512

Barkema HG, Bell JH, Pennings JM (1996) Foreign entry, cultural barriers, and learning. Strateg Manag J 17(2):151–166

Barney JB (1991) Firm resources and sustained competitive advantage. J Manag 17(1):99–120

Barron DN, West E, Hannan MT (1994) A time to grow and a time to die: growth and mortality of Credit Unions in New York City, 1914–1990. Am J Sociol 100(2):381–421

Baum JA, Calabrese T, Silverman BS (2000) Don't go it alone: alliance network composition and startup's performance in Canadian biotechnology. Strateg Manag J 21(3):267–294

Baum JA, Shipilov AW, Rowley TJ (2003) Where do small worlds come from? Ind Corp Chang 12 (4):697–725

Blossfeld H-P, Rohwer G (2002) Techniques of event history analysis – new approaches to causal analysis. Lawrence Erlbaum, London

Blossfeld H-P, Golsch K, Rohwer G (2007) Event history analysis with Stata. Lawrence Erlbaum, London

Boschma R (2005) Proximity and innovation: a critical assessment. Reg Stud 39(1):61–74

Boschma R, Frenken K (2010) The spatial evolution of innovation networks: a proximity perspective. In: Boschma R, Martin R (eds) The handbook of evolutionary economic geography. Edward Elgar, Cheltenham, pp 120–135

Boschma R, Martin R (2010) The aims and scope of evolutionary economic geography. In: Boschma R, Martin R (eds) The handbook of evolutionary economics geography. Edward Elgar, Cheltenham, pp 3–43

Brenner T, Cantner U, Graf H (2011) Innovation networks: measurement, performance and regional dimensions. Ind Innov 18(1):1–5

Broekel T, Graf H (2011) Public research intensity and the structure of German R&D networks: a comparison of ten technologies. Econ Innov New Technol 21(4):345–372

Buenstorf G (2007) Evolution on the shoulders of giants: entrepreneurship and firm survival in the German laser industry. Rev Ind Organ 30(3):179–202

Cassi L, Corrocher N, Malerba F, Vonortas N (2008) Research networks as infrastructure for knowledge diffusion in European regions. Econ Innov New Technol 17(7):665–678

Cleves MA, Gould WW, Gutierrez RG, Marchenko YU (2008) An introduction to survival analysis using Stata, 2nd edn. Stata Press, College Station

Coff RW (2003) The emergent knowledge-based theory of competitive advantage: an evolutionary approach to integrating economics and management. Manag Decis Econ 24(4):245–251

Cooke P (2001) Regional innovation systems, clusters, and the knowledge economy. Ind Corp Chang 10(4):945–974

Cowan R, Jonard N, Zimmermann J-B (2006) Evolving networks of inventors. J Evol Econ 16 (1):155–174

Dacin TM, Hitt MA, Levitas E (1997) Selecting partners for successful international alliances: examination of U.S. and Korean firms. J World Bus 32(1):3–16

David PA (1985) Clio and the economics of QWERTY. Am Econ Rev 75(2):332–337

De Rond M, Bouchiki H (2004) On the dialectics of strategic alliances. Organ Sci 15(1):56–69

Dierickx I, Cool K (1989) Asset stock accumulation and sustainability of competitive advantage. Manag Sci 35(12):1504–1511

Dopfer K (2005) The evolutionary foundation of economics. Cambridge University Press, Cambridge

Dopfer K, Foster J, Potts J (2004) Micro–meso–macro. J Evol Econ 14(3):263–279

Doreian P (2008) Actor utilities, strategic action and network evolution. In: Baum JA, Rowley TJ (eds) Advances in strategic management – network strategy, vol 25. Emerald Publishing, Bingley, pp 247–271

Doreian P, Stokman FN (2005) The dynamics and evolution of social networks. In: Doreian P, Stokman FN (eds) Evolution of social networks, 2nd edn. Gordon and Breach, New York, pp 1–17

Doz YL (1996) The evolution of cooperation in strategic alliances: initial conditions or learning processes? Strateg Manag J 17(1):55–83

Dwyer RF, Schurr PH, Oh S (1987) Developing buyer-seller relationships. J Mark 51(2):11–27

Elfring T, Hulsink W (2007) Networking by entrepreneurs: patterns of tie formation in emerging organizations. Organ Stud 28(12):1849–1872

Feldman MP (1999) The new economics of innovation, spillovers and agglomeration: a review of empirical studies. Econ Innov New Technol 8(1):5–25

Fornahl D, Broeckel T, Boschma R (2011) What drives patent performance of German biotech firms? The impact of R&D subsidies, knowledge networks and their location. Pap Reg Sci 90 (2):395–418

Forrest JE, Martin MJ (1992) Strategic alliances between large and small research intensive organizations: experiences in the biotechnology industry. R&D Manag 22(1):41–53

Freeman C (1991) Networks of innovators: a synthesis of research issues. Res Policy 20 (5):499–514

Glueckler J (2007) Economic geography and the evolution of networks. J Econ Geogr 7 (5):619–634

Goerzen A (2005) Managing alliance networks: emerging practices of multinational corporations. Acad Manag Exec 19(2):94–107

Grant RM, Baden-Fuller C (2004) A knowledge accessing theory of strategic alliances. J Manag Stud 41(1):61–84

Guimera R, Uzzi B, Spiro J, Armaral LA (2005) Team assembly mechanisms determine collaboration network structure and team performance. Science 308(29):697–702

Gulati R (1995) Social structure and alliance formation pattern: a longitudinal analysis. Adm Sci Q 40(4):619–652

Gulati R (2007) Managing network resources – alliances, affiliations and other relational assets. Oxford University Press, New York

Gulati R, Gargiulo M (1999) Where do interorganizational networks come from? Am J Sociol 104 (5):1439–1493

Gunasekaran A (1997) Essentials of international and joint R&D projects. Technovation 17 (11):637–647

Hagedoorn J (2006) Understanding the cross-level embeddedness of interfirm partnership formation. Acad Manag Rev 31(3):670–680

Hakansson H, Johanson J (1988) Formal and informal cooperation – strategies in international industrial networks. In: Contractor FJ, Lorange P (eds) Cooperative strategies in international business. Lexington Books, Lexington, pp 369–379

Hakansson H, Snehota I (1995) Stability and change in business networks. In: Hakansson H, Snetota I (eds) Developing relationships in business networks. Thomson, London, pp 24–49

Halinen A, Salmi A, Havila V (1999) From dyadic change to changing business networks: an analytical framework. J Manag Stud 36(6):779–794

Hamel G (1991) Competition for competence and inter-partner learning within international strategic alliances. Strateg Manag J 12(1):83–103

Hanusch H, Pyka A (2007a) Principles of neo-Schumpeterian economics. Camb J Econ 31 (2):275–289

Hanusch H, Pyka A (2007b) Elgar companion to neo-Schumpeterian economics. Edward Elgar, Cheltenham

Hite JM (2008) The role of dyadic multi-dimensionality in the evolution of strategic network ties. In: Baum JA, Rowley TJ (eds) Advances in strategic management – network strategy, vol 25. Emerald Publishing, Bingley, pp 133–170

Hofstede G (2001) Culture's consequences: comparing values, behaviors, institutions, and organizations across nations, 2nd edn. Sage, Thousand Oaks

Jackson MO, Watts A (2002) The evolution of social and economic networks. J Econ Theory 106 (2):265–295

Jeong H, Neda Z, Barabasi A-L (2003) Measuring preferential attachment in evolving networks. Europhys Lett 61(4):567–572

Jun T, Sethi R (2009) Reciprocity in evolving social networks. J Evol Econ 19(3):379–396

Kaplan EL, Meier P (1958) Nonparametric estimation from incomplete observations. J Am Stat Assoc 53:457–481

Kenis P, Knoke D (2002) How organizational field networks shape interorganizational tie-formation rates. Acad Manag Rev 27(2):275–293

Kenis P, Oerlmans L (2008) The social network perspective – understanding the structure of cooperation. In: Cropper S, Ebers M, Huxham C, Ring PS (eds) The Oxford handbook of inter-organizational relations. Oxford University Press, New York, pp 289–312

Kirman A (1993) Ants, rationality, and recruitment. Q J Econ 108(1):137–156

Klepper S (1997) Industry life cycles. Ind Corp Chang 6(1):145–181

Koka BR, Madhavan R, Prescott JE (2006) The evolution of interfirm networks: environmental effects on patterns of network change. Acad Manag Rev 31(3):721–737

Kudic M, Guhr K, Bullmer I, Guenther J (2011) Kooperationsintensität und Kooperations-förderung in der deutschen Laserindustrie. Wirtschaft im Wandel 17(3):121–129

Kudic M, Pyka A, Guenther J (2012) Determinants of evolutionary network change processes in innovation networks – empirical evidence from the German laser industry. In: Conference proceedings. The 14th international Schumpeter Society conference, Brisbane, pp 1–29

Kumar R, Nti KO (1998) Differential learning and interaction in alliance dynamics: a process and outcome discrepancy model. Organ Sci 9(3):356–367

Larson A (1992) Network dyads in entrepreneurial settings: a study of the governance of exchange relationships. Adm Sci Q 37(3):76–104

Lavie D, Miller SR (2008) Alliance portfolio internationalization and firm performance. Organ Sci 19(4):623–646

Levitt T (1965) Exploit the product life cycle. Harv Bus Rev 43(6):81–94

Lorenzoni G, Ornati OA (1988) Constellations of firms and new ventures. J Bus Ventur 3(1):41–57

Lu JW, Beamish PW (2006) Partnering strategies and performance of SMEs' international joint ventures. J Bus Ventur 21(4):461–486

Malerba F (2002) Sectoral systems of innovation and production. Res Policy 31(2):247–264

McPherson M, Smith-Lovin L, Cook JM (2001) Birds of a feather: homophily in social networks. Annu Rev Sociol 27(1):415–444

Menzel M-P, Fornahl D (2009) Cluster life cycles – dimensions and rationales of cluster evolution. Ind Corp Chang 19(1):205–238

Muldur U, Corvers F, Delanghe H, Dratwa J, Heimberge D, Sloan B, Vanslembrouck S (2006) A new deal for an effective European research policy: the design and impacts of the 7th Framework Programme. Springer Netherlands, Dordrecht

Murray EA, Mahon JF (1993) Strategic alliances: gateway to new Europe. Long Range Plan 26 (4):102–111

Nakamura M, Vertinsky I, Zietsam C (1997) Does culture matter in inter-firm cooperation? Research consortia in Japan and the USA. Manag Decis Econ 18:153–175

Nelson RR, Winter SG (2002) Evolutionary theorizing in economics. J Econ Perspect 16(2):23–46

Newman ME (2010) Networks – an introduction. Oxford University Press, New York

Nowak MA, Tarnita CE, Antal T (2010) Evolutionary dynamics in structured populations. Philos Trans R Soc B 365(1537):19–30

OECD (2008) OECD science, technology and industry outlook. OECD, Paris

Ostrom E (2009) Beyond markets and states: Polycentric Governance of Complex Economic Systems. Prize Lecture, December 8

Owen-Smith J, Riccaboni M, Pammolli F, Powell WW (2002) A comparison of U.S. and European University-Industry relations in the Life Sciences. Manag Sci 48(1):24–43

Ozman M (2009) Inter-firm networks and innovation: a survey of literature. Econ Innov New Technol 18(1):39–67

Parkhe A (1993) Strategic alliance structuring: a game theoretic and transaction cost examination of interfirm cooperation. Acad Manag J 36(4):794–829

Parkhe A, Wasserman S, Ralston DA (2006) New frontiers in network theory development. Acad Manag Rev 31(3):560–568

Peteraf MA (1993) The cornerstones of competitive advantage: a resource-based view. Strateg Manag J 14(3):179–191

Pittaway L, Robertson M, Munir K, Denyer D, Neely A (2004) Networking and innovation: a systematic review of the evidence. Int J Manag Rev 5(6):137–168

Powell WW, White DR, Koput KW, Owen-Smith J (2005) Network dynamics and field evolution: the growth of the interorganizational collaboration in the life sciences. Am J Sociol 110 (4):1132–1205

Pyka A (2002) Innovation networks in economics: from the incentive-based to the knowledge based approaches. Eur J Innov Manag 5(3):152–163

Ring PS, Van De Ven AH (1994) Developmental processes of cooperative interorganizational relationships. Acad Manag Rev 19(1):90–118

Rosenkopf L, Tushman ML (1998) The coevolution of community networks and technology: lessons from the flight simulation industry. Ind Corp Chang 7(2):311–346

Scherngell T, Barber MJ (2009) Spatial interaction modeling of cross-region R&D collaborations: empirical evidence from the 5th EU framework programme. Pap Reg Sci 88(3):531–546

Scherngell T, Barber MJ (2011) Distinct spatial characteristics of industrial and public research collaborations: evidence from the fifth EU framework programme. Ann Reg Sci 46 (2):247–266

Schilke O, Goerzen A (2010) Alliance management capability: an investigation of the construct and its measurement. J Manag 36(5):1192–1219

Schilling MA (2009) Understanding the alliance data. Strateg Manag J 30(3):233–260

Schwerk A (2000) Dynamik von Unternehmenskooperationen. Duncker & Humbolt, Berlin

Snijders TA (2004) Explained variation in dynamic network models. Math Soc Sci 42(168):5–15

Snijders TA, Van De Bunt GG, Steglich CE (2010) Introduction to actor-based models for network dynamics. Soc Networks 32(1):44–60

Soda G, Zaheer A (2004) Network memory: the influence of past and current networks on performance. Acad Manag J 47(6):893–906

Spence M (1976) Informational aspects of market structure: an introduction. Q J Econ 90 (4):591–597

Spence M (2002) Signaling in retrospect and the informational structure of markets. Am Econ Rev 92(3):434–459

Stata (2007) Stata statistical software: release 10. StataCorp LP, College Station

Stokman FN, Doreian P (2005) Evolution of social networks: processes and principles. In: Doreian P, Stockman FN (eds) Evolution of social networks, 2nd edn. Gordan Breach, New York, pp 233–251

Sydow J (2003) Dynamik von Netzwerkorganisationen – Entwicklung, Evolution, Strukturation. In: Hoffmann WH (ed) Die Gestaltung der Organisationsdynamik – Konfiguration und Evolution. Schäffer-Poeschel, Stuttgart, pp 327–357

Ter Wal AL, Boschma R (2011) Co-evolution of firms, industries and networks in space. Reg Stud 45(7):919–933

Tiberius V (2008) Prozesse und Dynamik des Netzwerkwandels. Gaber, Wiesbaden

Van De Ven AH, Poole MS (1995) Explaining development and change in organizations. Acad Manag Rev 20(3):510–540

Venkatraman S, Lee C-H (2004) Preferential linkage and network evolution: a conceptual model and empirical test in the U.S. video game sector. Acad Manag J 47(6):876–892

Walker G, Kogut B, Shan W (1997) Social capital, structural holes and the formation of an industry network. Organ Sci 8(2):109–125

Wasserman S, Faust K (1994) Social network analysis: methods and applications. Cambridge University Press, Cambridge

Wernerfelt B (1984) A resource based view of the firm. Strateg Manag J 5(2):171–180

Witt U (2008a) What is specific about evolutionary economics? J Evol Econ 18(5):547–575

Witt U (2008b) Recent developments in evolutionary economics. Edward Elgar, Cheltenham

Zaheer A, Soda G (2009) Network evolution: the origins of structural holes. Adm Sci Q 54 (1):1–31

Chapter 10
Ego Networks and Firm Innovativeness

*Innovation is all about people. Innovation thrives when the
population is diverse, accepting and willing to cooperate.*
(Vivek Wadhwa)

Abstract In this chapter we seek to analyze how firm innovativeness is related to
individual cooperation events and the structure and dynamics of firms' ego networks. On the one hand, we analyze to what extent individual cooperation events
have a direct effect on firm innovativeness. On the other hand, each cooperation
event changes the structural configuration of a firm's portfolio of cooperative
relationships. Evolutionary network change processes at the micro-level – i.e. tie-
formation as well as tie-termination – shape the structural configuration of firm-
specific ego networks which are assumed to have an indirect effect on innovation
output. Consequently, the aim of this second empirical section is to disentangle
these two cooperation-related innovation effects. To shed some light on the questions raised, we apply the longitudinal panel dataset described above (cf. Sect.
6.1.2). Network measures are calculated on the basis of 570 knowledge-related
publicly funded R&D cooperation projects. Firm innovativeness is measured by
patent grants with a 1 and 2 year time lag. Several robustness checks are performed
on the basis of patent application counts. The following empirical analysis is
organized as follows. We start in Sect. 10.1 with a short introduction. In Sect.
10.2 we provide a theoretical foundation, present our conceptual framework and
derive a set of testable hypotheses. A description of the data sources together with a
brief presentation of the variables used follow in Sect. 10.3. In Sect. 10.4 we discuss
some methodological issues, specify the econometric estimation approach and
present our empirical results. Finally the paper closes with a brief discussion of
our main findings.

© Springer International Publishing Switzerland 2015   217
M. Kudic, *Innovation Networks in the German Laser Industry*,
Economic Complexity and Evolution, DOI 10.1007/978-3-319-07935-6_10

10.1   Motivation and Research Questions

The very aim of this analysis is to investigate how firm innovativeness is related to individual cooperation events and the structure and dynamics of firms' ego networks.[1]

New knowledge in innovation processes is mainly generated through the exchange and recombination of existing knowledge content. From a firm's perspective, this recombination may be achieved either through internal learning processes within the boundaries of the firm or by interacting with other economic actors (Graf and Krueger 2011, p. 69). Long-term cooperation projects provide a particularly important way for firms to reach beyond their own corporate boundaries (Alic 1990). These projects often take the form of strategic alliances (Grunwald and Kieser 2007, p. 369) which can be defined as "[...] voluntary arrangements between firms involving exchange, sharing, or co-development of products, technologies, or services" (Gulati 1998, p. 293). Strategic alliances can be categorized based on the underlying motivation, goals or organizational forms (Osborn and Hagedoorn 1997; Mowery et al. 1996).[2] The number of R&D partnerships has increased considerably since the 1980s, especially in high-tech industries (Hagedoorn 2002). Thus, firms increasingly face the challenge of managing and controlling a portfolio of national and international alliances simultaneously.

In this analysis we apply an ego network perspective in order to capture the firm-specific cooperation patterns and subsequent innovation outcomes over time.[3] Ego networks are constructed on the basis of a specific type of cooperative relationship: knowledge-related publicly funded R&D alliances that aim to increase the innovativeness of the organizations involved. The subject of our analysis includes various types of individual cooperation events as well as firm-specific R&D cooperation project portfolios which are defined from the focal actor's perspective and consist of a set of direct, dyadic ties between the focal actor and its alters as well as indirect ties between the alters (Ahuja 2000). They do not include second-tier ties or second-step ties to which the focal actor is not directly connected (Hite and Hesterly 2001).

---

[1] This chapter draws upon a joint research project conducted together with Guido Buenstorf, University of Kassel, Institute of Economics and International Center of Higher Education Research (INCHER-Kassel) and Katja Guhr, Department for Structural Economics at the Halle Institute for Economic Research. We have greatly benefited from comments by the audience at the 7th EEMAE conference in 2011 in Pisa, Italy and the IIDEOS PhD colloquium in 2011 in Marburg. The latest draft of the paper was presented at the 5th EMNET Conference in 2011 in Limassol, Cyprus (Kudic et al. 2011a). I take full responsibility for any errors in this section.

[2] Section 2.5.3 provides a discussion on cooperation rationale and motives.

[3] These terms "ego network", "alliance portfolio" and "alliance constellation" are used in this paper interchangeably. For an overview and comparison of definitions and concepts, see Wassmer (2010).

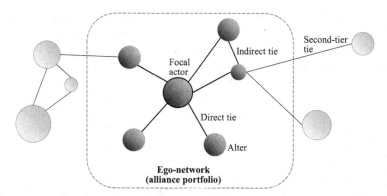

Fig. 10.1 Illustration of a typical ego network structure (Source: Kudic and Banaszak (2009, p. 9))

Figure 10.1 illustrates a typical ego network structure with one focal actor, five directly connected alters and one indirect connection (cf. Kudic and Banaszak 2009).

Earlier related work has analyzed the relationship between knowledge-intensive R&D alliances and firm innovativeness (Narula and Hagedoorn 1999; Stuart 1999; Stuart 2000) and introduced concepts explaining the identification and commercial utilization of knowledge (Cohen and Levinthal 1990) as well as disturbances in interorganizational knowledge transfer and learning processes (Simonin 1999). Moreover, scholars from various disciplines have analyzed how various dimensions of structural embeddedness in interorganizational networks (Powell et al. 1996; Rodan and Galunic 2004; Capaldo 2007) or the overall network structure itself (Schilling and Phelps 2007) affect innovativeness in the firms involved. In contrast, longitudinal empirical studies that explicitly analyze the relationship between ego network characteristics and firm innovativeness are comparably rare.[4]

One essential question that arises in this context is whether the innovativeness of firms in high-tech industries is directly affected by individual R&D cooperation events or more indirectly by structure and structural change in firm-specific ego network characteristics over time. In other words, through which transmission channels do cooperation events affect a firm's subsequent innovative performance? On the one hand it is plausible to assume that individual cooperation events directly affect firm innovativeness. On the other hand, past as well as present cooperation events determine the configuration of the focal actor's individual ego network structure over time which itself is likely to affect the firm's innovativeness. The explicit consideration of structural consequences of firm-level cooperation events raises the awareness of the existence of direct as well as indirect cooperation related innovation effects. Furthermore, Wassmer (2010, p. 162) concludes in his comprehensive review on alliance portfolios that further research based on longitudinal

---

[4] Most notable exceptions are: (Ahuja 2000; Baum et al. 2000; Wuyts et al. 2004).

studies is needed to understand how and why firms change the configuration of their alliance portfolios over time and how this affects a firm's performance. This dual character of individual cooperation events has been widely neglected in previous research on ego networks and constitutes the core of this investigation.

Consequently, we seek to answer the following research questions: **(I)** Do individual cooperation events (i.e. "direct effects") or rather structural ego network characteristics (i.e. "indirect effects") affect firm innovativeness over time? **(II)** How do individual cooperation events affect the structural configuration of the focal actor's ego network and which structural features affect its subsequent innovation output?

To answer these questions, we apply the longitudinal panel dataset introduced above (cf. Sect. 6.1.2). Information on type, content and funding of publicly funded R&D cooperation projects provides a solid basis for a fine-grained analysis of direct innovation effects. Structural ego network measures were calculated on a yearly basis by applying network data and quantitative network analysis methods (Wasserman and Faust 1994; Borgatti et al. 2002).[5]

10.2   Theoretical Reflections, Conceptual Framework and Hypotheses

Numerous theoretical contributions have sought to explain the nature of hybrid organizational forms and a firm's motives to cooperate in its innovation efforts (Hagedoorn 1993; Osborn and Hagedoorn 1997; Gulati 1998).[6] Some early explanations adopted the perspective of transaction cost economics (Jarillo 1988; Thorelli 1986; Williamson 1991). They interpret hybrid arrangements as strategic alliances (Borys and Jemison 1989) which are positioned between markets and hierarchies and reduce transaction costs under moderate asset specificity and frequency of disturbances (Williamson 1991, p. 292).

Other scholars have argued that hybrids have to be regarded as a unique organizational form that cannot be classified as an intermediate between markets and hierarchies (Powell 1990; Podolny and Page 1998). However, the structural forms behind these hybrids are manifold, ranging from short-term supply contracts, licensing and franchise agreements and consultancy contracts, to consortia, long-term partnerships and joint ventures (Podolny and Page 1998; Mowery et al. 1996). Previous studies on the motives for strategic alliances have shown that R&D alliances in particular provide significant cost saving potentials (Harrigan 1988; Hagedoorn 2002) and allow firms to reduce the risk inherent in R&D processes (Ohmae 1989; Hagedoorn 1993; Sivadas and Dwyer 2000). Furthermore, R&D

---

[5] We used standard ego network procedures implemented in UCI-Net 6.2 to calculate ego network measures (Borgatti et al. 2002).

[6] For an in-depth discussion on the motives for cooperating, see Sect. 2.5.3.

alliances provide access to new products and markets (Kogut 1991; Hagedoorn 1993), allow time to be saved by shortening the time-span between invention and market introduction (Mowery et al. 1996), and provide opportunities to internationalize business and penetrate markets abroad (Hakansson and Johanson 1988; Narula and Hagedoorn 1999). With the emergence of the knowledge-based approach in organization science (Kogut and Zander 1992; Spender and Grant 1996; Grant 1996), scholars realized the strategic importance of firm-specific knowledge resources for the competitive advantage of firms (Coff 2003). Knowledge related motives for interorganizational learning processes (Hamel et al. 1989; Hamel 1991; Khanna et al. 1998; Kale et al. 2000) as well as knowledge transfer processes (Rothaermel 2001; Grant and Baden-Fuller 2004; Buckley et al. 2009) have been analyzed from various angles in the field of alliance and network research. However, scholars have argued that "[...] among the various motivations for partnering, innovation is said to be a rationale of singular importance" (Bidault and Cummings 1994, p. 33).

10.2.1 R&D Alliances, Networks and Innovation Output

The relationship between knowledge transfer, R&D cooperation and firm innovativeness has been the subject of numerous case studies (Dyer and Nobeoka 2000; Ciesa and Toletti 2004; Eraydin and Aematli-Köroglu 2005; Capaldo 2007) as well as several survey-based empirical studies.

For instance, De Propris (2000) has studied the link between innovation performance and upstream as well as downstream interfirm partnerships drawing upon a unique dataset compromised of 435 firms located in the West Midlands, UK. Estimation results substantiate the importance of R&D cooperation as a driving force behind firm innovativeness. Harabi (2002) found statistically significant support for the impact of vertical R&D cooperation on firm-level innovation outcomes based on a sample of 370 small and medium sized German firms. The results indicate that informal modes of cooperation are apparently more important than formal modes. In a similar vein, Freel and Harrison (2006) investigated the impact of cooperation on firm-level innovation output. They conducted a survey-based study compromising 1,347 small firms from Northern Britain in both the manufacturing and service sectors. They report a positive correlation between product innovation success and cooperation with customers and public sector organizations.

Even though these studies provide us with important insights into the relationship between R&D partnerships and a firm's efforts to innovate, they suffer from at least three serious limitations. Firstly, the majority of survey-based cooperation studies focus on dyadic partnerships and neglect the structural dimension of the overall innovation network in which the firms under investigation are embedded. Secondly, network studies are quite sensitive with regard to network boundary misspecification and missing cooperation data. Empirical studies employing complete network data are quite rare. Finally, the majority of survey-based cooperation studies draw upon cross-sectional data and neglect the dynamic nature of cooperation activities and subsequent innovation consequences.

In response to these issues researchers have quite recently started to analyze the relationship between firm positioning in complex interorganizational networks and firm innovativeness based on longitudinal large-scale databases (Stuart 2000; Lee 2010; Fornahl et al. 2011).[7]

10.2.2   Ego Network Structure and Innovation Output

Over the past years the number of R&D collaborations has increased rapidly, especially in high-tech industries, (Hagedoorn 2002) and firms increasingly face the challenge of managing a portfolio of multiple collaborations simultaneously. This empirically observable fact places attention on firm-specific cooperation networks – so-called alliance portfolios or ego networks – (Wassmer 2010; Hite and Hesterly 2001) and raises several interesting and still widely unanswered research questions.

In the areas of economics, management and organization science, there are a number of excellent studies on "alliance network compositions" (Baum et al. 2000), "ego networks" (Ahuja 2000; Jarvenpaa and Majchrzak 2008; Hite and Hesterly 2001), "alliance constellations" (Das and Teng 2002; Gomes-Casseres 2003), "alliance portfolios" (George et al. 2001; Parise and Casher 2003; Hoffmann 2005, 2007; Lavie 2007; Lavie and Miller 2008) or "portfolios of interfirm agreements" (Wuyts et al. 2004). Our main interest is in the existence and the extent of additional ego network effects which are assumed to shape the focal actor's innovative performance over time. With few exceptions, previous studies have paid comparably less attention to links between the structural ego network configuration and firm innovativeness.

For instance, Ahuja (2000) has analyzed the relationship between three aspects of a firm's ego network characteristics – direct ties, indirect ties and structural holes – as well as subsequent firm-level innovation outcomes. The results confirm that direct and indirect ties positively affect innovation output, while also raising awareness for the negative innovation effects of structural holes. Baum and his colleagues (2000) have shown that the early innovative performance of Canadian biotech startups – measured by patent grant counts and R&D spending growth – is strongly affected by the alliance network composition of these firms at the time they are founded. Wuyts and his colleagues (2004) have analyzed the impact of different types of alliance portfolio descriptors on a firm's incremental and radical innovations as well as on firm profitability.

---

[7] Schilling (2009) provides a comprehensive overview of large-scale alliance and network data databases such as "SDC", "MERIT-CATI", "CORE", "RECAP", and "BIOSCAN".

Evidence that explains the overall advantages of alliance portfolios over dyadic cooperation linkages can be drawn from three lines of argument. Firstly, ego networks provide a risk reduction effect which goes beyond the dyadic level (Hoffmann 2007). By actively managing and controlling a portfolio of alliances, risk can be reduced by taking advantage of these risk diversification effects (Markowitz 1952). Given potentially high rates of failure in achieving risk reduction in dyadic alliances (Bleeke and Ernst 1991; Sivadas and Dwyer 2000), spreading risk over a portfolio of alliances helps firms reduce the variances in expected returns. Secondly, firms can gain cost savings by utilizing synergy effects in a portfolio of alliances (White 2005; Hoffmann 2005). Cooperation routines and standardized cooperation interfaces (Goerzen 2005), as well as alliance experience (Anand and Khanna 2000) and alliance management capabilities (Schilke and Goerzen 2010) save costs and increase the overall efficiency of a focal actor's ego network. For instance, Rothaermel and Deeds (2006) report a moderating effect of alliance experience on the relationship between a high-tech venture's R&D alliances and its new product development. Thirdly, an alliance portfolio enhances the scope of potential learning and knowledge access opportunities by providing access to multiple stocks of knowledge (Grant and Baden-Fuller 2004). Due to the heterogeneity of directly connected partners, the range of potentially accessible knowledge stocks increases. In addition, the interconnectedness of direct partners facilitates the flow of information in the narrower surroundings of the focal actor. The broader range of opportunities for knowledge access and learning, and the enhanced flow of information across partners are likely to have a positive impact on a firm's ability to innovate and gain competitive advantages (Gomes-Casseres 2003).

Most of the previously discussed arguments are directly reflected in the structural configuration of a focal actor's ego network. In other words, a focal actor's cooperative path is reflected in his past as well as present cooperation activities. Thus it is worthwhile taking a closer look at the structural features of firm-specific cooperation patterns over time in order to answer the research questions that were initially raised. Basically two distinct structural ego network dimensions can be identified in this context. On the one hand, we can analyze a firm's ego network structure with regard to features relating to the node level. This perspective refers, for instance, to the number of directly connected partners or to the heterogeneity of partners in an ego network. On the other hand, we can focus on the connectedness of partners in an ego network in order to characterize its structural features. From this point of view the various types and configurations of linkages between the actors in an ego network become relevant. In addition, ego networks are not static; they change continuously over time and shape the structural configuration of the focal actor's portfolio as well as the focal actor's subsequent innovative performance. This requires a dynamic view of networks which is provided in the following section.

10.2.3   An Evolutionary Perspective on Ego Networks

Recent reviews of overall interorganizational networks (Provan et al. 2007; Bergenholtz and Waldstrom 2011) and innovation networks (Pittaway et al. 2004; Ozman 2009) agree that the dynamic character of networks is still not understood sufficiently.[8] Changes in network structure are the result of events affecting two basic elements – nodes (i.e. organizations) and ties (i.e. R&D alliances) – of innovation networks (Doreian and Stokman 2005; Glueckler 2007). This means that an innovation network evolves as nodes enter and exit the population (i.e. changes in the number of organizations) and build and dissolve network relationships with other actors (i.e. changes in the number R&D partnerships). Structural network change can occur as a result of exogenous and endogenous factors. Determinants, mechanisms and structural change patterns as a consequence of micro-level network change processes are given a prominent role in evolutionary network studies (cf. Sects. 9.1 and 9.2). In comparison to the more general term "network dynamics" the concept of "network evolution" contains "[...] a stricter meaning that captures the idea of understanding change via some understood process" (Doreian and Stokman 2005, p. 5). However, the majority of previously conducted empirical studies on network evolution focus on the overall network level whereas research from the perspective of the focal actors is rare (Hite and Hesterly 2001). To date, only a small number of case studies (Dyer and Nobeoka 2000; Dittrich et al. 2007) have addressed the issue of how portfolios of collaborations change over time. Wassmer (2010, p. 165) concludes that "[...] little is still known on how alliance portfolio configurations change over time and what drives this evolution." In the present analysis we explicitly consider how tie formations and tie terminations of both the focal actors' cooperation activities as well as the network neighbors affect the structural configuration of ego networks and subsequent innovation outcomes.

10.2.4   Conceptual Framework: Direct and Indirect Innovation Effects

Our conceptual framework (cf. Fig. 10.2) draws upon the previously outlined theoretical considerations and seeks to substantiate the relationships between evolutionary micro-level network change processes, changes in ego network structure and firm-level innovation outcomes. The framework consists of four elements – (I) individual cooperation events, (II) ego network structure, (III) network environment, (IV) innovation outcomes – and illustrates four cooperation-related effects –

---

[8] Recently a number of excellent theoretical as well as empirical studies have addressed and analyzed the evolutionary change of networks. For an overview of contemporary research see Sect. 9.2.

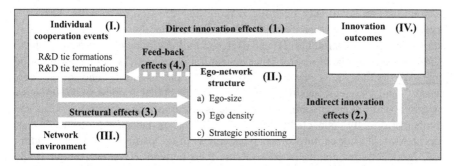

Fig. 10.2 Network change processes, ego network configuration and firm-level innovation output (Source: Author's own illustration)

(1) direct innovation effects, (2) indirect innovation effects, (3) structural effect, and (4) feed-back effects – all from a focal actor's perspective.

We start our argumentation by focusing on individual cooperation events (**I**). In this context, individual cooperation events encompass all tie formations and tie terminations on the micro-level which affect the structural configuration of the focal actor's ego network. These structural effects (3) can arise from the focal actor's own cooperation activities as well as from the cooperation activities of the focal actor's direct partners. In the first case, the size of the ego network is affected whereas in the second case the density of the focal actor's ego network is affected. In addition, the network environment (**III**) influences the ego network in at least two additional ways. Firstly, a focal actor's cooperation decisions are strongly influenced by the cooperation opportunities and restraints provided by the broader network environment. Secondly, even if an ego and its alters do not conduct any cooperation activities over a given period of time, the relative importance of its ego network changes continuously due to cooperation activities of other network actors in the broader network environment. This means that structural ego network features have to be analyzed in the context of the focal actor's broader network environment (**III**).

Now we turn our attention to the relationship between individual cooperation events (**I**) and firm-level innovation outcomes (**IV**). As outlined above, this direct innovation effect (1) has been the subject of a large number of empirical studies. The findings of these studies substantiate the assumption that cooperation events are positively related to firm-level innovation outcomes. However, especially in the case of publicly funded R&D cooperation projects, it is unclear whether it is the cooperation itself or, whether it is in fact the amount of funding received which affects firm innovativeness at a later point in time. To account for this issue we divide the direct cooperation-related drivers behind firm innovativeness into a "cooperation effect" and a "funding effect".

Firm-specific cooperation activities have an additional, more indirect innovation effect by shaping the focal actor's ego network structure. Theoretical arguments on risk diversification, synergy and cost-savings in alliance portfolios substantiate the assumption that an alliance portfolio is more than the sum of its parts. Thus, we argue that each cooperation event (**I**) affects the structural configuration of a focal actor's ego network structure (**II**) and exerts an indirect innovation effect (2) which is assumed to be related to firm-level innovation outcomes (**IV**) at a later point in time. We include three structural ego network dimensions – "ego size", "ego density" and "strategic positioning" – in our conceptual framework in order to capture a wide range of portfolio characteristics. Ego network size refers to the number of directly connected partners of a focal actor and the ego itself whereas ego network density captures the connectedness of the partners involved. In addition, firms act strategically in constructing their network (Dyer and Singh 1998; Gulati et al. 2000) and choose those network partners whose characteristics comply with their specific innovation process requirements. Consequently we include a structural component ("ego density") and strategic component ("strategic positioning") in our framework.

Finally, the dotted feedback line (4) illustrates the inter-temporal relationship between past and current cooperation events. The sum of all previously conducted tie formations and tie terminations of a focal actor itself and its closer network environment constitutes its individual ego network structure. New cooperation decisions are based on previous cooperation experiences and are determined by considerations of how new linkages fit into existing webs of linkages (Gulati and Gargiulo 1999). In other words, cooperation decisions are path-dependent. Some authors have argued that existing network structures are resistant to change. For instance, Kim and his colleagues (2006) have proposed a theoretical "network inertia" framework that explains the organizational resistance to changing interorganizational network ties as well as difficulties that an organization faces when it attempts to dissolve old relationships and form new network ties. In contrast, other authors have argued that firm strategies and actions can disrupt existing network paths (Glueckler 2007). Both, however, agree that a longitudinal setting is required to appropriately account for the inter-temporal dimension of structural ego network change patterns.

The deduction of testable hypotheses in the following section concentrates on the drivers as well as interrelationships between direct innovation effects (1) and indirect innovation effects (2) in our framework.

10.2.5   Hypotheses on Cooperation-Related Innovation Effects

Does R&D cooperation affect firm innovativeness, and if so, what are the rationales behind this assumption? The answer to at least the second part of this question was provided quite early by scholars (Alic 1990; Hagedoorn 1993). Due to the science-based character of the German laser industry (Grupp 2000) we refer to knowledge-related arguments to substantiate our first set of hypotheses. There are two streams of literature – the "knowledge acquiring approach" and the "knowledge accessing approach" which can be distinguished in this context (Al-Laham and Kudic 2008). The distinction is based on the underlying processes of knowledge generation (or "exploration") and knowledge application (or "exploitation") among partners in strategic alliances (Grant and Baden-Fuller 2004, p. 61).

According to the first approach, alliances can be regarded as "vehicles of learning" (Grant and Baden-Fuller 2004, p. 64) which allow a firm to share a particular part of its knowledge bases and exchange implicit stock of knowledge across firm boundaries. The firm's ability to "[...] recognize the value of new, external information, assimilate it, and apply it to commercial ends [...]" (Cohen and Levinthal 1990, p. 128) is of paramount importance for organizational as well as interorganizational learning processes. Since the introduction of the initial concept of "absorptive capacity", several scholars have contributed to a concretization of the concept itself (Van Den Bosch et al. 1999; Zahra and George 2002) and to a reconceptualization from a firm-level construct to a learning dyadic level concept (Lane and Lubatkin 1998; Lane et al. 2001). In addition, the establishment of mutual trust between partners (Lui 2009) has been recognized as a key factor in successful interorganizational learning processes in order to avoid learning races (Amburgey et al. 1996) or tensions between alliance partners (Das and Teng 2000) which can result in alliance instabilities or terminations (Park and Russo 1996; Inkpen and Beamish 1997).

The second approach suggests that firms cooperate in order to gain access to complementary stocks of knowledge (Grant and Baden-Fuller 2004) without necessarily internalizing the partner's skills (Doz and Hamel 1997). In other words, a knowledge accessing strategy focuses on the use of the partner's rich experience without acquiring any specific skills (Lui 2009). Grant and Baden-Fuller (2004, p. 69) argue in their "knowledge accessing" framework that the efficiency of knowledge integration through alliances can be superior compared to markets or hierarchies where products require a broad range of different types of knowledge. Firms do not necessarily have to generate new stocks of knowledge within the boundaries of the firm. Instead, they can collaborate with other firms or public research organizations to gain access to complementary stocks of explicit knowledge. However, several problems can occur during the interorganizational knowledge transfer processes. Simonin (1999) has introduced the concept of "causal ambiguity" and empirically analyzed the determinants affecting knowledge transfer processes in strategic alliances.

In summary, both knowledge acquiring as well as knowledge assessing strategies can significantly flexibilize and improve the firm's knowledge base – a necessary precondition for subsequent innovation processes. Broekel and Graf (2011, p. 6) argue that publicly funded R&D projects provide strong incentives for sharing knowledge and for innovating due to the regulative framework to which all cooperation partners involved have to agree. To test the empirical relationship between direct cooperation events and innovation output, we look at the two types of publicly funded R&D cooperation projects separately. Nationally funded cooperation projects predominantly address cooperation attempts among German firms and organizations. In contrast, supra-national cooperation projects are based on the notion of supporting pan-European research and development activities. Based on our previous considerations we can formulate the following two hypotheses:

H1a The annual number of nationally funded cooperation projects ("Foerderkatalog") is positively related to a firm's innovative performance at subsequent points in time.

H1b The annual number of supra-nationally funded cooperation projects ("CORDIS") is positively related to a firm's innovative performance at subsequent points in time.

Next we turn our attention to the structural dimension of individual cooperation events. The appropriate choice and establishment of R&D cooperation projects can increase the structural efficiency of an existing ego network. As outlined above, firms choose new partners based on strategic considerations (Dyer and Singh 1998; Gulati et al. 2000) which comply with their specific innovation process requirements. The rationale behind the establishment of a cooperative relationship is not necessarily direct access to the partner's resource pool. Instead the focal actor's intention may be to reduce its dependence on brokers by establishing alternative knowledge channels to strategically relevant actors or groups of actors. In other words, focal actors choose cooperation partners for strategic reasons in order to secure their network position, to complement their existing ego network structures and to increase efficiency. Consequently, tie formations and tie terminations may induce an additional structural effect (i.e. indirect innovation effect) by reshaping the configuration of the ego network. These individual cooperation events contribute to firm-specific innovation processes by filling "structural gaps" in existing ego networks. Thus, not only the "cooperation-specific" effect but also the superior "ego network-specific" effect is likely to determine firm innovativeness. In other words, it is plausible to assume that an additional innovation effect occurs which is caused by the focal actor's ego network structure. This implies that the several facets of the focal firm's ego network structure potentially affect the firm's innovativeness.

To test the empirical relationship between network structure and innovation output, we look separately at the distinct structural dimensions characterizing the ego network topology. The size of an ego network may affect the focal actor's innovativeness for a variety of reasons. As outlined above, collaborative arrangements provide access to new and complementary stocks of knowledge (Rothaermel 2001; Grant and Baden-Fuller 2004). This, however, is also of vital importance in portfolio settings. The more direct linkages there are in a portfolio, the broader the range of potentially accessible complementary knowledge stocks. Scholars have argued that a firm's ability to access new knowledge from external sources becomes itself a more relevant source for competitive success than the present stock of knowledge within the firm (Decarolis and Deeds 1999). Basically the same argument applies to knowledge-acquiring strategies. In addition, saving time, which can be achieved through cooperation, becomes increasingly important in science-based industries. Mowery and his colleagues (1996, p. 79) argue that the perceived shortening of product life-cycles increases the competitive pressure on firms in technology-intensive industries. They conclude that the rapid penetration of foreign markets becomes increasingly important, a goal which can be more easily achieved through alliances. These arguments become important, especially in an alliance portfolio context, as multiple collaborative R&D endeavors with diverse heterogeneous partners increase the accessibility to various types of knowledge stocks or learning opportunities and accelerate the development of new ideas and products. These arguments substantiate our next hypothesis:

H2a The greater the size of a focal actor's ego network, the higher its subsequent innovative performance.

As outlined above, in addition to node-related ego network features such as size we can distinguish between dimensions that are structurally and strategically oriented, i.e. degree of connectedness and brokerage positions. The degree of connectedness in an ego network is related to the extent to which firms gain innovation experience by being well connected to other firms or public research organizations. According to closure theory a high degree of connectedness increases the visibility of network actors (Coleman 1988). Furthermore, a high number of linkages in a densely connected ego network lower the risk of dependence on other organizations due to the existence of redundant ties and optional knowledge channels to relevant partners. Moreover, in highly connected networks, firms gain access to various types of potentially decisive stocks of explicit as well as implicit (or tacit) knowledge. This increases the scope of the firm's potentially available complementary knowledge stock and increases the firm's flexibility. These considerations lead to the following prediction[9]:

H2b The higher the degree of connectedness in a focal actor's ego network, the greater its subsequent innovative performance.

A central debate in alliance and network literature occurs around Coleman's "closure theory". Burt's (1992) "structural hole" theory highlights the importance of strategic positions and brokerage activities of actors in sparsely connected networks. Recent theoretical and empirical studies (Rowley et al. 2000; Burt 2005) indicate that these two perspectives are not mutually exclusive. We follow Burt's line of argument with regard to our last hypothesis. According to this perspective it is not so much a high degree of connectedness but rather the occupation of strategically relevant network positions that is decisive. Actors

---

[9] Even though we argue in this paper that the connectedness of an actor exerts a positive effect on innovation output, one has to keep contrary lines of argument in mind. For instance, Uzzi (1997) proposes that the effects of network embeddedness may become negative with an increasing level of connectedness.

connecting a large number of otherwise unconnected actors – so-called "brokers" – occupy such positions. Referring to this argument and keeping in mind our ego network perspective, we put forward the following argument: like brokers in overall networks, we can identify strategically decisive actors in ego networks who mediate the majority of the relationships between the other ego network actors. "When 'ego' is tied to a large number of 'alters' who themselves are not tied to one another, then ego has a network rich in structural holes" (Podolny 2001, p. 34). These positions are beneficial for several reasons. Brokers can facilitate, control or prevent the flow of knowledge into an ego network to a large extent by bridging structural holes in existing network structures. They are in a position that allows them to bring together firms as well as other organizations. Consequently we formulate our last hypothesis as follows:

H-2c Focal actors that occupy a brokerage position show a higher innovative performance at a later point in time.

10.3   Data, Methods and Variable Specification

10.3.1   Applied Data Sources

The analytical part of this book is based on three main data sources: patent data, industry data and network data. [10]

We use patent data to construct indicators reflecting the innovative performance at the firm level. A lot has been written about the empirical challenges of measuring innovation processes. Despite the methodological constraints related to the use of patents to measure innovation performance (Patel and Pavitt 1995), patent indicators are commonly used in analyzing innovation processes (Jaffe 1989; Jaffe et al. 1993). Raw data was taken from the EPO Worldwide Statistical Database. DEPATISnet (the German Patent and Trade Mark Office's online database) and ESPACEnet (European Patent Office database) were used to check results for integrity and consistency. Our database includes patent applications as well as patents granted by the German Patent Office and by the European Patent Office. [11]

Industry data came from a proprietary dataset containing the entire population of German laser source manufacturers between 1969 and 2005 (Buenstorf 2007). Based on this initial dataset we used additional data sources to gather information about firm entries and exits after 2005. For the purpose of this paper we chose the

---

[10] Fo an in-depth description of applied data sources and data gathering procedures, see Sect. 4.2.

[11] Identifying patent grants is a difficult task. We used the "patent first granted" flag (PatStat) in combination with the variable "publn_kind" to identify all granted patents.

business-unit or firm level. We ended up with an industry dataset encompassing 233 laser source manufacturers over the entire period under observation. In addition, we identified 145 universities and public research organizations with laser related activities by using the methodical procedure described below.

Network data came from two official databases on publicly funded R&D collaboration projects. The first source was the Foerderkatalog database provided by the German Federal State, which contains information on a total of more than 110,000 completed or ongoing subsidized research projects and provides detailed information on the starting point, duration, funding and characteristic features of the project partners involved. This data source has quite recently been used by other researchers to gather network data (Fornahl et al. 2011; Broekel and Graf 2011). The publicly funded research projects are subsidized by five German federal ministries. In total, we were able to identify, for the entire population of 233 German laser source manufacturers, 417 R&D projects with up to 33 project partners from various industry sectors, non-profit research organizations and universities. The second raw data source was an extract from the *CORDIS* project database which includes a complete collection of R&D projects for all German companies which were funded by the European Commission between 1990 and 2010. Data on EU Framework programs has also been used by other researchers to construct R&D networks (Cassi et al. 2008; Protogerou et al. 2010; Scherngell and Barber 2011). In total, this database extract consisted of a project dataset with over 31,000 project files and an organization dataset with over 57,100 German organizations and roughly 194,000 international project partners. Based on this raw data, we identified 155 R&D projects with up to 53 project partners for the entire sample of German laser source manufacturers. Finally, both cooperation data sources were used to construct interorganizational innovation networks on a yearly basis.

We used both data sources on publicly funded projects because the German national funding paradigm differs in several ways from the supra-nationally oriented funding paradigm of the European Union. For instance, a comparison of *Foerderkatalog and CORDIS* data shows a much higher heterogeneity of projects in terms of partner nationality, number of project partners and funding received (Kudic et al. 2011b). In addition, other researchers have pointed out that supranational projects have a much higher involvement of public research organizations (Scherngell and Barber 2011; Broekel and Graf 2011, p. 5).

Using information about publicly funded research projects to construct R&D networks raises potentially grave selectivity concerns. It is conceivable – and indeed desirable from a societal perspective – that funding decisions reflect the heterogeneous quality of applicants. In our empirical setting, this concern seems to be of limited salience for several reasons.[12] Another potential concern is that publicly funded R&D projects primarily affect innovation outcomes through their resource effects. We checked for the resource effects by including funding as a control variable in our empirical analysis.

---

[12] A detail discussion of potential selection biases is provided in Sect. 4.2.3.

10.3.2   The Data Preparation Process

The empirical analysis is based on the full population of German laser source manufacturers between 1990 and 2010 – an unbalanced panel of 233 firms with a total of 2,645 firm years. Over the entire observation period we had an average of 11.08 observations per firm. Annual counts of patent grants and applications were used as the measure of innovation output, with a 2 year lag structure accounting for the time required to arrive at patentable innovations.

To construct the R&D network we had to identify all laser-related public research organizations (PROs). Two complementary methods were applied to obtain a complete list of all PROs involved (cf. Sect. 4.2.1). We started with the "expanding selection method" according to Doreian and Woodard (1992). Using the initial list of 233 laser source manufacturers we added to our extended ID-list all non-profit research organizations and universities active in the field of laser research as long as these organizations established two or more links to at least one firm on our initial list. In contrast to the "snowball sampling method" (Frank 2005) we did not immediately include organizations with just one link in our sample. Instead, we checked in each of these cases whether the identified public research organization was active in the field of laser research or not. In total we identified 138 laser-related public research organizations. This procedure, however, has a serious limitation. All laser-related PROs that did not cooperate with LSMs in the period under observation were systematically ignored. Thus, we applied a second methodological approach to complement our sample. Based on a bibliometric analysis we identified all of the organizations that published laser papers in conference proceedings or academic journals over the past two decades. Raw data for this analysis, provided by the LASSSIE project consortium (Albrecht et al. 2011), was used and supplemented by searches for laser-related publications listed in the ISI Web of Knowledge database. Thus we were able to generate a complete list of all PROs that have published at least one paper in the field of laser research. By comparing and consolidating the results of these two data gathering methods we ended up with a final list of 145 laser active PROs for the time spanning between 1990 and 2010. Finally, entry and exit dates and addresses were retrieved for all identified PROs in the dataset.

In a second step we broke down the overall network into 21 time-distinctive network layers, one network for each year. Each network layer is based on a symmetric undirected and binary adjacency matrix (Wasserman and Faust 1994) whereas the number of rows or columns was determined by the number of active laser source manufacturing firms in a given year. The decomposition of multi-partner R&D cooperation projects into dyadic network linkages is based on the assumption that all partners involved have linkages to one another (cf. Sect. 5.2).

This converted dataset allowed us to capture and quantify structural network characteristics over time and to account for several key network variables – especially ego network measures – that may influence the innovative performance of laser source manufacturing firms during the period under observation. We used standard ego network procedures implemented in UCI-Net 6.2 to calculate ego network measures (Borgatti et al. 2002).

For the patent data gathering process we used the names of the firms in the sample and assigned a patent to a firm if its name appeared as an applicant and if either applicant or inventor had a German address. We also traced changes in corporate names and legal status, as well as organizational changes and the establishment of spin-offs to allocate annual patent counts to each company.

10.3.3 Variable Specification

In previous studies, both patent applications and grants were used as innovation proxies (Powell et al. 1996; Ahuja 2000; Jaffe et al. 1993). We decided in favor of patent grants *[pgcnt]* because they indicate the actual securitization of a patent. In other words, we chose a more restrictive innovation indicator for the purpose of this empirical section. In addition, we used patent application *[pacnt]* as an additional innovation proxy to cross-check our results and ensure robustness of our findings. Application counts are frequently used in innovation studies as this reflects the earliest point in time that research was completed (Jaffe et al. 1993). A 1 and 2 year time lag structure was applied in line with previous research in this area.

The key explanatory variables are two types of cooperation counts and three basic ego network measures (cf. Sect. 5.2.2). On the one hand, we measured firm-specific cooperation propensity with two cooperation count measures based on the *Foerderkatalog* data *[coopcnt_fk]* and *CORDIS* data *[coopcnt_c]*, respectively, as well as a combined cooperation count indicator *[coopcnt_fkc]* consisting of the sum of both. On the other hand we applied three structural ego network indicators. We used procedures implemented in UCI-Net 6.2 (Borgatti et al. 2002) to generate our ego network variables. We repeated this procedure for each year under observation. The first measure is a size variable *[ego_size]*. It is defined by the number of actors (alters) that are directly connected to the focal actor (ego). The second ego network measure is a density variable *[ego_density]*. This variable is defined as the number of de facto ties at a given point in time divided by the number of pairs, multiplied by a factor of 100.[13] The third ego network variable is a normalized ego network

---

[13] The number of pairs of alters in an ego network is a measure for the maximum connectedness, i.e. potential ties that can be realized, of the ego network.

brokerage indicator *[ego_nbroke]*. This measure captures the number of times a focal actor of an ego network lies on the shortest path between two alters, normalized by the number of brokerage opportunities, which is a function of ego network size (Borgatti et al. 2002).

For firm-level control variables, we include a linear firm age measure *[firmage]* as well as a squared term *[firmage_sq]*. To account for overall network effects we include two types of network level control variables. The first variable captures the size of the overall network *[nw_size]* defined as the proportion of firms with at least one dyadic partnership in a given year. The second variable measures the connectedness of the overall network *[nw_density]* calculated by using the standard network density procedure implemented in UCI-Net 6.2 (Borgatti et al. 2002). In addition, we include annual time-dummies to control for inter-temporal effects. We included a set of year dummies *[yr97-yr08]* to account for year-specific effects in our estimations. Finally, we include a cooperation funding *[coopfund_fkc]* variable in our model. The funding received is measured in 1,000 euros.

Table 10.1 provides an overview of the variables and corresponding definitions on the left-hand side. Summary statistics for the dependent and independent variables are displayed on the right. Table 10.2 presents the correlation matrix for all variables used.

Table 10.1 Descriptive statistics – cooperation events and ego network characteristics

Variable	Variable definition	Summary statistics				
		Obs.	Mean	Std. dev.	Min	Max
Endogenous variables						
pacnt	Patent applications (annual count)	2,645	2.662004	17.43323	0	366
pgcnt	Patent grants (annual count)	2,645	0.339130	1.635554	0	28
Control variables						
firmage	Age of the firm	2,645	8.055955	6.800477	0	43
firmage_sq	Age of the firm, squared	2,645	111.1274	177.8146	0	1,849
nw_size	Network size	2,645	0.381853	0.060200	0.240506	0.472393
nw_density	Network density	2,645	0.088119	0.069955	0.037300	0.440500
yr97	=1, if year = 1997	2,645	0.04272	0.20227	0	1
yr98	=1, if year = 1998	2,645	0.04499	0.27322	0	1
yr99	=1, if year = 1999	2,645	0.05217	0.22241	0	1
yr00	=1, if year = 2000	2,645	0.05252	0.22318	0	1
yr01	=1, if year = 2001	2,645	0.05861	0.23492	0	1
yr02	=1, if year = 2002	2,645	0.06049	0.23844	0	1
yr03	=1, if year = 2003	2,645	0.06049	0.23844	0	1
yr04	=1, if year = 2004	2,645	0.06201	0.24131	0	1
yr05	=1, if year = 2005	2,645	0.06351	0.24393	0	1
yr06	=1, if year = 2006	2,645	0.06162	0.24052	0	1
yr07	=1, if year = 2007	2,645	0.06049	0.22384	0	1
yr08	=1, if year = 2008	2,645	0.06163	0.24051	0	1
Cooperation count variables						
coopcnt_c	Count of CORDIS cooperation projects	2,645	0.06578	0.322242	0	4
coopcnt_fk	Count of Foerderkatalog cooperation projects	2,645	0.21921	0.632911	0	8
coopcnt_fkc	Count of both types of cooperation projects	2,645	0.27599	0.774139	0	8

(continued)

Table 10.1 (continued)

Variable	Variable definition	Summary statistics				
		Obs.	Mean	Std. dev.	Min	Max
Ego network variables						
ego_size	Ego network size	2,645	2.46992	5.25304	0	41
ego_density	Ego network density	2,645	13.89213	25.73445	0	100
ego_nbroke	Normalized ego network brokerage	2,645	24.8449	67.4958	0	65.09

Source: Author's own calculations

Table 10.2 Correlation matrix – cooperation events and ego network characteristics

	pgcnt	pacnt	nw_size	nw_density	firm_age	firmage_sq	coopfund_fkc	coopcnt_c	coopcnt_fk	coopcnt_fkc	ego_size	ego_density	ego_nbroke
pgcnt	1.0000												
pacnt	0.6506	1.0000											
nw_size	0.0670	0.0448	1.0000										
nw_density	−0.0831	−0.0529	−0.6576	1.0000									
firmage	0.0106	−0.0566	0.2138	−0.2007	1.0000								
firmage_sq	−0.0046	−0.0455	0.1609	−0.1451	0.9275	1.0000							
coopfund_fkc	0.3114	0.3922	0.0146	−0.0144	−0.0279	−0.0100	1.0000						
coopcnt_c	0.2310	0.1754	0.0234	−0.0175	0.0193	0.0266	0.3144	1.0000					
coopcnt_fk	0.2157	0.2208	0.0420	−0.0132	0.0027	0.0200	0.4651	0.2326	1.0000				
coopcnt_fkc	0.2725	0.2535	0.0441	−0.0181	0.0102	0.0274	0.5112	0.6064	0.9144	1.0000			
ego_size	0.3170	0.2732	0.0313	−0.0071	0.1101	0.1306	0.3393	0.3217	0.6128	0.6349	1.0000		
ego_density	0.0542	0.0908	0.1265	−0.0724	0.0654	0.0754	0.0783	0.0661	0.2496	0.2316	0.4129	1.0000	
ego_nbroke	0.2733	0.2352	0.0959	−0.1032	0.1149	0.1244	0.2875	0.2442	0.4740	0.4892	0.6607	0.2035	1.0000

Source: Author's own calculations

10.4   Empirical Analysis: Model Specification and Results

In this paper we use panel count data techniques to test our hypotheses.[14] In general, the use of fixed effects models provides some important advantages. Most importantly, the fixed effects estimator is unbiased as it includes dummy variables for the different intercepts and is more robust against selection bias problems than the random effects estimator (Kennedy 2003, p. 304). However, fixed effects models also have two considerable drawbacks. Firstly, all time-invariant explanatory variables are thrown out because the estimation procedure fails to estimate a slope coefficient for variables that do not vary within an individual unit (Kennedy 2003, p. 304). Secondly, using only within-variation leads to less efficient estimates and the model loses its explanatory power (Cameron and Trivedi 2009, p. 259). The random effects model compensates for some of these disadvantages. On the one hand random effects estimators make better use of the information values of patent data and generate efficient estimates with higher explanatory power. In addition, random effects estimators can generate coefficient estimates of both time-variant as well as time-invariant explanatory variables (Kennedy 2003, p. 307). The major drawback of the random effects model is that correlations between the error term and the explanatory variables generate biased estimates (Kennedy 2003, p. 306). In other words, the random effects estimator generates potentially inconsistent results when the model assumptions are violated.

10.4.1   Empirical Model Specification

As our endogenous variable accepts only nonnegative integer values, we chose a count data model specification for the purpose of this analysis.[15] Following Ahuja (2000) and Stuart (2000) we estimated panel count models and adopted the following estimation strategy to test our hypotheses. First we estimated panel Poisson models in order to obtain an initial idea of the relationship between cooperation counts, network positioning measures and firm-specific patenting activity. As our endogenous variables exhibited strong over-dispersion, we then turned to a negative binomial model specification with random effects. This generalization of the Poisson model allows for overdispersion by including an individual, unobserved effect into the conditional mean (Schilling and Phelps 2007, p. 1119). In the next step we estimated both fixed effects and random effects models.[16] We

---

[14] We used STATA 10.1 (Stata 2007), a standard software package for statistical data analysis.

[15] For an in-depth discussion of panel data count models, see Sect. 6.1.2.

[16] The main difference between the estimation techniques is that fixed effects models allows for correlations to be made between the unobserved individual effect and the included explanatory variables whereas random effects models require the unobserved individual effect and the explanatory variables to be uncorrelated (Greene 2003, p. 293).

used the Standard Hausman Test (1978) to decide which results to interpret.[17] Finally, we ran several consistency checks to ensure robustness of the reported results. We used several time lags for the estimations. Additionally, we used patent applications in some cases as an additional innovation measure to ensure the results.

10.4.2   Estimation Results

Tables 10.3, 10.4, 10.5, and 10.6 report the estimation results for patent grants based on a panel negative binomial model with both fixed effects and random effects estimation techniques. The tables are organized as follows. The baseline model (i.e. BL Model) consists of a set of time dummies, two firm age variables, two network control variables and a funding variable. Models I–III address direct cooperation effects and Models IV–VI report ego network effects. The last three models (i.e. Model VII–IX) provide the results for the fully specified models. Fixed effects as well as random effects estimates are reported for both patent grants with a lag of $t = 1$ and patent grants with a lag of $t = 2$. Results are reported under consideration of Standard Hausman Test results and interpreted on the basis of the fully specified models.

We start the discussion with Tables 10.3 and 10.4 which illustrate the estimation results for patent grants with a time lag of 2 years. The baseline model (cf. Table 10.3, BL Model) provides fixed effects estimation results for a set of time dummies, two firm age variables, two network control variables and a funding variable. The time dummies show positive and significant effects for the time period from 1998 to 2007. Models I–III (Table 10.3) address direct cooperation effects. The fixed effects model reveals no significant effects for *CORDIS [coopcnt_c]* or *Foerderkatalog [coopcnt_fk]*. The last cooperation count model (cf. Table 10.3, Model III) addresses combined cooperation counts *[coopcnt_fkc]*. Fixed effects estimates are significant at the 0.1 level indicating a moderate relatedness between combined cooperation counts and firm innovativeness. Models IV–VI (Table 10.3) address structural ego network effects. The ego size variable *[ego_size]* as well as the ego brokerage variable *[ego_nbroke]* show highly significant and positive coefficients at the 0.01 level for the fixed effects model. Surprisingly, network density *[ego_density]* shows no significant effect (cf. Table 10.3, Model V). Finally, we turn our attention to the fully specified models (cf. Table 10.3, Models VII–IX). The results are consistent with the previously reported findings on

---

[17] The basic idea of the Standard Hausman specification test is to test the null hypothesis that the unobserved effect is uncorrelated with the explanatory variables (Greene 2003, p. 301). If the null hypothesis cannot be rejected, both fixed effects estimates as well as random effects estimates are consistent and the model of choice is the random effects model due to its higher explanatory power. Under the alternative, random effects and fixed effects estimators diverge and it is argued that the latter model is the appropriate choice (Cameron and Trivedi 2009, p. 260).

Table 10.3 Estimation results – cooperation events and ego network characteristics; panel data count model, patent grants, time lag (t − 2): fixed effects

Variable	Estimation results									
	BL model	Model I	Model II	Model III	Model IV	Model V	Model VI	Model VII	Model VIII	Model IX
yr97	.28301388	.27134911	.33328969	.33802164	.23259319	.25011823	.25915583	.24418028	.2869446	.3046047
yr98	.63349169	.63238627	.69985957	.7092334	.56353661	.65428546	.55159163	.56265345	.69367349	.62658734
yr99	.82586662**	.88061534***	.86092359***	.90019169***	.87293139***	.82059957**	.75845013**	.9056236***	.88917565**	.85159569**
yr00	.95342567***	.99923104***	.98786659***	1.0216749***	1.0312264***	.96580641***	.90642038***	1.0555268***	1.0232238***	.99003889***
yr01	.92004125**	.98326257***	.93150456**	.97397866***	1.0491278***	.92587801**	.82002215**	1.0825001***	.98419502**	.90722633**
yr02	.84997207**	.93087144*	.8529483**	.90519472**	1.0329185**	.84339954**	.85076682**	1.0781388***	.91179799**	.94296864**
yr03	.97198301**	1.0752515***	.98363511**	1.0516739***	1.1652561***	.97673664**	.90963022**	1.2198901***	1.0686577**	1.0209909**
yr04	1.0875851**	1.1864142**	1.073296**	1.1348294**	1.2989609**	1.0902405**	1.0001675**	1.3523733***	1.1584861**	1.0958454**
yr05	1.347883***	1.4177135***	1.32162201***	1.3624504***	1.5210474***	1.3623725***	1.1419374**	1.5569011***	1.3886375***	1.2125097***
yr06	1.1220418**	1.205549***	1.165116**	1.2248344***	1.3246384***	1.1130364**	.93178546*	1.3671364***	1.2250648**	1.0651463**
yr07	.95981367*	1.0862809**	.99180778**	1.0777645**	1.1370854**	.95285036*	.76052249	1.2115581**	1.0747211**	.93667485*
yr08	.67159305	.77197143	.70824405	.77595561	.83052698	.68624225	.50677561	.8836538	.78164193	.64574211
nw_size	.90263709	.98554392	.32529304	.30542586	1.045863	.62072613	1.0811379	1.0960311	.33799293	.66379718
nw_density	−3.001253	−2.735748	−3.321225	−3.172295	−2.579785	−3.150222	−2.721877	−2.428528	−3.096249	−2.75881
firmage	.01604764	.01039181	.02262534	.02010696	.00761643	.01785	.03356662	.00449849	.01887192	.03180792
firmage_sq	−.0015182	−.00142676	−.00164884	−.00161259	−.00092361	−.00148828	−.00169563	−.00090271	−.001519	−.00170399
coopfund_fkc	−5.63E-06	−.00001184	−.00001427	−.00001984	−.0000115	−1.10E-06	−.00001693	−.00001494	−.0000143	−.00003011
coopcnt_c		.14257143						.07254601	.13064184	.11665961
coopcnt_fk			.086562					−.00420153	.06514561	.06454571
coopcnt_fkc				.0953059*				.03404157***		
ego_size					.03505361				.00377074	
ego_density						.00430862				
ego_nbroke							.01321065***			.01233842***
_cons	−.01708228	−.10815625	.10310738	.05881304	−.44521632	−.02884002	−.2722868	−.4810172	−.03998001	−.22026722
chi²	52.509824	55.411193	54.147803	56.250292	66.074274	54.534457	61.149145	67.531997	58.406049	64.025971
ll	−642.8041	−641.8139	−641.9164	−641.1722	−635.854	−641.6564	−639.0199	−635.575	−640.2322	−637.7467
aic	1,321.6081	1,321.6279	1,321.8327	1,320.3444	1,309.708	1,321.3127	1,316.0398	1,313.15	1,322.4643	1,317.4935
bic	1,409.1377	1,414.0203	1,414.2251	1,412.7368	1,402.1004	1,413.7051	1,408.4322	1,415.2679	1,424.5822	1,419.6114
N	956	956	956	956	956	956	956	956	956	956

Legend: * $p < .1$; ** $p < .05$; *** $p < .01$
Source: Author's own calculations

Table 10.4 Estimation results – cooperation events and ego network characteristics; panel data count model, patent grants, time lag (t − 2): random effects

Estimation results

Variable	BL model	Model I	Model II	Model III	Model IV	Model V	Model VI	Model VII	Model VIII	Model IX
yr97	.28958433	.26156408	.40949314	.40041785	.22462128	.24614093	.27302751	.26437514	.35062709	.37382097
yr98	.72302556	.71322182	.89548873*	.8820191*	.65197964	.74708038	.63789276	.71806691	.864066**	.81086855
yr99	.82907701***	.88984488***	.92798811***	.96839459***	.91596182***	.82506891***	.78335293**	.97467178***	.94276367***	.93589088***
yr00	.95077403***	.99707042***	1.0482623***	1.0775272***	1.0974957***	.97550928***	.94377903***	1.1409193***	1.0774213***	1.0793283***
yr01	.83318421**	.88915129***	.88611249***	.9367763***	1.0347635***	.85001325***	.76243324**	1.071034***	.93900536***	.89332907***
yr02	.72281924**	.81511353**	.77725397**	.8452026**	1.0035318***	.72687333**	.78666605**	1.0530985***	.83581661**	.92121655***
yr03	.89798708**	1.0097849***	.9719612***	1.0539452***	1.1863358***	.91524701**	.87845921**	1.2431772***	1.0524818***	1.0358243***
yr04	.95081279**	1.0549921**	.97754627**	1.0552204**	1.2689355**	.97050824**	.91898499**	1.3064586***	1.062301**	1.042154**
yr05	1.240967***	1.2965524***	1.2479777***	1.2807483***	1.5089173***	1.2696625***	1.0600288***	1.5215934***	1.2907343***	1.1484343***
yr06	1.0399992***	1.116053***	1.1898283***	1.2422235***	1.3446564***	1.064106***	.87002448**	1.401828***	1.2267438***	1.0893737***
yr07	.89268594**	1.0224223**	1.0264416**	1.120329*	1.1708172***	.89241509**	.7213689**	1.2676747***	1.0863207**	.99479006**
yr08	.62559058	.7195015	.7772816	.83709645	.88454814*	.64516148	.51333118	.95571482*	.81283373	.73222948
nw_size	-.12371274	-.01476782	-1.3205184	-1.1774041	.09392536	-.41114889	.24700993	-.22673423	-1.174295	-.73962207
nw_density	-3.1861801	-2.8161108	-3.7008682	-3.3468854	-2.7453617	-3.3526623	-2.9368624	-2.7026842	-3.4065215	-3.0772957
firmage	.01972481	.01581655	.02832724	.02586692	.00514208	.02206826	.03293266	.00686375	.02719964	.03414089
firmage_sq	-.00094217	-.00084522	-.00114572	-.00108107	-.0002293	-.00098127	-.00121826	-.00031047	-.00108934	-.00124884
coopfund_fkc	.0000103	2.89E-06	-5.78E-06	-.00001274	1.25E-06	.00001542	-5.25E-06	-6.41E-06	-6.53E-06	-.00002515
coopcnt_c		.20296094*								
coopcnt_fk			.17515551***					.08820682	.17173138*	.14649198
coopcnt_fkc				.16547723***				.03174681	.14507879**	.12989886**
ego_size					.04775071***			.04409843***		
ego_density						.00596513**	.0181826***		.00490688*	.01599729***
ego_nbroke										
_cons	.20849708	.07200637	.43883649	.31596308	-.3642751	.17140244	-.14703416	-.29658238	.22780358	.02877153
ln_r_cons	.79957613***	.82964915***	.82943021***	.84928918***	.91174454***	.80721387***	.83048072***	.92259594***	.85348863***	.87566941***
ln_s_cons	-1.4847918***	-1.4648625***	-1.4145574***	-1.4039554***	-1.3271212**	-1.4583187***	-1.3972224***	-1.3214073***	-1.3894995***	-1.3526565*
chi²	54.440312	60.435554	61.91168	67.27798	84.718806	59.268746	72.699885	86.973403	70.893299	81.056876

(continued)

Table 10.4 (continued)

Variable	Estimation results									
	BL model	Model I	Model II	Model III	Model IV	Model V	Model VI	Model VII	Model VIII	Model IX
ll	−1,037.63	−1,035.5521	−1,033.8359	−1,032.3531	−1,023.0145	−1,035.117	−1,030.0798	−1,022.4587	−1,030.6775	−1,026.5078
aic	2,115.2599	2,113.1042	2,109.6717	2,106.7062	2,088.0291	2,112.234	2,102.1595	2,090.9174	2,107.3551	2,099.0156
bic	2,228.9923	2,232.5233	2,229.0908	2,226.1253	2,207.4289	2,231.6337	2,221.5593	2,221.6886	2,238.1263	2,229.7867
N	2,179	2,179	2,179	2,179	2,177	2,177	2,177	2,177	2,177	2,177

Legend: $^{*} p < .1$; $^{**} p < .05$; $^{***} p < .01$
Source: Author's own calculations

Table 10.5 Robustness check – cooperation events and ego network characteristics; panel data count model, patent grants, time lag (t − 1): fixed effects

Variable	Estimation results									
	BL model	Model I	Model II	Model III	Model IV	Model V	Model VI	Model VII	Model VIII	Model IX
yr97	.68785935**	.66294073	.74171482*	.72638156*	.6754754*	.69253539*	.87113862**	.67944759*	.7380029*	.89218256**
yr98	1.1279063**	1.090248**	1.2345816**	1.197073**	1.1529394*	1.1258483**	1.2607989**	1.1585452**	1.2073965**	1.341176***
yr99	.61629188**	.63616725**	.62699145**	.64986356**	.61403569*	.61773981**	.65599545**	.62534616**	.64955633**	.69013901***
yr00	.84741821***	.85623162***	.87221792***	.87942742***	.88632774*	.84600484***	.85838016***	.89262562***	.87780939***	.89305845***
yr01	.22683027	.24861631	.19232501	.21743036	.24204418	.22795257	.17979738	.24470296	.2124664	.18003816
yr02	-.22577387	-.18664191	-.28803614	-.24488059	-.14145878	-.22494196	-.15822143	-.13762631	-.2565136	-.18553153
yr03	-.40210607	-.35731528	-.46965221	-.41903848	-.33126601	-.40147874	-.35362227	-.3263149	-.43293813	-.38143208
yr04	-.69302719	-.65073261	-.82785045	-.77720869	-.68068334	-.69169004	-.69602363	-.68326859	-.79496528	-.77552661
yr05	-.17803578	-.15739976	-.29019569	-.26237468	-.11706538	-.17914557	-.24835379	-.1233949	-.27781723	-.31295593
yr06	.32909289	.33850187	.30519773	.31727216	.38580388	.32540284	.2603632	.38326797	.3063161	.25760356
yr07	.15369194	.18051145	.13103061	.16074346	.20540014	.15516037	.0998801	.21096708	.15682981	.11545088
yr08	.43233178*	.42533845*	.42750285*	.42036388*	.43745154*	.43104303*	.42295142*	.43239495*	.41925132*	.41762267*
nw_size	-10.819289**	-10.505499**	-12.204879**	-11.830165**	-12.057715**	-10.771181**	-11.51977**	-12.107176**	-11.912055**	-12.478111***
nw_density	-17.409863**	-17.033061**	-18.769641***	-18.244603***	-16.773439**	-17.384798*	-15.70962*	-16.789017**	-18.424683***	-16.611578**
firmage	.10819177***	.10748294***	.11637332***	.11580378***	.12233203***	.10790828***	.13292805***	.12373452***	.1158048***	.13944415***
firmage_sq	-.00237513	-.00226284	-.00234645	-.00222562	-.00110621	-.00238232	-.00211179	-.00111539	-.00226306	-.00206684
coopfund_fkc	-.00001606	-.00001958	-.00003609	-.0000395	-.00003263	-.00001662	-.00003245	-.00003508	-.00004154	-.00005128
coopcnt_c		.10273775								
coopcnt_fk			.12046422*					.03039635	.09064787	.07093859
coopcnt_fkc				.11098304**				.00820148	.12122404*	.08743728
ego_size					.04274911***			.04191475***		
ego_density						-.00068882			-.0013467	
ego_nbroke							.01687165***			.01573293***
_cons	5.5180797**	5.3313742**	6.0586559**	5.8412712**	5.5225411**	5.5160388**	5.3467067**	5.5313789**	5.9226843**	5.7184145**
chi²	67.746283	68.216716	71.267474	71.333738	87.241417	67.821444	81.283363	86.98889	71.599599	82.397879

(continued)

Table 10.5 (continued)

Variable	Estimation results									
	BL model	Model I	Model II	Model III	Model IV	Model V	Model VI	Model VII	Model VIII	Model IX
ll	−687.9207	−687.4949	−686.126	−685.8197	−676.8627	−687.8914	−681.5634	−676.8159	−685.6867	−680.3578
aic	1,411.8413	1,412.9898	1,410.2521	1,409.6394	1,391.7255	1,413.7827	1,401.1267	1,395.6319	1,413.3734	1,402.7156
bic	1,501.2807	1,507.3979	1,504.6602	1,504.0476	1,486.1336	1,508.1909	1,495.5349	1,499.9777	1,517.7193	1,507.0615
N	1,063	1,063	1,063	1,063	1,063	1,063	1,063	1,063	1,063	1,063

Legend: $^{*}p<.1$; $^{**}p<.05$; $^{***}p<.01$
Source: Author's own calculations

Table 10.6 Robustness check – cooperation events and ego network characteristics; panel data count model, patent grants, time lag $(t-1)$: random effects

Estimation results

Variable	BL model	Model I	Model II	Model III	Model IV	Model V	Model VI	Model VII	Model VIII	Model IX
yr97	.4632924	.40007221	.50604133	.46998198	.29400179	.4505134	.52573381	.2957748	.4713754	.52550231
yr98	.89284237*	.81495246*	1.0094595*	.9408987*	.74367303	.89233824*	.87920164*	.72602959	.94854238*	.94793599**
yr99	.61120188**	.64617483**	.6327672**	.66529358**	.62902407**	.60877572**	.64562722**	.64497679**	.6617576**	.69362867***
yr00	.90011859***	.91932571***	.96036612***	.97340323***	1.0223323***	.9054814***	.93522689***	1.0306877***	.97565071***	.9972492***
yr01	.30508782	.35439181	.27663257	.32587166	.40339144	.30743574	.31199899	.41252192	.32292195	.3340893
yr02	-.21379731	-.13330299	-.26707914	-.19457355	-.04157813	-.211843	-.08333328	-.02591321	-.20106005	-.08751152
yr03	-.35717163	-.27470625	-.42681579	-.34932546	-.23750035	-.35553922	-.2721566	-.22154065	-.35612416	-.28789802
yr04	-.61165042	-.52957056	-.77196361	-.69117413	-.52427483	-.60949472	-.5452333	-.51478462	-.69970397	-.62491829
yr05	-.13659284	-.10159446	-.27286589	-.23641331	-.02795895	-.13150452	-.16901561	-.03261082	-.23939201	-.23869546
yr06	.37005431	.38796475	.35983868	.38037001	.47571513	.37911672	.31452759	.47093128	.38169832	.33173044
yr07	.18871255	.23819118	.18055139	.22660797	.27929432	.18763892	.14666464	.29181355	.2220858	.19573344
yr08	.43894939*	.42518311*	.44609622*	.43230737*	.44698795*	.44038243*	.42527008*	.4338907*	.43374551*	.42197885*
nw_size	-9.6061159*	-8.9198244*	-11.258832**	-10.572557**	-9.6574063*	-9.659645*	-9.2507536*	-9.5466639**	-10.674452**	-10.214994*
nw_density	-18.879345**	-18.373738***	-21.149769***	-20.536506***	-20.458121***	-18.934135***	-17.853806**	-20.43898**	-20.642559***	-19.378541***
firmage	.0768271**	.07463565**	.08392102**	.08225298**	.0637425*	.07727288*	.08744924**	.0645004	.08266315**	.09054553***
firmage_sq	-.00215862*	-.00206355	-.00225022*	-.00218002*	-.00112079	-.00216901*	-.0022373*	-.00114604	-.00219021*	-.00224674*
coopfund_fkc	-2.43E-07	-6.56E-06	-.0000241	-.00003033	-.00002091	8.53E-07	-.00002042	-.00002358	-.00002977	-.00004263
coopcnt_c		.17767547*						.05606854	.1482401	.10678382
coopcnt_fk			.17407617***					.00336744	.16471489***	.11765017*
coopcnt_fkc				.160997***						
ego_size					.05132971***			.05022702***		
ego_density						0.00144519			.00051321	
ego_nbroke							.01955928***			.01717148***
_cons	5.226859*	4.8656738*	5.8961892**	5.542724**	4.974592*	5.2151074*	4.729544*	4.919327*	5.5786573**	5.1408288*
ln_r_cons	.77643775***	.79111548***	.79492294***	.80543569***	.87561503***	.77952532***	.8029451***	.87810292***	.80648061***	.82384172***
ln_s_cons	-1.4661876***	-1.4447477***	-1.4163338***	-1.4030847***	-1.3109938***	-1.4584033***	-1.385512***	-1.3080685***	-1.4001969***	-1.3599306***
chi²	65.795584	68.225196	72.497399	73.98586	97.423975	66.038451	87.174111	97.482609	74.049408	89.786435

(continued)

Table 10.6 (continued)

Estimation results

Variable	BL model	Model I	Model II	Model III	Model IV	Model V	Model VI	Model VII	Model VIII	Model IX
ll	−1,097.0172	−1,095.6135	−1,093.0796	−1,092.1249	−1,079.1377	−1,096.8302	−1,087.6593	−1,078.9803	−1,092.0537	−1,085.2273
aic	2,234.0344	2,233.2271	2,228.1592	2,226.2498	2,200.2755	2,235.6604	2,217.3186	2,203.9606	2,230.1073	2,216.4545
bic	2,349.7986	2,354.7795	2,349.7116	2,347.8022	2,321.8105	2,357.1954	2,338.8536	2,337.0704	2,363.2171	2,349.5643
N	2,412	2,412	2,412	2,412	2,410	2,410	2,410	2,410	2,410	2,410

Legend: $^*p < .1$; $^{**}p < .05$; $^{***}p < .01$
Source: Author's own calculations

cooperation count (Models I–III) and ego networks (Model IV–VI). The effects for ego network size and ego network brokerage remain robust in Models VII and IX (Table 10.3) whereas no effect could be identified for ego network density in Model VIII (Table 10.3) based on fixed effect estimation.

However, a look at the results of the random effects model (cf. Table 10.4) reveals a slightly different picture. Estimation results for both cooperation count measures as well as ego network measures are positive and highly significant in nearly all model specifications. In other words, the previously reported findings are supported by random effects models. These estimation results, however, have to be interpreted with caution bearing in mind the results of the Hausman Test.

In order to check the robustness and consistency of these initial findings we estimated all previously discussed models again with a time lag of 1 year (cf. Tables 10.5 and 10.6).[18] Table 10.5 reports results for fixed effects estimation techniques whereas Table 10.6 provides results based on random effects estimators. Just as before, Models I–III (cf. Table 10.6) address direct cooperation effects. This specification confirms the previously reported combined cooperation count effect *[coopcnt_fkc]* with an increased 0.05 significance level. Moreover, we can now observe an additional direct cooperation for nationally funded cooperation projects *[coopcnt_fk]* at the 0.1 significance level.

The results for the ego network effects (cf. Table 10.5, Model IV–VI) are fully consistent with our previous findings (cf. Table 10.3, Model IV–VI). Again, ego size (cf. Table 10.5, Model IV) as well as the ego brokerage variable (cf. Table 10.5, Model VI) show highly positive and significant coefficients at the 0.01 level and no network-density effects (cf. Table 10.5, Model V). The fully specified models (cf. Table 10.5, Model VII–IX) reconfirm our previous ego network results and reveal at the same time some interesting additional insights with regard to individual cooperation effects. The effects for ego network size *[ego_size]*, and ego network brokerage *[ego_nbroke]* remain robust (cf. Table 10.5, Model VII and IX) and ego network density *[ego_density]* still shows no significant effect (cf. Table 10.6, Model VIII). Surprisingly, now the nationally funded cooperation counts *[coopcnt_fk]* are directly related to firm-level innovation output, but the estimates are only marginally significant at the 10 % level (cf. Table 10.6, Model VII). A look at the fully specified random effects model (cf. Table 10.6, Model VII–IX) confirms this finding. Model VII (Table 10.6) reports a highly significant coefficient for nationally funded cooperation counts at the 0.01 significance level and no effect for ego network density.

What do these results tell us about our previously formulated hypotheses? Hypotheses H1a and H1b suggest that both nationally (i.e. Foerderkatalog counts) and supra-nationally funded (i.e. CORDIS counts) collaborations are positively related to firm innovativeness. Our results show that nationally funded cooperation projects are positively related to innovation output in three out of four fully

---

[18] Additional robustness checks have been conducted by using patent application data. Most of the results confirm the reported findings. All estimations are available upon request.

specified models (Model VII, in: Tables 10.4, 10.5, and 10.6). Thus we find at least modest support for Hypothesis H1a. In addition, these findings support our initial conjecture that individual cooperation effects diminish at least partially when considering structural ego network effects at the same time.

Now we turn to Hypothesis H1b. Based on our previously discussed estimation results we have to reject Hypothesis H1b. Moreover, it is interesting to note that none of the models (cf. Tables 10.3, 10.4, 10.5, and 10.6) reveal significant coefficient estimates for funding. In other words, it is not the funding effect but rather the cooperation itself that is related to firm-level innovativeness. Hypothesis H2a suggests that the size of an ego network is positively related to firm-level innovation output. Estimation results provide strong support for Hypothesis H2a, predicting that innovation output is positively related to a firm's number of direct linkages' to other laser source manufacturers or public research organizations with laser-related activities. Likewise our estimation results provide strong support for Hypothesis H2c suggesting that brokerage positions in ego networks are positively related to subsequent firm-level innovation outcomes. Surprisingly, estimation results provide no support for Hypothesis H2b.

In summary, it turns out that the estimation models confirm the existence of direct innovation effects of individual cooperation events as long as portfolio characteristics are ignored. These effects partially diminish when ego network characteristics are taken into consideration at the same time (cf. comparison of Model VIII, Tables 10.3 and 10.5). Funding plays a subordinate role in the innovative performance of the firms under investigation. In contrast to the ego network size and brokerage, the ego network density proves to be of subordinate importance for firms in their attempts to innovate.

10.5 Discussion and Implications

This analysis was motivated by a goal to broaden our understanding of the relationship between individual cooperation events, ego network structures and firm level innovation output in the German laser industry. Our research in this area is still in an early stage. We started the analysis by taking a closer look at individual cooperation events of laser manufacturing firms.

The results of our analysis imply that the initialization of new collaborative arrangements seems to be an important driver behind a firm's innovation performance. Participation in new R&D projects with multiple profit and non-profit organizations broadens the scope of potentially accessible knowledge stocks. At the same time this increases the diversity of the knowledge base of focal firms. The subsequent impact of newly initialized R&D collaboration projects on innovation output is in line with theoretical reasoning from a knowledge-based perspective as outlined above. Surprisingly, this result only applies to nationally funded projects whereas the supra-nationally funded cooperation projects end up showing no significant effects. Furthermore, our findings relativize the argument that a firm's innovative performance is affected more by public funding than the cooperation activities themselves. With regard to the structural configuration of a firm's ego network it becomes obvious that the size of the ego network does matter. The findings for ego size suggest that the number of direct connections between the focal actor and ego network alters are especially decisive in terms of innovation output. This result is consistent with the initial findings as the diversity of potentially accessible knowledge stocks increases with the size of the ego network. Surprisingly, we found no support for ego network density. In other words, the existence of ties among alters seems to be less important for firm-level innovation outcome in the German laser industry innovation network. Finally, it turns out that the ego network brokerage has significant coefficient estimates. In other words, there is a positive and significant relationship between ego network brokerage and a firm's patenting activity. Thus, the strategic positioning of focal actors and their ability to mediate and control knowledge flows between other pairs of ego network actors appears to be of vital importance for their innovative performance.

The limitations of our empirical analysis (cf. Sect. 13.2) and our strategy to solve these issues (cf. Sect. 14.2) is subject to discussion in the final chapter of this study.

References

Ahuja G (2000) Collaboration networks, structural hole, and innovation: a longitudinal study. Adm Sci Q 45(3):425–455

Albrecht H, Buenstorf G, Fritsch M (2011) System? What system? The (co-) evolution of laser research and laser innovation in Germany since 1960. Working paper, pp 1–38

Alic JA (1990) Cooperation in R&D. Technovation 10(5):319–332

Al-Laham A, Kudic M (2008) Strategische Allianzen. In: Corsten H, Goessinger R (eds) Lexikon der Betriebswirtschaftslehre, 5th edn. Oldenbourg Verlag, München, pp 39–41

Amburgey TL, Dacin T, Singh JV (1996) Learning races, patent races, and capital races: strategic interaction and embeddedness within organizational fields. In: Baum JA (ed) Advances in strategic management. Elsevier, New York, pp 303–322

Anand BN, Khanna T (2000) Do firms learn to create value? The case of alliances. Strateg Manag J 21(3):295–315

Baum JA, Calabrese T, Silverman BS (2000) Don't go it alone: alliance network composition and startup's performance in Canadian biotechnology. Strateg Manag J 21(3):267–294

Bergenholtz C, Waldstrom C (2011) Inter-organizational network studies – a literature review. Ind Innov 18(6):539–562

Bidault F, Cummings T (1994) Innovating through alliances: expectations and limitations. R&D Manag 24(1):33–45

Bleeke J, Ernst D (1991) The way to win in cross-border alliances. Harv Bus Rev 69(6):127–135

Borgatti SP, Everett MG, Freeman LC (2002) Ucinet for windows: software for social network analysis. Analytic Technologies, Harvard

Borys B, Jemison DB (1989) Hybrid arrangements as strategic alliances: theoretical issues in organizational combinations. Acad Manag Rev 14(2):234–249

Broekel T, Graf H (2011) Public research intensity and the structure of German R&D networks: a comparison of ten technologies. Econ Innov New Technol 21(4):345–372

Buckley PJ, Glaister KW, Klijn E, Tan H (2009) Knowledge accession and knowledge acquisition in strategic alliances: the impact of supplementary and complementary dimensions. Br J Manag 20(4):598–609

Buenstorf G (2007) Evolution on the shoulders of giants: entrepreneurship and firm survival in the German laser industry. Rev Ind Organ 30(3):179–202

Burt RS (1992) Structural holes: the social structure of competition. Harvard University Press, Cambridge

Burt RS (2005) Brokerage & closure – an introduction to social capital. Oxford University Press, New York

Cameron CA, Trivedi PK (2009) Microeconometrics using Stata. Stata Press, College Station

Capaldo A (2007) Network structure and innovation: the leveraging of a dual network as a distinctive relational capability. Strateg Manag J 28(6):585–608

Cassi L, Corrocher N, Malerba F, Vonortas N (2008) Research networks as infrastructure for knowledge diffusion in European regions. Econ Innov New Technol 17(7):665–678

Ciesa V, Toletti G (2004) Network of collaborations for innovation: the case of biotechnology. Tech Anal Strat Manag 16(1):73–96

Coff RW (2003) The emergent knowledge-based theory of competitive advantage: an evolutionary approach to integrating economics and management. Manag Decis Econ 24(4):245–251

Cohen WM, Levinthal DA (1990) Absorptive capacity: a new perspective on learning and innovation. Adm Sci Q 35(3):128–152

Coleman JS (1988) Social capital in the creation of human capital. Am J Sociol 94:95–120

Das TK, Teng B-S (2000) Instabilities of strategic alliances: an internal tensions perspective. Organ Sci 11(1):77–101

Das TK, Teng B-S (2002) Alliance constellations: a social exchange perspective. Acad Manag J 27 (3):445–456

De Propris L (2000) Innovation and inter-firm co-operation: the case of the West Midlands. Econ Innov New Technol 9(5):421–446

Decarolis DM, Deeds DL (1999) The impact of stocks and flows of organizational knowledge on firm performance: an empirical investigation of the biotechnology industry. Strateg Manag J 20 (10):953–968

Dittrich K, Duysters G, De Man A-P (2007) Strategic repositioning by means of alliance networks: the case of IBM. Res Policy 36(10):1496–1511

Doreian P, Stokman FN (2005) The dynamics and evolution of social networks. In: Doreian P, Stokman FN (eds) Evolution of social networks, 2nd edn. Gordon and Breach, New York, pp 1–17

Doreian P, Woodard KL (1992) Fixed list versus snowball selection of social networks. Soc Networks 21(2):216–233

Doz Y, Hamel G (1997) The use of alliances in implementing technology strategies. In: Tushman MT, Anderson P (eds) Managing strategic innovation and change. Oxford University Press, New York, pp 556–580

Dyer JH, Nobeoka K (2000) Creating and managing a high-performance knowledge-sharing network: the Toyota case. Strateg Manag J 21(3):345–367

Dyer JH, Singh H (1998) The relational view: cooperative strategy and sources of international competitive advantage. Acad Manag Rev 23(4):660–680

Eraydin A, Aematli-Köroglu B (2005) Innovation, networking and the new industrial clusters: the characteristics of networks and local innovation capabilities in the Turkish industrial clusters. Entrepren Reg Dev 17(4):237–266

Fornahl D, Broeckel T, Boschma R (2011) What drives patent performance of German biotech firms? The impact of R&D subsidies, knowledge networks and their location. Pap Reg Sci 90 (2):395–418

Frank O (2005) Network sampling and model fitting. In: Carrington PJ, Scott J, Wasserman S (eds) Models and methods in social network analysis. Cambridge University Press, Cambridge, pp 31–56

Freel MS, Harrison RT (2006) Innovation and cooperation in the small firm sector: evidence from 'Northern Britain'. Reg Stud 40(4):289–305

George G, Zahra SA, Wheatley KK, Khan R (2001) The effects of alliance portfolio characteristics and absorptive capacity on performance: a study of biotechnology firms. J High Technol Manag Res 12(2):205–226

Glueckler J (2007) Economic geography and the evolution of networks. J Econ Geogr 7(5):619–634

Goerzen A (2005) Managing alliance networks: emerging practices of multinational corporations. Acad Manag Exec 19(2):94–107

Gomes-Casseres B (2003) Competitive advantage in alliance constellations. Strateg Organ 1(3):327–335

Graf H, Krueger JJ (2011) The performance of gatekeepers in innovator networks. Ind Innov 18(1):69–88

Grant RM (1996) Towards a knowledge based theory of the firm. Strateg Manag J 17(2):109–122

Grant RM, Baden-Fuller C (2004) A knowledge accessing theory of strategic alliances. J Manag Stud 41(1):61–84

Greene WH (2003) Econometric analysis, 5th edn. Prentice Hall, Upper Saddle River

Grunwald R, Kieser A (2007) Learning to reduce interorganizational learning: an analysis of architectural product innovation in strategic alliances. J Prod Innov Manag 24(4):369–391

Grupp H (2000) Learning in a science driven market: the case of lasers. Ind Corp Chang 9(1):143–172

Gulati R (1998) Alliances and networks. Strateg Manag J 19(4):293–317

Gulati R, Gargiulo M (1999) Where do interorganizational networks come from? Am J Sociol 104(5):1439–1493

Gulati R, Nohria N, Zaheer A (2000) Strategic networks. Strateg Manag J 21(3):203–215

Hagedoorn J (1993) Understanding the rational of strategic technology partnering – organizational modes of cooperation and sectoral differences. Strateg Manag J 14(5):371–385

Hagedoorn J (2002) Inter-firm R&D partnership: an overview of major trends and patterns since 1960. Res Policy 31(4):477–492

Hakansson H, Johanson J (1988) Formal and informal cooperation – strategies in international industrial networks. In: Contractor FJ, Lorange P (eds) Cooperative strategies in international business. Lexington Books, Lexington, pp 369–379

Hamel G (1991) Competition for competence and inter-partner learning within international strategic alliances. Strateg Manag J 12(1):83–103

Hamel G, Doz YL, Prahalad CK (1989) Collaborate with your competitors – and win. Harv Bus Rev 67(1):133–139

Harabi N (2002) The impact of vertical R&D cooperation on firm innovation: an empirical investigation. Econ Innov New Technol 11(2):93–108

Harrigan KR (1988) Joint ventures and competitive strategy. Strateg Manag J 9(2):141–158

Hausman JA (1978) Specification tests in econometrics. Econometrica 46(6):1251–1271

Hite JM, Hesterly WS (2001) The evolution of firm networks: from emergence to early growth of the firm. Strateg Manag J 22(3):275–286

Hoffmann WH (2005) How to manage a portfolio of alliances. Long Range Plan 38(2):121–143

Hoffmann WH (2007) Strategies for managing alliance portfolios. Strateg Manag J 28(8):827–856

Inkpen AC, Beamish PW (1997) Knowledge, bargaining power, and the instability of international joint ventures. Acad Manag Rev 22(1):177–202

Jaffe AB (1989) Real effects of academic research. Am Econ Rev 79(5):957–970

Jaffe AB, Trajtenberg M, Henderson R (1993) Geographic localization of knowledge spillovers as evidenced by patent citations. Q J Econ 108(3):577–598

Jarillo CJ (1988) On strategic networks. Strateg Manag J 9(1):31–41

Jarvenpaa SL, Majchrzak A (2008) Knowledge collaboration among professionals protecting national security: role of transactive memories in ego-centered knowledge networks. Organ Sci 19(2):260–276

Kale P, Singh H, Perlmutter H (2000) Learning and protection of proprietary assets in strategic alliances: building relational capital. Strateg Manag J 21(3):217–237

Kennedy P (2003) A guide to econometrics. Blackwell, Oxford

Khanna T, Gulati R, Nohria N (1998) The dynamics of learning alliances: competition, cooperation, and relative scope. Strateg Manag J 19(3):193–210

Kim T-Y, Oh H, Swaminathan A (2006) Framing interorganizational network change: a network inertia perspective. Acad Manag Rev 31(3):704–720

Kogut B (1991) Joint ventures and the option to expand and acquire. Manag Sci 37(1):19–33

Kogut B, Zander U (1992) Knowledge of the firm, combinative capabilities, and the replication of technology. Organ Sci 3(3):383–397

Kudic M, Banaszak M (2009) The economic optimality of sanction mechanisms in interorganizational ego networks – a game theoretical analysis. In: 35th European International Business Academy conference, Valencia, pp 1–40

Kudic M, Buenstorf G, Guhr K (2011a) Analyzing the relationship between cooperation events, ego-networks and firm innovativeness – empirical evidence from the German laser industry. In: Conference proceedings. The 5th international EMNet conference, Limassol, pp 1–42

Kudic M, Guhr K, Bullmer I, Guenther J (2011b) Kooperationsintensität und Kooperationsförderung in der deutschen Laserindustrie. Wirtschaft im Wandel 17(3):121–129

Lane PJ, Lubatkin MH (1998) Relative absorptive capacity and interorganizational learning. Strateg Manag J 19(5):461–477

Lane PJ, Salk JE, Lyles MA (2001) Absorptive capacity, learning, and performance in international joint ventures. Strateg Manag J 22(12):1139–1161

Lavie D (2007) Alliance portfolios and firm performance: a study of value creation and appropriation in the U.S. software industry. Strateg Manag J 28(12):1187–1212

Lavie D, Miller SR (2008) Alliance portfolio internationalization and firm performance. Organ Sci 19(4):623–646

Lee JJ (2010) Heterogeneity, brokerage, and innovative performance: endogenous formation of collaborative inventor networks. Organ Sci 21(4):804–822

Lui SS (2009) Interorganizational learning the roles of competence trust, formal contract, and time horizon in interorganizational learning. Organ Stud 30(4):333–353

Markowitz H (1952) Portfolio selection. J Financ 7(1):77–91

Mowery DC, Oxley JE, Silverman BS (1996) Strategic alliances and interfirm knowledge transfer. Strateg Manag J 17(2):77–92

Narula R, Hagedoorn J (1999) Innovating through strategic alliances: moving towards international partnerships and contractual agreements. Technovation 19(5):283–294

Ohmae K (1989) The global logic of strategic alliances. Harv Bus Rev 67(3/4):143–154

Osborn RN, Hagedoorn J (1997) The institutionalization and evolutionary dynamics of interorganizational alliances and networks. Acad Manag J 40(2):261–278

Ozman M (2009) Inter-firm networks and innovation: a survey of literature. Econ Innov New Technol 18(1):39–67

Parise S, Casher A (2003) Alliance portfolios: designing and managing your network of business-partner relationships. Acad Manag Exec 17(4):25–39

Park SH, Russo MV (1996) When competition eclipses cooperation: an event history analysis of joint venture failure. Manag Sci 42(6):875–890

Patel P, Pavitt K (1995) Patterns of technological activity: their measurement and interpretation. In: Stoneman P (ed) Handbook of the economics of innovation and technological change. Blackwell, Oxford, UK, pp 14–51

Pittaway L, Robertson M, Munir K, Denyer D, Neely A (2004) Networking and innovation: a systematic review of the evidence. Int J Manag Rev 5(6):137–168

Podolny JM (2001) Networks as the pipes and prisms of the market. Am J Sociol 7(1):33–60

Podolny JM, Page KL (1998) Network forms of organization. Annu Rev Sociol 24(1):57–76

Powell WW (1990) Neither market nor hierarchy: networks forms of organization. Res Organ Behav 12(1):295–336

Powell WW, Koput KW, Smith-Doerr L (1996) Interorganizational collaboration and the locus of innovation – networks of learning in biotechnology. Adm Sci Q 41(1):116–145

Protogerou A, Caloghirou Y, Siokas E (2010) Policy-driven collaborative research networks in Europe. Econ Innov New Technol 19(4):349–372

Provan KG, Fish A, Sydow J (2007) Interorganizational networks at the network level: a review of the empirical literature on whole networks. J Manag 33(3):479–516

Rodan S, Galunic C (2004) More than network structure: how knowledge heterogeneity influences managerial performance and innovativeness. Strateg Manag J 25(6):541–562

Rothaermel FT (2001) Incumbent's advantage through exploiting complementary assets via interfirm cooperation. Strateg Manag J 22(6):687–699

Rothaermel FT, Deeds DL (2006) Alliance type, alliance experience and alliance management capability in high-technology ventures. J Bus Ventur 21(4):429–460

Rowley TJ, Behrens D, Krackhardt D (2000) Redundant governance structures: an analysis of structural and relational embeddedness in the steel and semiconductor industries. Strateg Manag J 21(3):369–386

Scherngell T, Barber MJ (2011) Distinct spatial characteristics of industrial and public research collaborations: evidence from the fifth EU framework programme. Ann Reg Sci 46 (2):247–266

Schilke O, Goerzen A (2010) Alliance management capability: an investigation of the construct and its measurement. J Manag 36(5):1192–1219

Schilling MA (2009) Understanding the alliance data. Strateg Manag J 30(3):233–260

Schilling MA, Phelps CC (2007) Interfirm collaboration networks: the impact of large-scale network structure on firm innovation. Manag Sci 53(7):1113–1126

Simonin BL (1999) Ambiguity and the process of knowledge transfer in strategic alliances. Strateg Manag J 20(1):595–623

Sivadas E, Dwyer RF (2000) An examination of organizational factors influencing new product success in internal and alliance-based processes. J Mark 64(1):31–49

Spender JC, Grant RM (1996) Knowledge and the firm: overview. Strateg Manag J 17(2):5–10

Stata (2007) Stata statistical software: release 10. StataCorp LP, College Station

Stuart TE (1999) A structural perspective on organizational performance. Ind Corp Chang 8 (4):745–775

Stuart TE (2000) Interorganizational alliances and the performance of firms: a study of growth and innovational rates in a high-technology industry. Strateg Manag J 21(8):791–811

Thorelli HB (1986) Networks: between markets and hierarchies. Strateg Manag J 7(1):37–51

Uzzi B (1997) Social structure and competition in interfirm networks : the paradox of embeddedness. Adm Sci Q 42(1):35–67

Van Den Bosch FA, Volberda HW, De Boer M (1999) Coevolution of firm absorptive capacity and knowledge environment: organizational forms and combinative capabilities. Organ Sci 10 (5):551–568

Wasserman S, Faust K (1994) Social network analysis: methods and applications. Cambridge University Press, Cambridge

Wassmer U (2010) Alliance portfolios: a review and research agenda. J Manag 36(1):141–171

White S (2005) Cooperation costs, governance choice and alliance evolution. J Manag Stud 42 (7):1383–1413

Williamson OE (1991) Comparative economic organization: the analysis of discrete structural alternatives. Adm Sci Q 36(2):269–296

Wuyts S, Dutta S, Stremersch S (2004) Portfolios of interfirm agreements in technology-intensive markets: consequences for innovation and profitability. J Mark 68(2):88–100

Zahra SA, George G (2002) Absorptive capacity: a review, reconceptualization, and extension. Acad Manag Rev 27(2):185–203

Chapter 11
Small World Patterns and Firm Innovativeness

Innovation is the central issue in economic prosperity.
(Michael E. Porter 1980)

Abstract In this section we switch analytical perspective and take a closer look at the systemic or overall network level. As outlined before (cf. Sect. 8.3), an in-depth understanding of large-scale network patterns is important for several reasons. On the one hand, previous studies have demonstrated that networks with comparably short path lengths and a high level of clustering – so-called "small-world" networks – can facilitate the exchange of information, ideas and knowledge in networks (Fleming, L., C. King, A. I. Juda. 2007. Small worlds and regional innovation. *Organ. Sci.* 18(6) 938-954). This, however, substantiates the assumption that systemic level network properties are likely to affect the embedded firms in their efforts to innovate. On the other hand, systemic level studies have some far-ranging implications, not only for firms but also for policy makers, by providing an informative basis for the evaluation of cooperation-related innovation policies at the national and supra-national level. In a nutshell, the aim of the third empirical part of this study is to shed light on the relationship between specific types of large-scale network properties at the macro-level and firm-level innovation outcomes at the micro-level. This investigation is organized as follows. After a short introduction in Sect. 11.1 we outline selected theoretical concepts. Next, we continue by providing the graph theoretical underpinnings of small-world properties in Sect. 11.2. Then, we introduce our conceptual framework and derive a set of testable hypotheses. In Sect. 11.3 we provide a short overview of data and methods used for this analysis. After these preparatory steps, we continue with a description of the empirical model and present our estimation results in Sect. 11.4. Finally, after a brief discussion of our main findings we conclude with some critical remarks.

© Springer International Publishing Switzerland 2015   255
M. Kudic, *Innovation Networks in the German Laser Industry*,
Economic Complexity and Evolution, DOI 10.1007/978-3-319-07935-6_11

11.1   On Small-World Characteristics in Innovation Networks

In the late 1960s Stanley Milgram conducted an experiment that is still highly topical, especially in the field of network research. The specific concern of his research project was to understand how communication processes work in social systems (Uzzi and Spiro 2005, p. 450). The constellation of his so-called "letter-passing" experiment was quite simple. He sent letters to a randomly chosen set of participants who were scattered throughout the United States. Written instructions were included asking the recipients to pass the letter on to a pre-specified target individual (Newman 2010, p. 55). It turned out that almost one third of the letters sent even reached far away targets after an average of around six distinct steps. Milgram's (1967) groundbreaking experiment demonstrated that people in the United States are separated by more or less six degrees of separation.

Only recently have economists, sociologists and management scholars started to address the "small-world" phenomenon (for a comprehensive review see: Uzzi et al. 2007). Milgram's findings have some far-reaching implications for innovation networks. The experiment implies that the network topology itself is likely to affect the exchange of knowledge in innovation networks. This, however, substantiates the assumption that large-scale properties at the overall network level affect the innovative performance of network actors at the micro-level. It is all the more astonishing that large-scale network properties have been widely neglected in the field of interorganizational alliance and network research over the past decades.[1] One possible explanation is that it took scholars about 30 years to quantify Milgram's initial idea. Watts and Strogatz (1998) have shown that the "small-world" phenomenon can be empirically analyzed by using relatively simple network measures which were originally designed for unipartite networks (cf. Sect. 8.3.2). Some years later a reconceptualization for bipartite networks was proposed by Newman and colleagues (2001). Quite recently, a few excellent empirical studies were conducted which explicitly analyzed the relationship between "small-world" properties and the creation of novelty and innovation (Uzzi and Spiro 2005; Fleming et al. 2007; Schilling and Phelps 2007).

One of the first studies on collaboration, creativity and small worlds was conducted by Uzzi and Spiro (2005). The authors analyzed the relationship between small-world properties in the Broadway musical industry and creativity in terms of the financial and artistic performance of musicals produced from 1945 to 1989. This setting is remarkable for two reasons. Firstly, the network measures were constructed based on bipartite network data. In other words, groups of artists were treated as fully connected cliques. To handle the data properly, novel statistical techniques (Newman et al. 2001) were applied to detect and interpret small-world properties which were explicitly designed for the analysis of bipartite

---

[1] Most notable exceptions are the studies by Baum et al. (2003), Corrado and Zollo (2006), Uzzi and Spiro (2005), Fleming et al. (2007), Schilling and Phelps (2007) and Cassi and Zirulia (2008).

networks. Finally, it is interesting to note that Uzzi and Spiro (2005) measured performance outcomes at the team level and not the actor level. They reported a parabolic small-world network effect in a sense that performance increased initially and then decreased after a certain point.

In a similar vein, Fleming and colleagues (2007) raised the question of why some regions outperform others in terms of innovativeness. Like Uzzi and Spiro (2005) they focused explicitly on small-world networks. However, both "small-world" properties and innovative performance were measured at the regional level. Based on patent co-authorship data they showed that comparably short path lengths and larger connected components are positively correlated with increased innovation. Nonetheless, they failed to find empirical evidence that the small-world properties of the regional innovation network enhanced firm innovativeness.

The most comprehensive study on small worlds and firm innovativeness was provided by Schilling and Phelps (2007). They analyzed the patent performance of 1,106 firms in 11 industry-level alliance networks based on a comprehensive panel dataset. The findings of the study provide support for the small-world hypothesis by showing that networks with comparably short path lengths and high clustering have a significant impact on the innovativeness of the firms involved. The authors came to the conclusion that local density and global efficiency can exist simultaneously, and in particular, the combination of these two network characteristics enhances innovation (Schilling and Phelps 2007, p. 1124). Despite these interesting findings the study has some limitations. The most notable is that the authors had to make assumptions about alliance duration due to a lack of information on alliance termination dates. They assumed that alliance relationships last for 3 years on average. In the worst case, this could result either in a systematic underestimation or overestimation of small-world network properties.

All of these studies provide us with valuable insights into the small-world phenomenon. However, this discussion also reveals that recent empirical findings have so far been rather mixed and inconclusive. In addition, we still lack an in-depth understanding of how large-scale network properties affect firm innovativeness. In other words, we have to open up the black box in order to understand the mechanisms or transmission channels through which firm innovativeness is affected by systemic-level network properties. Thus, the aim of this investigation is twofold. From a theoretical point of view, we draw upon a reconceptualization of the absorptive capacity concept proposed by Zahra and George (2002) to provide the missing link between overall network characteristics and a firm's innovative performance. From an empirical point of view, we put the "small-world" hypothesis to the test according to which small-world networks are assumed to enhance an embedded firm's creativity and its ability to create novelty in terms of innovation. More precisely, we analyze the relationship between distinct large-scale patterns (i.e. "weighted clustering coefficient" *or* "avg. path-length") and firm innovativeness on the one hand, and small-world properties (i.e. "weighted clustering coefficient" *and* "avg. path-length") and firm innovativeness on the other.

11.2   Small-World Networks and Absorptive Capacity

11.2.1   Graph Theoretical Basis of the "Small-World" Phenomenon

Small-world networks are characterized by two structural particularities: a high level of clustering and short average path lengths. The theoretical conceptualization and quantification of the small-world phenomenon can be traced back to the pioneering work of Watts and Strogatz (1998). The authors argued that a compression of real-world networks and randomly generated networks should reveal some systematic differences with regard to network clustering and reachability. They proposed using two simple graph theoretical concepts – "cluster coefficient" and "average distance" – and calculating two ratios – "clustering coefficient ratio" (CC ratio) and "path length ratio" (PL ratio) – in order to check for the existence of small-world properties (cf. Sect. 8.3.2). Quantitative network analysis methods provide a rich toolbox for calculating these indicators (cf. Wasserman and Faust 1994; Borgatti et al. 2013).

The actor-specific clustering coefficient varies from 0 to 1.0 whereby high values indicate that many of the actor's direct contacts are connected to each other (Wasserman and Faust 1994). The overall clustering coefficient is an indicator that allows the connectedness and crowding in a network to be quantified. This measure is simply defined as the average of all individual clustering coefficients for a well-specified population of network actors. In contrast, the weighted overall clustering coefficient is defined as the weighted mean of the clustering coefficient of all the actors, each one weighted by its degree (Borgatti et al. 2002). The shortest path between two network actors is referred to as a geodesic whereas the length of the geodesic between a pair of network actors is referred to as the geodesic distance (Wasserman and Faust 1994, p. 110). The average path length captures the reachability among all network actors in a connected graph or subgraph. The measure can be defined as "[...] the average number of intermediaries, that is, the degrees of separation between any two actors in the network along their shortest path of intermediaries" (Uzzi et al. 2007, p. 78).[2]

Watts and Strogatz (1998) concluded that real-world networks with a CC ratio much higher than 1.0 and a PL ratio of about 1.0 have a small-world character. A related indicator is the so-called "small-world Q" (defined as: the CC ratio divided by the PL ratio), where Q values that are much greater than 1.0 indicate the small-world nature of a real-world network (Uzzi et al. 2007, p. 79). Newman et al. (2001, 2002) have shown that the "path length ratio" in bipartite networks has basically the same interpretation as in unipartite networks (Uzzi and Spiro 2005, p. 454). In contrast, the "clustering coefficient ratio" has to be interpreted differently in the

---

[2] For further details on the calculation and interpretation of these two measures, see Sects. 5.2.3 and 8.3.2.

sense that a coefficient ratio of about 1.0 indicates within-team clustering whereas a higher clustering coefficient ratio indicates an increase in between-team clustering (Uzzi and Spiro 2005, pp. 454–455).

What do these graph theoretical considerations tell us with regard to firm innovativeness? Or to put it another way, what is the theoretical explanation that substantiates the assumption that small-world properties at the systemic level enhance a firm's ability to innovate? Earlier researchers have argued as follows (Schilling and Phelps 2007, pp. 1114–1115): On the one hand, a high level of clustering increases the network's information transmission rate, enhances a firm's willingness and ability to exchange knowledge and enables richer and greater amounts of information and knowledge to be integrated. On the other hand, networks with short average path lengths enhance reachability among actors and generally improve information accessibility at the systemic level. There is no doubt that these arguments provide an intuitive reasoning behind the consequences of potential firm-level innovation outcomes caused by increased information permeability in a small-world network. However, these arguments do not directly address what is happening at the firm level during the firm's efforts to innovate.

11.2.2   Potential and Realized Absorptive Capacity: The Missing Link

We argue that Zahra and George's (2002) reconceptualization of Cohen and Levinthal's (1990) initially proposed "absorptive capacity" concept provides the missing link in understanding the interrelationship between systemic network level properties and firm-level innovation outcomes.

The originally proposed "absorptive capacity" concept by Cohen and Levinthal (1989, 1990) has significantly enhanced our understanding of a firm's ability to identify, exploit and assimilate external knowledge and apply it for commercial ends. Cohen and Levinthal (1989) focused initially on the costs of acquiring new technological knowledge and on the incentives for learning that determine the firm's willingness to invest in creating and establishing absorptive capacity. Later the authors enriched the construct by emphasizing the relevance of individual learning processes and incorporating the notion that learning is a cumulative process (Cohen and Levinthal 1990). Furthermore, they adapted insights from research on individual cognitive structures and individual learning processes. They applied these findings to the organizational level and emphasized that an organization's absorptive capacity is path-dependent, builds on prior investments in individual absorptive capacity and depends on an organization's internal communication processes and its ability to share knowledge (Lane et al. 2006, p. 838). In addition, they pointed to the fact that previously accumulated knowledge enables the firm to predict and appraise new technological trends and developments in a timely way. Since then the concept has attracted a great deal of attention.[3] Several scholars have proposed insightful reconceptualizations of Cohen and Levinthal's original concept (Lane and Lubatkin 1998; Van Den Bosch et al. 1999; Zahra and George 2002).

For the purpose of this analysis we draw upon the concept proposed by Zahra and George (2002). This reconceptualization builds upon the distinction between "capabilities" and "dynamic capabilities". By starting from the dynamic capability perspective (Teece et al. 1997; Katkalo et al. 2010) they suggest a separation of the original absorptive capacity concept into potential absorptive capacity and realized absorptive capacity and introduce an efficiency factor η that captures the interrelationship between these two constructs (Zahra and George 2002, p. 194). They argue that four capabilities[4]– i.e. knowledge acquisition, assimilation, transformation and exploitation – are combinative in nature and build upon each other. These four capabilities make up a firm's absorptive capacity that has to be regarded as a dynamic capability pertaining to knowledge creation and utilization that enhances a firm's innovative performance and ability to gain and sustain a knowledge-based competitive advantage (Zahra and George 2002, p. 185). They define absorptive capacity as "[...] a set of organizational routines and processes by which firms acquire, assimilate, transform, and exploit knowledge to produce a dynamic organizational capability" (Zahra and George 2002, p. 186).

Figure 11.1 illustrates a slightly refined version of Zahra and George's (2002) model. The absorptive capacity construct, at the core of the model (cf. Fig. 11.1 center), is divided into potential absorptive capacity (PACAP), which includes knowledge acquisition and assimilation, and realized absorptive capacity (RACAP), that consists of knowledge transformation and exploitation capabilities. This absorptive capacity construct connects the antecedents, i.e. external knowledge sources, knowledge complementarities and experiences (cf. Fig. 11.1, left) with firm-level outcomes, i.e. firm innovativeness and sustainable competitive advantages (cf. Fig. 11.1, right). In addition, the model accounts for several moderating effects: "activation triggers", "social integration mechanisms", and "regimes of appropriability". An efficiency factor η is integrated into the model that captures a firm's ability to transform and exploit external knowledge sources in order to gain a sustainable competitive advantage. This factor reflects the extent to which a firm can make commercial use of potentially available knowledge. In other words, RACAP approaches PACAP in firms with a high efficiency factor (Zahra and George 2002, p. 191). This model paves the way for a dynamic conceptualization of absorptive capacity and provides several interesting implications for systemic level network studies. Below we argue that a simple extension of the model

---

[3] Lane et al. (2006) identified a total of 289 papers in 14 academic journals between July 1991 and June 2002 that cite Cohen and Levinthal's (1990) "absorptive capacity" concept.

[4] Zahra and George (2002) draw upon Winter (2000, p. 983) who defines capabilities as "[...] a high-level routine that, together with its implementing input flows, confers upon an organization's management a set of decision options for producing significant outputs of a particular type."

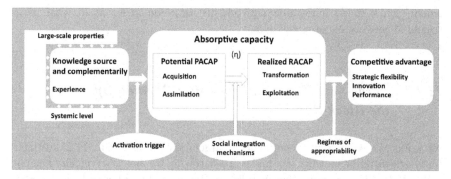

Fig. 11.1 Conceptual framework – an adapted model of potential and realized absorptive capacity (Source: Zahra and George (2002, p. 192), extended and modified)

provides the missing link for understanding how large-scale properties at the overall network level affect innovation outcomes at the firm level.

In doing so, we have to take a closer look at the first element of the framework (cf. Fig. 11.1, left). According to the model originally proposed by Zahra and George (2002, p. 191) there is a direct link between external knowledge sources and complementarities and a firm's PACAP. These external knowledge sources encompass, among other things, various structural forms of interorganizational relationships such as R&D consortia, alliances, or joint ventures.[5] Thus cooperative relationships to external partners can serve as a vehicle for accessing new information and knowledge. However, it is important to note that not only direct but also indirect interorganizational linkages have to be considered in this context (Gulati 1998). As a consequence, we apply here not a relational but rather a structural network embeddedness perspective (cf. Sect. 2.5.4). One particular feature of a network is that a particular firm can even reach far distant organizations that are spread throughout the entire network space by second or third tier ties. This means that a firm that is a part of the industry's innovation network has potential access to an extensive pool of external technological knowledge sources spread throughout the entire network. Thus, in line with previous systemic-level studies (Uzzi and Spiro 2005; Fleming, et al. 2007; Schilling and Phelps 2007), we argue that actual access to information and the knowledge stocks of other firms is likely to be affected by the structure of the network in question. The network topology itself plays a key role in the permeability of the network. In contrast to previous research, we believe that an extension of the absorptive capacity concept outlined above and an in-depth exploration of structural network characteristics adds extra value to our understanding of how large-scale properties at the systemic level affect a firm's efforts to innovate (cf. Fig. 11.1, left). Or to put it differently, given that network topologies can facilitate but also hamper the flow of information and knowledge

---

[5] Due to the purpose of this study we focus explicitly on the innovation network as one particular type of external knowledge source that can be tapped by the firms.

among actors in an innovation network, the question arises as to what these structural network patterns look like.

11.2.3   Large-Scale Network Properties: Opening Up the Black Box

Networks can exhibit quite heterogeneous structural patterns. Figure 11.2 illustrates four fairly different network topologies. To start with, we look at a typically random network. It is important to note that the emergence of these networks is not very likely under realistic conditions. Nonetheless, we explicitly consider and discuss all four cases in order to develop our theoretical arguments.

The first network example is characterized by a rather fragmented network structure that consists of five components (cf. Fig. 11.2, I). The structural config- uration of the network shows no significant peaks in term of the actors' nodal degrees. The minimum degree is one and the maximum degree is two. Network actors within a component are not directly but rather are indirectly connected to other actors in the same component. The benefits of a firm in participating in such a fragmented, randomly distributed network are rather limited. The reasons for this are straightforward. Firstly, the pool of potentially accessible knowledge sources is limited by the size of the component in which the firm is embedded. Secondly, the geodesic distances to most other actors are infinite due to the high degree of fragmentation. Thus, knowledge transfer processes are likely to be hampered by the component's size or even entirely prevented by the overall network structure.

These issues lead to our second network example. Figure 11.2 (II) illustrates a fully connected but randomly distributed network structure. Like before there are no systematic biases in the degree distribution at the overall network level. The main difference is that the network consists of only one large component. This, however, has some important implications with regard to knowledge diffusion processes. Theoretically, we would expect that a firm's participation in such a

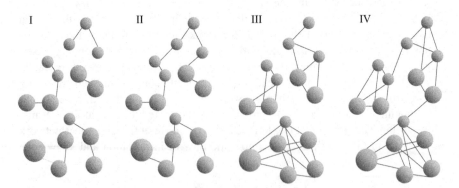

Fig. 11.2 Illustration of network topologies (Source: Author's own illustration)

network broadens the scope and variety of potentially accessible information and knowledge sources. One could argue that the firm's chance of identifying and actually accessing external knowledge sources that fit with its own set of capabilities increases with the number of potentially accessible knowledge sources. The crucial point is that such an increased set of opportunities would allow a firm to make better use of its knowledge exploitation capabilities. According to Zahra and George (2002) this would be reflected in a higher efficiency factor η and lead to a higher firm-level innovation outcome at subsequent points in time. In fact the actual situation looks somewhat different. The likelihood of successfully exchanging knowledge between two indirectly connected network actors decreases with the number of other actors that lie on the geodesic between them. A closer look at our network example illustrates this point (Fig. 11.2, II). In this case we have up to 11 intermediates between the most distant actors in the network.

Next, we turn our attention to a somewhat more realistic network structure. By now, it is well-recognized that some nodes attract ties at a higher rate than others. This is reflected in real-world networks by the emergence of a strongly biased degree distribution at the overall network level. These types of networks are also known as power law distributed or scale free networks (cf. Sect. 8.3.1). Real-world network topologies can differ significantly in terms of their structural features.

Our third network example consists of three components (two peripheral and one main component) and the nodal degrees range from one to five (cf. Fig. 11.2, III). The network is disconnected and clustered. The nodes within these components are well-connected themselves but have no linkages to actors in other areas of the network. We start our line of argument by focusing on the network's main component (cf. Fig. 11.2, III, bottom). A firm's involvement in a highly interconnected main component of a disconnected network has some considerable advantages. Firstly, all main component firms are connected to one another. A main component firm can reach most other actors in the same component in only a few steps. Short paths are likely to facilitate potential knowledge transfer and learning processes. Most innovation researchers would agree that a decreasing path length is positively related to firm innovativeness (Fleming et al. 2007, p. 941).

Secondly we turn our attention to clustering within connected network components. A high degree of interconnectedness allows a focal firm to achieve cooperation-related synergy effects. These effects can result from direct but also from indirect linkages among a focal actor's directly connected partners (White 2005; Hoffmann 2005). Redundant knowledge transfer channels allow firms to circumvent potentially emerging knowledge transfer barriers. It has been argued that clustering promotes collaboration, resource pooling and risk sharing (Fleming et al. 2007, p. 940).[6]

---

[6] It is important to note that these considerations only hold true as long as the number of disconnected network components is comparably small. The benefits diminish with an increasing number of disconnected subgroups in the network. Or to put it another way, increasing fragmentation disestablishes the benefits described above.

In summary, the previously outlined arguments substantiate the assumption that a firm's embeddedness in the main component of a highly clustered but disconnected innovation network enhances a firm's scope and variety of accessible knowledge sources. Two structural characteristics, i.e. short path lengths and a high level of clustering are considered to be important in this context. Keeping the extension of Zahra and George's (2002) absorptive capacity model in mind, it is plausible to assume that these structural features enhance a firm's efficiency factor η. This, in turn, is likely to be positively related to firm-level innovation outcomes at later points in time. The arguments above form our first two hypotheses:

H1 Short average path length in the overall network level is positively related to its innovative performance at later points in time.

H2 A high degree of clustering at the overall network level is positively related to its innovative performance at later points in time.

Last but not least, we address small-world properties of innovation networks. It becomes apparent that the previously discussed real-world network in itself encounters barriers in information and knowledge transfer. As already stated above, the network consists of several densely interconnected components which are not connected to one another. This leads us to take a look at the last network example. Figure 11.2 (IV) illustrates a highly clustered but fully connected real-world network. The simultaneous occurrence of cohesive subgroups and short paths in a network has some interesting implications.

Firstly, such a network is rich in structural holes and the cohesive subgroups are interconnected through network brokers (Burt 1992). They bridge structural gaps in a network and establish important connections between otherwise unconnected or at least loosely connected network subgroups (ibid). This, however, significantly decreases the average path lengths at the overall network level and increases, at the same time, information permeability. Secondly, the benefits of cohesive subgroups in a firm's close network surroundings are maintained. The simultaneous occurrence of clustering and short average path length indicate the small-world nature of a network (Watts and Strogatz 1998).

In line with previous research (Schilling and Phelps 2007) we argue that small-world network properties are accompanied by some extra additive effects which are assumed to enhance a firm's efficiency factor η. The simultaneous occurrence of both high clustering and short average path lengths is likely to catalyze and foster local cooperation effectiveness and enhance global information transmission efficiency (Schilling and Phelps 2007, p. 1116). These considerations substantiate our last hypothesis:

H3 A firm's participation in a small-world network (characterized by short average path lengths and a high level of clustering) is positively related to its innovative performance at later points in time.

11.3 Data, Variables and Descriptive Statistics

11.3.1 Data Sources

Four main data sources were used to construct a longitudinal panel dataset: patent data, industry data, geographical data and network data (cf. Sect. 4.2).

Patent data was used to measure innovative performance at the firm level. We are not the first to use patent data as an innovation proxy (Jaffe 1989; Jaffe et al. 1993). Previous studies provide us with important insights into the pros and cons of using patents to measure innovation performance.[7] In accordance with contemporary research (Schilling and Phelps 2007), we used annual patent counts as a proxy for innovation output. Our database (cf. Sect. 6.1.2) includes patent applications as well as patents granted by the German Patent Office and by the European Patent Office. DEPATISnet (the German Patent and Trade Mark Office's online database) and *ESPACEnet* (the European Patent Office database) were employed to cross check the results from our initial data gathering procedure. For the purpose of this analysis we used the annual count of patent applications [*pacnt*] as an endogenous variable.

Industry data came from a proprietary dataset containing the entire population of German laser source manufacturers between 1969 and 2005 (Buenstorf 2007). Based on this initial dataset we used additional data sources to gather information about firm entries and exits after 2005. We chose the business unit or firm level for the purpose of this analysis.[8] In addition, we identified 145 universities and public research organizations with laser-related activities by using two complementary methods – the expanding selection method and the bibliometric approach.[9]

Network data was gathered from two official databases on publicly funded R&D collaboration projects – the *Foerderkatalog* database and *CODRIS* database.[10] The first database contains information on more than 110,000 ongoing or completed subsidized research projects. The second raw data source was an extract from the *CORDIS* project database which includes a complete collection of R&D projects for all of the German companies funded by the European Commission. This database extract encompasses a project dataset with over 31,000 project files and an organization dataset with over 57,100 German organizations and roughly 194,000 international project partners. In total, we were able to identify, for the entire population of 233 German laser source manufacturers, 570 R&D projects with up to 33 project partners from various industry sectors, non-profit research organizations and universities.

---

[7] Section 4.2.4 provides a detailed discussion on the measurement of innovation and describes the patent data sources and data gathering procedure used for the purpose of this study.

[8] For a detailed description of industry data used for this study, see Sect. 4.2.1.

[9] Both methods are described in detail in Sect. 4.2.

[10] For a detailed description of both cooperation data sources (*CORDIS* and *Foerderkatalog*) and the methods used to construct annual networks, see Sect. 4.2.3.

11.3.2 Variable Specification

The data sources described above were used to construct interorganizational innovation networks and calculate network indicators on a yearly basis. We calculated
weighted clustering coefficients *[nw_wclust]* and average path length *[nw_areach]*
on an annual basis (cf. Sect. 5.2.3, Eqs. 5.10 and 5.11). An interaction term was
calculated to capture the small-world properties of the network *[inter_sw]*. Several
additional control variables were calculated. We measured firm-specific cooperation activities with two cooperation count measures based on the *Foerderkatalog*
data *[coopcnt_fk]* and *CORDIS* data *[coopcnt_c]* respectively, as well as a combined cooperation count indicator *[coopcnt_fkc]* consisting of the sum of both.
Moreover, we accounted for cooperation funding by including a variable that
measures the firm's amount of cooperation funding received annually
[coopfund_fkc] in 1,000 euros. We also included a linear firm age measure
[firmage] as well as a squared term *[firmage_sq]* to account for firm maturity. In
addition, two network level variables were included to control for the structural
network characteristics at the overall network level. The first variable captured
the size of the overall network *[nw_size]* defined as the proportion of firms with
at least one dyadic partnership in a given year. The second variable measured
the connectedness of the overall network *[nw_density]*. Standard algorithms
implemented in UCI-Net 6.2 were used to calculate the network measures (Borgatti
et al. 2002).

11.3.3 Descriptive Statistics

Next, we take a brief look at the variable description and basic summary statistics
(cf. Table 11.1). In total, we have 2,645 firm-year observations in the time between
1990 and 2010. The average number of observations per firm amounts to 11.35.
Table 11.2 reports the correlation coefficients for all variables in our empirical
models.

Based on the data sources described above we conducted an initial exploratory
analysis to get an idea of what the overall network topology looks like. Figure 11.3
(top) displays the weighted overall clustering coefficients and the average overall
path length for both the German laser industry innovation network and a randomly
generated Erdös-Renyi network that is comparable in terms of size and density.[11]
Network measures are calculated on an annual basis and the period under observation is from 1990 to 2010. All measures are calculated using UCI-Net 6.2 (Borgatti
et al. 2002). The corresponding CC ratios, the PL ratios and the small-world

---

[11] The construction of the reference network is described in Sect. 8.3.2.

Table 11.1 Descriptive statistics – clustering, reach and small-world properties

Variable	Variable definition	Summary statistics				
		Obs.	Mean	Std. dev.	Min	Max
Endogenous variables						
papcount	Patent applications (annual count)	2,645	2.662004	17.43323	0	366
pgrcount	Patent grants (annual count)	2,645	0.339130	1.635554	0	28
Control variables						
firmage	Age of the firm	2,645	8.055955	6.800477	0	43
firmage_sq	Age of the firm, squared	2,645	111.1274	177.8146	0	1,849
coopcount	Count of cooperation events (annual)	2,645	0.275992	0.774138	0	8
coopfund	Annual cooperation funding received (in k€)	2,645	132.299	851.8748	0	31.863
nw_size	Network size (overall network level)	2,645	0.381853	0.060200	0.240506	0.472393
nw_density	Network density (overall network level)	2,645	0.088119	0.069955	0.037300	0.440500
Network level properties						
nw_wclust	Weighted clustering coefficient	2,645	0.58152	0.161069	0.345	0.906
nw_areach	Average distance based reach measure	2,645	3.09431	0.504183	2.075	3.786
inter_sw	"Small world" indicator (nw_wclust × nw_areach)	2,645	1.7324	0.298921	1.14021	2.18748

Source: Author's own calculations

Table 11.2 Correlation matrix – clustering, reach and small-world properties

	Pgr count	Pap count	Firm age	Firm age_sq	Coopcount	Coopfund	Nw_size	Nw_density	Nw_wclust	Nw_areach	Inter_sw
Papcount	1.0000										
Pgrcount	0.6506	1.0000									
Firmage	-0.0566	0.0105	1.0000								
Firmage_sq	-0.0455	-0.0047	0.9276	1.0000							
Coopcount	0.2535	0.2726	0.0101	0.0272	1.0000						
Coopfund	0.3923	0.3114	-0.0279	-0.0100	0.5112	1.0000					
Nw_size	0.0448	0.0670	0.2131	0.1603	0.0442	0.0147	1.0000				
Nw_density	-0.0529	-0.0832	-0.2006	-0.1450	-0.1450	-0.0145	-0.6576	1.0000			
Nw_wclust	-0.0529	-0.0835	-0.2530	-0.1901	-0.0074	-0.0153	-0.4684	0.6934	1.0000		
Nw_areach	0.0637	0.0965	0.2761	0.2062	0.0164	0.0100	0.7154	-0.7499	-0.8255	1.0000	
Inter_sw	-0.0219	-0.0414	-0.1639	-0.1275	0.0026	-0.0141	-0.0656	0.3186	0.8357	-0.4020	1.0000

Source: Author's own calculations

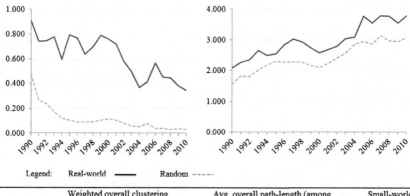

Legend: Real-world ——   Random -----

| Year | Weighted overall clustering coefficient | | | Avg. overall path-length (among reachable pairs) | | | Small-world properties |
	Real-world	Random	CC	Real-world	Random	PC	(Q)
1990	0.906	0.477	1.899	2.075	1.560	1.330	1.4279609
1991	0.743	0.263	2.825	2.268	1.820	1.246	2.267051589
1992	0.746	0.240	3.108	2.351	1.810	1.299	2.393059691
1993	0.777	0.175	4.440	2.658	2.020	1.316	3.374266366
1994	0.595	0.120	4.958	2.501	2.180	1.147	4.321937892
1995	0.793	0.105	7.552	2.553	2.300	1.110	6.803946804
1996	0.767	0.088	8.716	2.852	2.280	1.251	6.967837562
1997	0.638	0.093	6.860	3.027	2.280	1.328	5.167258118
1998	0.701	0.093	7.538	2.929	2.270	1.290	5.841731003
1999	0.791	0.104	7.606	2.735	2.160	1.266	6.006750105
2000	0.761	0.115	6.617	2.579	2.090	1.234	5.36267849
2001	0.720	0.105	6.857	2.685	2.220	1.209	5.669592977
2002	0.579	0.082	7.061	2.789	2.410	1.157	6.101452571
2003	0.499	0.059	8.458	3.040	2.560	1.188	7.12221231
2004	0.369	0.053	6.962	3.090	2.860	1.080	6.444037369
2005	0.413	0.078	5.295	3.768	2.950	1.277	4.145401219
2006	0.567	0.042	13.500	3.548	2.850	1.245	10.84413754
2007	0.452	0.039	11.590	3.786	3.140	1.206	9.612201498
2008	0.446	0.032	13.938	3.768	2.970	1.269	10.98576831
2009	0.380	0.036	10.556	3.548	2.950	1.203	8.776462483
2010	0.345	0.031	11.129	3.786	3.080	1.229	9.053729359
Mean	0.6185	0.1157		2.9684	2.4171		
Std. dev.	0.1692	0.1037		0.5386	0.4582		

Fig. 11.3 Weighted overall clustering coefficient and average overall path length

Q values are reported in the illustration below (cf. Fig. 11.3, bottom). The following structural patterns are noteworthy.[12]

Firstly, the German laser industry innovation network shows a relatively high level of clustering and rather short average path lengths overall. Secondly, over time we can observe decreasing weighted clustering coefficients and increasing average path length. This is primarily due to the fact that the German laser industry network has demonstrated a pronounced growth tendency over time. In other

---

[12] Note that the calculations are based on bipartite network data. This is in line with the study by Uzzi and Spiro (2005). However, the use of bipartite network data generates relatively high clustering coefficients. This should be kept in mind when interpreting the results. For an in-depth discussion on the differences between unipartite and bipartite network data, see Sect. 8.3.2. To ensure robustness of the reported results we calculated both small-world indicators based on an alternative network data decomposition assumption. Additional calculations confirm the small world character of the network (cf. Appendix 3).

words, the number of laser-related organizations that actively participate in the industry's innovation network increases over time. Thirdly, small-world measures indicate the emergence and consolidation of the network's small-world nature. More precisely, a comparison of the real-world network with a randomly generated reference network reveals that the German laser industry innovation network exhibits both higher overall clustering coefficients and longer average path lengths for each year throughout the entire observation period. The annually calculated CC ratios are clearly above 1.0 and increase over time. PC ratios do not exceed the value range between 1.0 and 1.35 and the small-world Q ratio lies significantly above 1.0 and demonstrates, like the CC ratio, a pronounced tendency towards increasing values over time.

In summary, the results of the exploratory analysis of large-scale network properties for the German laser industry are suggestive of an increasing emergence and solidification of small-world properties over time.

11.4   Estimation Results and Empirical Findings

11.4.1   Model Specification and Estimation Strategy

As our endogenous variable – annual patent application counts – only accepts nonnegative integer values, we choose a count data model specification for the purpose of this analysis.[13] Following Ahuja (2000), Stuart (2000) and Schilling and Phelps (2007), we estimated panel data count models.[14] Basically, two estimation techniques can be distinguished: the fixed effects and random effects methods. In general, the use of fixed effects models provides some important advantages. The fixed effects estimator is unbiased as it includes dummy variables for the different intercepts and is more robust against selection bias problems than the random effects estimator (Kennedy 2003, p. 304). The problem that occurs with fixed effects models is that all time-invariant explanatory variables are thrown out because the estimation procedure fails to estimate a slope coefficient for variables that do not vary within an individual unit (Kennedy 2003, p. 304). In addition, using only within-variation leads to less efficient estimates and the model loses its explanatory power (Cameron and Trivedi 2009, p. 259). In contrast, random effects estimators make better use of the information values of patent data and generate efficient estimates with higher explanatory power. In addition, random effects estimators can generate coefficient estimates of both time-variant as well as time-invariant explanatory variables (Kennedy 2003, p. 307). The major drawback of the random effects model is that correlations between the error term and the explanatory variables generate biased estimates and thus inconsistent estimation results (Kennedy 2003, p. 306).

---

[13] For an in-depth discussion on the use of panel data count models, see Sect. 6.1.2.

[14] We used STATA 10.1 (Stata 2007), a standard software package for statistical data analysis.

We adopted the following estimation strategy to test our hypotheses. First, we implemented a 2-year time lag structure in our empirical setting. Then, we estimated panel Poisson models in order to obtain an initial idea of the relationship between cooperation counts, network positioning measures and firm-specific patenting activity. As our endogenous variables exhibited strong overdispersion, we then turned to a Negative Binomial model specification with random effects (cf. Sect. 6.2.2). This generalization of the Poisson model allows for overdispersion by including an individual, unobserved effect in the conditional mean (Schilling and Phelps 2007, p. 1119). In the next step, we estimated both fixed effects and random effects models. Usually the Standard Hausman Test (1978) is used to decide which results to interpret. In this analysis, most fixed effects and random effects estimates are consistent. In a final step, we ran consistency checks to ensure the robustness of our results by using a 1-year time lag structure.

11.4.2   Estimation Results

The presentation and discussion of our empirical findings is centered on the Negative Binomial model for panel count data reported in Table 11.3. Robustness of our findings is ensured by additional estimation results reported in Table 11.4. Results from both estimation techniques (fixed effects and random effects) are reported in the tables below.

Table 11.3 includes information on the total of four models. In addition to a baseline model (i.e. BL Model), there is one model that includes the network clustering coefficient (i.e. Model I), one model that comprises the overall average path length indicator (i.e. Model II), and one model that accounts for small-world properties of networks (i.e. Model III). We did not specify a full model that incorporates path-length, clustering and small-world indicators simultaneously because we are primarily interested in testing the relatedness between three distinct and structurally quite different network topologies and firm-level innovativeness. At the same time we face the risk of running into methodological problems when including all three variables in one estimation model. Potential methodological extensions and refinements of the empirical setting are discussed in Sect. 14.2.

The baseline model (cf. Table 11.3, BL Model) provides results for firm-level controls (i.e. firm age & firm age squared), cooperation-related controls (i.e. cooperation counts & cooperation funding) and overall network level control variables (i.e. network size & network density). Results from a random effects specification (time lag, t-2) reveal a positive and significant coefficient for cooperation counts (cf. Table 11.3). This should be viewed with great caution because the fixed effects specification fails to show a positive and significant relationship between cooperation counts and firm innovativeness. The same is true for both the fixed effects and the random effects model with a time lag t-1 (cf. Table 11.4). The situation looks fairly different for overall network control variables, especially in terms of network size. Estimation results (cf. Table 11.3, FE & RE; Table 11.4,

Table 11.3 Estimation results – clustering, reach and small-world properties; patent applications, time lag (t-2)

Variables	Estimation results							
	Baseline model		Model (I)		Model (II)		Model (III)	
	Fixed effects	Random effects	Fixed effects	Random effects	Fixed effects	Random effects	Fixed effects	Random effects
firmage	.01022012	.01024719	.02827257	.02714577	.02608898	.02586384	.02002634	.01947403
firmage_sq	-.00073156	-.00052259	-.00118264	-.00095319	-.00111437	-.00091046	-.00098806	-.00076586
coopcnt	.05714882	.09544829**	.05419512	.09080471**	.05701504	.09396949**	.05424240	.09156302**
coopfund	.00000789	.00001272	.00000904	.00001399	.00000728	.00001199	.00000962	.00001464
nw_size	-3.8560791***	-4.0591897***	-4.2459032***	-4.4569542***	-3.0717744***	-3.1756426***	-4.8507463***	-5.0871176***
nw_density	-3.9025440***	-3.7342787***	-5.4885633***	-5.4673971***	-4.5450498***	-4.522689***	-4.8823746***	-4.7819077***
nw_wclust			1.1088603***	1.1611018***				
nw_areach					-.28236740*	-.31326415*		
inter_sw							.44901422***	.46640724***
_cons	1.4738306***	1.4692506***	1.0165371**	1.0050505**	2.0200482***	2.0895598***	1.100405**	1.0870732**
ln_r_cons	-.13420195	-.13420195	-.13538645	-.13538645	-.1364538	-.1364538		-.1329913
ln_s_cons	-.97386831***	-.97386831***	-.97281046***	-.97281046***	-.96994511***	-.96994511***		-.97476673***
chi²	24.34534	30.054709	29.330156	35.970943	26.690456	33.130402	30.034183	36.51918
ll	-1,470.5418	-2,093.1967	-1,467.0863	-2,089.2416	-1,469.0602	-2,091.2694	-1,466.9632	-2,089.2587
aic	2,955.0837	4,204.3934	2,950.1725	4,198.4833	2,954.1204	4,202.5388	2,949.9263	4,198.5174
bic	2,991.3389	4,255.573	2,991.607	4,255.3495	2,995.5549	4,259.405	2,991.3608	4,255.3836
N	1,312	2,179	1,312	2,179	1,312	2,179	1,312	2,179

Legend: * $p < .1$; ** $p < .05$; *** $p < .01$
Source: Author's own calculations

Table 11.4 Robustness check – clustering, reach and small-world properties; patent applications, time lag (t-1)

Variables	Estimation results							
	Baseline model		Model (I)		Model (II)		Model (III)	
	Fixed effects	Random effects	Fixed effects	Random effects	Fixed effects	Random effects	Fixed effects	Random effects
firmage	.00994378	.00829882	.03733312 *	.03220746	.00423757	.00460507	.03259915	.02808188
firmage_sq	−.00053489	−.00034804	−.00105358	−.00081466	−.00041663	−.00026863	−.00089849	−.0006686
coopcnt	−.01811569	.01934418	−.01818331	.01694328	−.01846864	.01938771	−.01811078	.01728223
coopfund	.000009877	.00001396	.00001412	.00001792	.00001004	.0000141	.000017	.00002073
nw_size	−2.9475613***	−3.1684482***	−3.3421082***	−3.5298766***	−3.2089678***	−3.3685447***	−4.2628185***	−4.468513***
nw_density	−2.8696261***	−2.88051***	−4.8546***	−5.0429677***	−2.6009588***	−2.6603955***	−4.1461309***	−4.3348133***
nw_wclust			1.5707897***	1.5695806***				
nw_areach					.11411269	.08427262		
inter_sw							.81154585***	.81613743***
_cons	1.0926888***	1.1293445***	.35436373	.4183161	.85300282 *	.94940078 *	.18453871	.2419253
ln_r_cons		−.09498949		−.09417647		−0.0937929		−.08599318
ln_s_cons		−.91609186***		−.91730297***		−.91830232***		−.92419009***
chi²	15.736744	17.259548	30.991406	33.142983	16.399322	17.688214	44.277213	46.67405
ll	−1,674.1401	−2,339.2755	−1,665.8197	−2,330.532	−1,673.8832	−2,339.1265	−1,659.2927	−2,323.8767
aic	3,362.2801	4,696.5511	3,347.6395	4,681.064	3,363.7664	4,698.2529	3,334.5855	4,667.7534
bic	3,399.6523	4,748.645	3,390.3506	4,738.9461	3,406.4775	4,756.135	3,377.2966	4,725.6355
N	1,539	2,412	1,539	2,412	1,539	2,412	1,539	2,412

Legend: * $p < .1$; ** $p < .05$; *** $p < .01$
Source: Author's own calculations

FE & RE) provide empirical evidence for a negative relatedness between network size and firm innovativeness.

To start with, the estimation results are robust for both time lags (Table 11.3, time lag t-2; Table 11.4, time lag t-1) and for both estimation techniques (i.e. random effects & fixed effects models). Coefficient estimates for network clustering are positive and highly significant at the 0.01 level (cf. Table 11.3, Model I; Table 11.4, Model I). Estimation results for average path length are negative and show a minor significance at the 0.1 level (cf. Table 11.3, Model II) and no significance in the robustness check (cf. Table 11.4, Model II). Finally, coefficient estimates for the small-world indicator are positive, consistent over all specifications and highly significant at the 0.01 level (cf. Table 11.3, Model III; Table 11.4, Model III). In summary, our estimation results provide strong empirical support for Hypotheses H2 & H3 but only minor support for Hypothesis H1.

11.5   Discussion and Implications

In this section we were primarily interested in testing the relatedness between three distinct, structurally different, network topologies and firm-level innovativeness.

Our results for the overall average path lengths (Hypothesis H1) are as expected and in line with previous empirical findings (Schilling and Phelps 2007; Fleming et al. 2007). Both studies report a negative[15] and, in most cases, highly significant correlation between the average path length at the overall network level and firm innovativeness. Schilling and Phelps (2007) pay little attention to these individual effects. Fleming et al. (2007, p. 949) conclude: "Shorter path length [...] correlates with an increase in subsequent patenting." However, in our setting the significance-level for this coefficient is fairly low and a robustness check did not support the initially identified effect.

Our results for the clustering coefficient are in line with our theoretical expectations (Hypothesis H2), however, it is interesting to note that the findings for the individual clustering of the German laser industry innovation network are not in line with previous empirical findings in several respects. Schilling and Phelps (2007, p. 1122) report in four out of six empirical settings a negative but insignificant effect. Similarly, the results of Fleming and colleagues (2007, p. 948) reveal negative and significant coefficient estimates. This is an issue that clearly calls for clarification and further research.

Last but not least, we take a look at a network's small-world properties. Firstly, the descriptive analysis shows that the German laser industry network clearly fulfills the small-world criteria according to Watts and Strogatz (1998). Moreover, results are suggestive of an increasing solidification of small-world properties over

---

[15] Note that Fleming and colleagues (2007) use an inverse path length measure. Thus, the coefficient estimates are positive.

time. Secondly, in our estimation, results clearly support Hypothesis H3 and provide empirical evidence for a positive relatedness between a network's small-world nature and a firm's subsequent innovativeness. This is in sharp contrast to the findings of Fleming et al. (2007, p. 949); the authors conclude: "The small world effect is not observed in our data." However, our results are in line with previous findings by Schilling and Phelps (2007) who summarize their findings as follows: "[. . .] networks that have both the high information transmission capacity enabled by clustering, and the high quantity and diversity of information provided by reach, should facilitate greater innovation by firms that are members of the network" (Schilling and Phelps 2007, p. 1124).

This empirical analysis has several important implications for both managers and policy makers. Most noteworthy is the recognition that the network topology itself seems to affect the innovative performance of firms at the micro-level in multiple ways. In other words, analyzing firm-specific cooperation patterns is necessary but not sufficient for a comprehensive understanding of a firm's innovative performance. Another important implication is that regional innovation networks can significantly gain in effectiveness when they concurrently show high clustering and short average path lengths. Moreover, regional networks should have a certain degree of openness in a sense that trans-regional linkages should be established and maintained.

The limitations of this analysis are the subject of a discussion in Sect. 13.2. In addition, we outline some fruitful avenues for further research into large-scale networks in Sect. 14.2.

References

Ahuja G (2000) Collaboration networks, structural hole, and innovation: a longitudinal study. Adm Sci Q 45(3):425–455

Baum JA, Shipilov AW, Rowley TJ (2003) Where do small worlds come from? Ind Corp Chang 12 (4):697–725

Borgatti SP, Everett MG, Freeman LC (2002) Ucinet for windows: software for social network analysis. Analytic Technologies, Harvard

Borgatti SP, Everett MG, Johnson JC (2013) Analyzing social networks. Sage, London

Buenstorf G (2007) Evolution on the shoulders of giants: entrepreneurship and firm survival in the German laser industry. Rev Ind Organ 30(3):179–202

Burt RS (1992) Structural holes: the social structure of competition. Harvard University Press, Cambridge

Cameron CA, Trivedi PK (2009) Microeconometrics using Stata. Stata Press, College Station

Cassi L, Zirulia L (2008) The opportunity cost of social relations: on the effectiveness of small worlds. J Evol Econ 18(1):77–101

Cohen WM, Levinthal DA (1989) Innovation and learning: the two faces of R&D. Econ J 99 (397):569–596

Cohen WM, Levinthal DA (1990) Absorptive capacity: a new perspective on learning and innovation. Adm Sci Q 35(3):128–152

Corrado R, Zollo M (2006) Small worlds evolving: governance reforms, privatizations, and ownership networks in Italy. Ind Corp Chang 15(2):319–352

Fleming L, King C, Juda AI (2007) Small worlds and regional innovation. Organ Sci 18(6):938–954

Gulati R (1998) Alliances and networks. Strateg Manag J 19(4):293–317

Hausman JA (1978) Specification tests in econometrics. Econometrica 46(6):1251–1271

Hoffmann WH (2005) How to manage a portfolio of alliances. Long Range Plan 38(2):121–143

Jaffe AB (1989) Real effects of academic research. Am Econ Rev 79(5):957–970

Jaffe AB, Trajtenberg M, Henderson R (1993) Geographic localization of knowledge spillovers as evidenced by patent citations. Q J Econ 108(3):577–598

Katkalo VS, Pitelis CN, Teece DJ (2010) Introduction: on the nature and scope of the dynamic capabilities. Ind Corp Chang 19(4):1175–1186

Kennedy P (2003) A guide to econometrics. Blackwell, Oxford

Lane PJ, Lubatkin MH (1998) Relative absorptive capacity and interorganizational learning. Strateg Manag J 19(5):461–477

Lane PJ, Koka BR, Pathak S (2006) The reification of absorptive capacity: a critical review and rejuvenation of the construct. Acad Manag Rev 31(4):833–863

Milgram S (1967) The small-world problem. Psychol Today 1(1):60–67

Newman ME (2010) Networks – an introduction. Oxford University Press, New York

Newman ME, Strogatz S, Watts D (2001) Random graphs with arbitrary degree distributions and their applications. Phys Rev E 64:1–17

Schilling MA, Phelps CC (2007) Interfirm collaboration networks: the impact of large-scale network structure on firm innovation. Manag Sci 53(7):1113–1126

Stata (2007) Stata statistical software: release 10. StataCorp LP, College Station

Stuart TE (2000) Interorganizational alliances and the performance of firms: a study of growth and innovational rates in a high-technology industry. Strateg Manag J 21(8):791–811

Teece DJ, Pisano GP, Shuen A (1997) Dynamic capabilities and strategic management. Strateg Manag J 18(7):509–533

Uzzi B, Spiro J (2005) Collaboration and creativity: the small world problem. Am J Sociol 111 (2):447–504

Uzzi B, Amaral LA, Reed-Tsochas F (2007) Small-world networks and management science research: a review. Eur Manag Rev 4(2):77–91

Van Den Bosch FA, Volberda HW, De Boer M (1999) Coevolution of firm absorptive capacity and knowledge environment: organizational forms and combinative capabilities. Organ Sci 10 (5):551–568

Wasserman S, Faust K (1994) Social network analysis: methods and applications. Cambridge University Press, Cambridge

Watts DJ, Strogatz SH (1998) Collective dynamics of 'small-world' networks. Nature 393 (6684):440–442

White S (2005) Cooperation costs, governance choice and alliance evolution. J Manag Stud 42 (7):1383–1413

Winter S (2000) The satisficing principle in capability learning. Strateg Manag J 21:981–996

Zahra SA, George G (2002) Absorptive capacity: a review, reconceptualization, and extension. Acad Manag Rev 27(2):185–203

Chapter 12
Network Positioning, Co-Location or Both?

An invention is a major one if it provides the basis for
extensive applications and improvements [...]
(Simon Kuznetz 1971)

Abstract Previous research indicates that firm innovativeness can either be determined by a firm's position within the network dimension or by its position within the geographical dimension. Integrative studies addressing both distinct and combined proximity effects remains rare (cf. Whittington et al. 2009). Thus, we address in this Chapter the following research question: Are firm-level innovation outcomes positively or negatively related to network positioning effects, geographical co-location effects or combined proximity effects; and if the latter case is true, are the combined effects substitutional or complementary in nature? Panel data count models with fixed and random effects were used to analyze a firm's innovative performance as measured by patent application counts. This last empirical analysis is organized as follows: We start with a short introduction in Sect. 12.1. Next, we provide a brief discussion of theoretical background in Sect. 12.2. In Sect. 12.3 we introduce our conceptual framework and derive our hypotheses. In Sect. 12.4 we introduce the data and methods used. Next, we outline the estimation strategy and report our empirical results in Sect. 12.5. Finally, we discuss our findings and conclude with a number of critical remarks in Sect. 12.6.

12.1   Introduction: Proximity and Firm Innovativeness

In this analysis we seek to disentangle the relationship between network positioning, geographical co-location and firm innovativeness.[1] The knowledge-based approach started to thrive in economics in the early 1980s. The neo-Schumpeterian school of thought explicitly emphasized the central role of knowledge and innovation as drivers of economic change and prosperity (Hanusch and Pyka 2007b). From this perspective, knowledge is no longer considered purely a public good but rather as a scarce and highly valuable firm-specific resource. Firms have to take action, spend time and resources, and develop routines and capabilities to successfully tap external stocks of knowledge. The creation of novelty is regarded as a collective process that involves knowledge exchange and learning processes among mutually interconnected economic actors, each of whom seeks to improve their own imperfect knowledge base and accomplish their very individual goals (Hanusch and Pyka 2007a). These actors are embedded in socio-economic systems of innovation (Lundvall 1992; Nelson 1992) that are complex and dynamic in nature (Fagerberg 2005). Quite recently it has been argued that the proximity concept originally proposed by Boschma (2005b) can be regarded as an integral part and extension of the evolutionary economic approach (Boschma and Frenken 2010).

The knowledge-based view in management science emerged just a few years later. Scholars started to emphasize the strategic nature of knowledge and pointed to the fact that both knowledge accessing and learning processes are crucial for firms to gain competitive advantages and outperform competitors (Nonaka 1991; Kogut and Zander 1992; Grant 1996; Coff 2003). According to this view, firms have to build up a certain degree of absorptive capacity (Cohen and Levinthal 1990), avoid learning races (Amburgey et al. 1996) and reduce causal ambiguities (Simonin 1999) to make successful use of external knowledge sources. These challenges are essentially attributable to the tacit nature of non-codified and context-specific knowledge (Polanyi 1958; Polanyi 1967) that underlies innovation processes at the micro-level. With the emergence of the proximity concept it has been argued that proximity in all its facets can enhance a firm's ability to access new stocks of knowledge and generate novelty in the form of innovation (Amin and Wilkinson 1999).

---

[1] This section is based on a joint research project conducted together with Dr. Peter Boenisch, chair for Statistics and Econometrics at Martin Luther University in Halle and Dr. Iciar Dominguez Lacasa, Department of Structural Economics at the Halle Institute for Economics (Kudic et al. 2010). Moreover we thank Dr. Michael Schwartz and Dr. Marco Sunder for reviewing the paper and providing critical comments and helpful suggestions. We have benefited from comments from the audience at the 13th International Schumpeter Society Conference in 2010 in Aalborg, Denmark and the 36th EIBA Annual Conference in 2010 in Porto, Portugal. I take full responsibility for the content or any errors in this completely revised version of the initial paper.

We are surely not the first to address the direct relationship between proximity and firm innovativeness. An excellent overview of contemporary research in this area is provided by the study carried out by Knoben and Oerlemans (2006). Quite recently, a few excellent theoretical (e.g. Boschma 2005b; Torre and Rallet 2005; Visser 2009; Boschma and Frenken 2010) and empirical studies (e.g. Oerlemans et al. 2001; Oerlemans and Meeus 2005; Owen-Smith and Powell 2004; Whittington et al. 2009) have started to address both distinct and combined prox-imity affects. Nonetheless, we still face more questions than answers.

For instance, Boschma (2005a) calls for a clear analytical separation of distinct proximity dimensions, a more dynamically oriented proximity perspective and stronger recognition of both positive and negative effects of proximity on innova-tion. However, these issues have several far-reaching implications. Due to the conceptual ambiguity of the proximity concept, the underlying mechanisms are not clearly assigned to one specific proximity dimension. Thus, we are still lacking an in-depth understanding and a clear separation of the mechanisms that foster or hamper innovation processes at the firm level. In addition, we have a rather vague idea of how and why one proximity dimension affects another and the underlying logic of combined proximity effects is not yet sufficiently understood. Quite recently some pioneering studies have started to address these questions empirically (most notably: Owen-Smith and Powell 2004; Whittington et al. 2009). Nonethe-less, it should be noted that most previous studies are cross-sectional in nature. This neglects the dynamic nature of the mechanisms that underlie the various types of proximity dimensions. Moreover, previous studies inherently imply that proximity is positive per se; the "dark side" of proximity is widely ignored. Finally, the majority of previous empirical studies are based on data from the biotech industry. However, knowledge exchange, learning and innovation processes can significantly differ across industries due to differences in the degree of the industry's techno-logical maturity, different industry life-cycle stages and differences in firm size distribution.

Consequently, this study contributes to the existing body of literature in several ways. Firstly, by focusing on two proximity dimensions – network proximity and geographical proximity – we seek to deepen our understanding as to how distinct, or potentially, combined proximity effects relate to firm-level innovation outcomes. Secondly, in response to Boschma's critique we provide an evolutionary proximity framework and apply longitudinal data and panel estimation techniques to account for the dynamic nature of proximity in all its facets. In doing so, we seek to understand the underlying mechanisms that determine both distinct as well as combined proximity effects. Thirdly, we supplement existing research by providing new empirical evidence from a unique panel dataset for the entire population of 233 German laser source manufacturers between 1990 and 2010.

In a nutshell, inspired by the conceptual framework of Boschma (2005a) and the empirical study of Whittington et al. (2009) and supplemented by our own consid-erations, we raise the following research question: Are firm-level innovation out-comes of German laser source manufacturers related to network proximity effects, to geographical co-location effects or to combined proximity effects; and if so, are the effects positively or negatively related to firm innovativeness at later points in time?

12.2   Theoretical Background

12.2.1   The Multifaceted Character of Proximity

Over the past few years, scholars in the field of economics, sociology, geography and management science have significantly improved our understanding of how proximity can improve a firm's ability to tap into new knowledge sources, learn to recombine existing knowledge stocks and finally generate new and commercializable goods and services. Firms are simultaneously exposed to a variety of proximity dimensions such as institutional proximity, organizational proximity, cultural proximity, technological proximity, network proximity and geographical proximity (cf. Knoben and Oerlemans 2006, p. 71). In the most general sense, proximity can be defined as "[...] being close to something measured on a certain dimension" (Knoben and Oerlemans 2006, pp. 71–72). One of the main issues that is common to all literature on proximity is the conceptual ambiguity of previous approaches. In this context Knoben and Oerlemans (2006, p. 71) criticize the fact that previous research has failed to provide a clear separation of proximity dimensions which is still reflected in conceptual overlaps across many proximity dimensions.

We follow the proximity framework proposed by Boschma (2005b) for several reasons. Firstly, the framework allows five proximity dimensions to be clearly defined and separated: cognitive, organizational, social, institutional and geographical proximity (Boschma 2005b, p. 62). He stresses the lack in understanding combined proximity effects and emphasizes that proximity in all its facets can both facilitate and impede knowledge access and learning processes over time. Thus the framework enables both the positive as well as the negative impact of proximity on firm-level innovation outcomes to be explained. Secondly, the proposed proximity framework can be regarded as an integral part of the evolutionary economic approach (Boschma and Frenken 2010, p. 121). The integration of the concept broadens the analytical scope of the evolutionary approach by explicitly considering several types of proximity dimensions (ibid). Moreover, it also paves the way for a more dynamic and process-oriented understanding of the proximity concept. Thirdly, proximity as an analytical concept allows multiple proximity dimensions to be incorporated into an explanatory framework (Boschma and Frenken 2010, p. 124). The proximity dimensions are clearly separated and thus independent of each other. This implies, however, that one can reduce as well as extend the list of relevant proximity dimensions without changing the meaning of each dimension (ibid).

In summary, the analytical proximity framework originally proposed by Boschma (2005b) and then extended upon by Boschma and Frenken (2010), clearly promotes a more process-oriented understanding of how changes in proximity affect innovation outcomes over time. It allows the interplay between selected proximity dimensions to be analyzed and provides potential explanations for both complementary and substitutional effects. Finally, the framework offers the possibility of analyzing the relatedness between firm innovativeness and individual proximity dimensions and provides a solid basis for examining whether combined proximity effects are positively or negatively related to firm innovation outcomes at later points in time.

12.2.2 Network Proximity and Firm Innovativeness

Now we take a closer look at the network proximity dimension. This type of proximity is frequently referred to as relational or social proximity (Coenen et al. 2004). The social proximity concept is strongly influenced by social capital and embeddedness literature (Laumann et al. 1978; Granovetter 1985; Uzzi 1996; Uzzi 1997; Granovetter 2005). According to this perspective, economic actions and outcomes are influenced by the context in which they occur (Uzzi 1996; Gulati 2007). Boschma (2005b) defines social proximity "[...] in terms of socially embedded relations between agents at the micro-level." The use of this proximity dimension requires an in-depth specification of at least three constituent features: the agents, the type of relations that connect these agents and the system boundaries that define the scope and size of the overall network.

For the purpose of this paper we focus on interorganizational innovation networks (Pyka 2002) consisting of all German laser source manufacturers (LSMs) and laser-related public research organizations (PROs) that were actively operating in the field of laser source research and production between 1990 and 2010. In the most general sense, these actors can be interconnected in multiple ways and can exchange knowledge either through informal or formal relationships. According to Pyka (1997, p. 210) the former encompasses "[...] any action that can contribute to disclosure, dissemination, transmission and communication of knowledge." The latter addresses a broad variety of structural forms ranging from short term contractual alliances and minority alliances, characterized by an intermediate degree of hieratical control, to long-term equity alliances such as joint ventures (Gulati and Singh 1998). All formalized partnerships, however, stipulate that all parties involved agree upon more or less formalized obligations, rights and common goals. We focus on a specific type of innovation network that is constructed on the basis of nationally or supra-nationally funded R&D cooperation projects.[2] The

---

[2] We are not the first to use data on nationally or supra-nationally R&D cooperation projects to construct knowledge-related innovation networks (cf. Broekel and Graf 2011; Fornahl et al. 2011; Scherngell and Barber 2009, 2011; Cassi et al. 2008).

reasoning for this is straightforward. All project partners have to agree on contract clauses that aim to improve knowledge exchange among project partners and initiate innovation activities (Fornahl et al. 2011; Scherngell and Barber 2009). At the same time the concretization of node and tie dimensions outlined above specifies the boundaries of the network under investigation.

Previous research provides strong evidence that not only structural network characteristics such as network density, structural holes, or structural equivalence (Gulati et al. 2000, p. 205) but also a firm's structural position within the overall industry network can significantly affect various dimensions of firm-level performance (Baum et al. 2000; Stuart et al. 1999; Zaheer and Bell 2005). It has been demonstrated that a firm's occupation of strategically important network positions can improve its ability to access external knowledge sources (Grant and Baden-Fuller 2004; Buckley et al. 2009) and facilitate interorganizational learning processes (Hamel 1991; Schoenmakers and Duysters 2006; Nooteboom 2008). In a similar vein, previous studies have explored the importance of structural network characteristics and various types of network positions in a firm's innovative performance (Powell et al. 1996; Stuart 1999; Stuart 2000; Fornahl et al. 2011). The potential benefits of a firm's network position are closely related to the overall network topology. Some scholars have argued that brokerage positions in sparsely connected networks are the most beneficial – the "structural hole theory" (Burt 1992) – whereas others have stressed the importance of high nodal degrees at the actor level and closely interconnected overall network structures – the "closure theory" (Coleman 1988).[3] For the purpose of this study we focus particularly on the latter stream of research.

However, not only the positive but also the negative impact of a firm's network embeddedness on performance outcomes has been the subject of debate over the past few years (Boschma 2005b). By now it is well-recognized that a firm's position within the innovation network can have a positive impact on its innovative performance at subsequent points in time (Uzzi 1997). However, after a certain point the positive effects of social proximity may move in the opposite direction and have an adverse effect on learning and innovation (Boschma 2005b, p. 66). This phenomenon is referred to as "overembeddedness" (Uzzi 1997). The main argument behind this concept is straightforward. Too much social proximity can cause a lock-in effect in a sense that actors remain in an established web of habitual partnerships. Such a closed network system generates opportunity costs because the actors involved isolate themselves from other firms and organizations with fresh and novel ideas (Boschma 2005b, p. 66). Others have pointed to the fact that some organizations face considerable difficulties in dissolving old relationships and forming new network ties (Kim et al. 2006). These authors coined the term "network inertia" to address the persistent organizational resistance to changing its interorganizational network. Kim and colleagues (2006, p. 706) argue that

---

[3] By now, both theoretical (Burt 2000, 2005) and empirical studies (Rowley et al. 2000) accept the partial compatibility of both theories (cf. Sect. 2.5.4).

change is mainly influenced by four types of constraints: internal constraints, network tie-specific constraints, network position-specific constraints, and external or environmental constraints. This concept has its intellectual roots in the structural inertia theory in organizational ecology (Hannan and Freeman 1984) and is clearly evolutionary in nature. The process-oriented network inertia concept explains negative innovation outcomes of firms not primarily because of overembeddedness issues but rather because some firms cannot react and adapt fast enough to new conditions and needs. In other words, a firm may face a situation in which the formation of connections to new cooperation partners that would bring in innovation stimuli is likely to be seriously delayed or even entirely impaired whereas obsolete linkages simply cannot be resolved. Following this reasoning one could argue that a firm can face a situation in which it is not necessarily overembedded but rather missembedded.

12.2.3 Geographical Proximity and Firm Innovativeness

Next, we focus on the spatial or geographical dimension of proximity. Oerlemans and Meeus (2005, p. 94) point out that this body of research can be grouped into two categories: one which focuses on spatial (or face-to-face) interaction and interactive learning (Saxanian 1990; Maskell and Malmberg 1999) and one focusing on spatially mediated knowledge spillovers (Feldman 1993; Audretsch and Feldman 1996, 2003). For the purpose of this paper we stick to the latter perspective for two reasons. On the one hand, the knowledge spillovers perspective allows us to define geographical proximity in a quite restrictive manner and isolate the spatial proximity dimension from other proximity dimensions (Boschma 2005b, p. 69). On the other hand, the knowledge spillover perspective acknowledges the partially non-rival, dynamic and cumulative character of knowledge (Oerlemans and Meeus 2005, p. 94) and puts forward the argument that knowledge tends to spill over locally between firms of the same industry – so-called intra-industry or MAR externalities (Marschall 1890; Arrow 1962; Romer 1986) – or between firms of different industries – so-called inter-industry or Jacobs externalities (Jacobs 1969). Due to the aim and the scope of this study we focus specifically on intra-industry knowledge spillover.

   This perspective stresses that proximity influences a company's ability to benefit from knowledge spillover stemming from research and development activities taking place outside the boundaries of the firm (Audretsch 1998). In the early 1990s a vibrant field of research started to address the spatial dimension of knowledge and innovation by introducing novel methods for measuring the extent of local knowledge spillovers and innovative activities (Audretsch 1998, p. 22). Empirical studies from this strand of research suggest that physical proximity of firms to external knowledge sources enhances innovative and economic performance (Jaffe 1989; Audretsch and Feldman 1996, 2003; Audretsch and Dohse 2007). However, they have also demonstrated that spatial knowledge accumulation effects and innovation activities can be strongly determined by the knowledge intensity of the industry and stage of the industry life cycle (Audretsch 1998).

Keeping in mind the previous considerations, we define geographical proximity as a concept that "[...] refers to the spatial or physical distance between economic actors, both in its absolute and relative meaning" (Boschma 2005b, p. 69). This perspective is consistent with a related concretization that defines geographical proximity as "[...] kilometric distance that separates units (e.g. individuals, organizations, towns) in geographical space" (Torre and Rallet 2005, p. 49) and, at the same time, provides a solid basis for taking a closer look at the link between geographical co-location, knowledge and innovation. The main argument is that any firm located in an agglomeration area can benefit from local knowledge spillovers as long as geographical openness of the agglomeration is ensured (Boschma 2005b, p. 69). The mechanisms that generate the knowledge spillover are not relational in nature. In other words, knowledge spillovers can occur regardless of whether firms in a region are interconnected by a formal relationship or not (Boschma 2005b, p. 69). Geographical proximity has to be defined in such a restrictive manner as to allow a clear separation of other proximity dimensions (ibid), especially relational knowledge transfer mechanisms such as informal social relationships at an interpersonal level.

However, geographical proximity does not have a positive effect on knowledge transfer and learning processes per se. Boschma (2005b, p. 70) stresses the risk that spatial lock-in effects and a lack of openness to the outside world can result in situations in which local knowledge quickly becomes outdated and knowledge-based agglomeration effects become increasingly eroded over time. Thus, firms in closed agglomeration areas become increasingly inward-looking and isolate themselves from the other actors in the industry. In summary, the geographical proximity dimension defined in this way addresses positive but also negative consequences of local knowledge spillovers on firm innovativeness due to a firm's co-location to other organizations.

12.2.4 Addressing Combined Proximity Effects and Firm Innovativeness

In the real world, it is very unlikely that only one proximity dimension affects a firm's innovative performance in isolation. Instead, firms are simultaneously exposed to multiple mutually interdependent proximity dimensions. Below we focus exclusively on combined proximity effects between two proximity dimensions: network proximity and geographical proximity.

According to Whittington et al. (2009, pp. 97–98), there are theoretically three ways in which combined proximity effects can affect a firm's innovation outcome. Firstly, independent effects of network and geographical proximity on innovation would imply that both proximity dimensions influence innovation through autonomous mechanisms. Secondly, substitutional effects of network and geographical proximity on innovation are based on the notion that one proximity dimension can compensate for a lack of another proximity dimension. Finally, complementary effects of network and geographical proximity on innovation imply that these two dimensions are supplementary in nature due to extra-additive effects. Unfortunately, Whittington and colleagues (2009) do not clearly address and separate the mechanisms that underlie these combined proximity effects. This is precisely the point at which our conceptual framework comes into play.

12.3 Conceptual Framework and Hypotheses Development

12.3.1 Specifying Distinct and Combined Proximity

Our conceptual framework (cf. Fig. 12.1) builds upon the theoretical considerations outlined above and aims to contribute to an in-depth understanding of the mechanisms that underlie both distinct and, especially, combined proximity effects. The first requires a concretization of the elementary building blocks in our framework. The framework consists of four elements – (**I**) network proximity (**II**) geographical proximity, (**III**) combined proximity, and (**IV**) innovation outcomes.

To start with, we outline our notion of network proximity. We argue that a firm with a high number of direct partners, irrespective of whether these direct partners are themselves well-connected or not, has a high level of network proximity. We focus on a firm's degree of connectedness for the following reasons. Firstly, a firm's nodal degree is quite a simple and straightforward network concept that reflects the full range of its external knowledge channels. Secondly, densely connected actors are highly visible and well-recognized by other actors in the network (Wasserman and Faust 1994, p. 179).

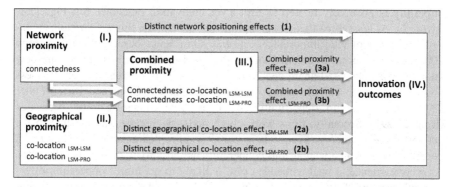

Fig. 12.1 Conceptual framework – distinct and combined proximity and firm-level innovation output (Source: Author's own illustration)

Similarly, geographical proximity has to be specified. One way to accomplish this task is to focus on the physical distances between firms in a well-specified population. Thus, a firm with a short average distance to all other firms at the same stage of the industry value chain has a high level of geographical proximity (Whittington et al. 2009). However, not only firms but also universities or other public research organizations (PROs) are an important source of new technological knowledge (Agrawal 2001, p. 285). It has been argued that PROs follow quite different rules for the dissemination and use of scientific findings than profit-oriented firms (Owen-Smith and Powell 2004, p. 7). In other words, there is a qualitative difference as to whether an LSM is co-located to other firms at the same stage of the industry value chain, or to other laser-related public research organizations carrying out either basic or applied research. In line with Whittington et al. (2009) we consider a second type of geographical proximity that reflects a firm's average distance to all laser related PROs in the sample. The combined proximity dimension captures a firm's simultaneous positioning in both the network space and the geographical space.

Finally, a firm's innovation output has to be clarified. The Oslo Manual (OECD 2005) differentiates between four types of innovation: "product innovation", "process innovation", "organizational innovation", and "marketing innovation". We focus here on all kinds of novel ideas generated by laser-related firms that are truly new to the market and thus at least theoretically patentable.

12.3.2   The Link Between Proximity Effects and Firm Innovativeness

Previous research has significantly contributed to our understanding of how a firm's structural embeddedness and network positioning affects the innovation generating process (Shan et al. 1994; Powell et al. 1996; Ahuja 2000; Owen-Smith and Powell 2004; Gilsing et al. 2008). These findings leave us to suppose that a firm's network proximity is positively related to firm-level innovation outcomes (cf. Fig. 12.1, Arrow 1). At least three theoretical arguments substantiate this assumption.

Firstly, it is of vital importance, especially in science-based industries (Grupp 2000), to have access to external knowledge stocks and to be able to acquire new knowledge stocks (Al-Laham and Kudic 2008). A high degree of connectedness provides access to complementary knowledge sources (Grant and Baden-Fuller 2004) and opens up opportunities of interorganizational learning processes (Hamel 1991). Secondly, firms with a high nodal degree are most visible in the network. Agency theory (Spence 1976, 2002) implies that an above-average nodal degree signalizes an advantageous cooperation opportunity to other network actors. As a consequence, a well-embedded firm is likely to get more cooperation offers than other actors (Hanneman and Riddle 2005). A broad opportunity-set of potential cooperation partners increases the probability of finding the right partner when required. Finally, well-embedded firms are characterized by a comparably high level of alliance experience that allows alliance capabilities to be built up over time (Kale et al. 2000, p. 750; Schilke and Goerzen 2010). The implementation of cooperation routines saves costs (Zollo et al. 2002) and increases managerial efficiency over time (Goerzen 2005). As a consequence, well-experienced firms have a higher chance of completing innovation projects successfully than firms that are cooperating for the first time. In summary, these considerations result in the formulation of our first hypothesis:

H1   A firm's network proximity is positively related to its innovative performance at later points in time.

Next, we turn our attention to the relationship between a firm's geographical proximity to other firms at the same stage of the industry's value chain and its subsequent innovation outcomes (cf. Fig. 12.1, Arrow 2a). The following theoretical arguments substantiate the assumption that geographical co-location can generate localized knowledge spillovers. Knowledge spillovers provide valuable information and increase a firm's awareness of new industrial and technological trends. Consequently, firms who benefit from knowledge spillovers increase their general technological understanding. According to Breschi and Lissoni (2001) at least three underlying mechanisms are responsible for knowledge spillovers at the local level: local labor markets (Almeida and Kogut 1999; Zucker et al. 1998), local technology markets (Lamoreaux and Sokoloff 1999) and the low propensity of skilled workers to relocate in space (Breschi and Lissoni 2009). In particular, it has been argued that these mechanisms enable knowledge transmission and facilitate knowledge reuse among firms of the same industry at the local level. These considerations substantiate the assumption that a firm can benefit from regional knowledge spillovers due to its geographical closeness to other firms at the same stage of the industry value chain. Consequently we elaborate on our hypothesis:

H2a   A firm's geographical proximity to other laser source manufacturers is positively related to its innovative performance at later points in time.

Now we address a firm's proximity to other laser-related public research organizations (cf. Fig. 12.1, Arrow 2b). Jaffe (1989) and Acs et al. (1992) provide interesting empirical results concerning the contribution of knowledge spillovers from public research organizations (PROs). Using patents (Jaffe 1989) and direct counts of innovation outputs (Acs et al. 1992), these studies provide evidence for the positive impact of knowledge spillovers from universities on corporate innovation activity. According to Jaffe (1989), this effect is particularly significant in the areas of drugs and medical technology, electronics, optics and nuclear technology. At least two theoretical arguments come into play in this context. Firstly, knowledge flows out of PROs, especially universities, much more readily than it does from privately owned firms (Jaffe et al. 1993; Owen-Smith and Powell 2004, p. 7). Firms tend to retain information or implement secrecy strategies to protect their knowledge stock whereas universities and PROs tend to disseminate research findings instantly. Secondly, shortages in qualified personnel can become an existential problem for a firm. These issues become all the more significant for firms operating in science-driven and highly interdisciplinary fields of research. A firm's geographical co-location to technical universities and universities of applied science increases the firm's chance of hiring well-qualified graduates such as engineers, scientists and other experts. Drawing on these considerations we formulate the following hypothesis:

H2b A firm's geographic proximity to laser-related public research organizations is positively related to its innovative performance at later points in time.

On the one hand, the independent effects of network proximity and geographical proximity on innovation imply that both proximity dimensions influence innovation through completely unrelated mechanisms and transmission channels. In other words, the effects of positioning in the geographical and network space do not influence each other. On the other hand, it is plausible to assume that these two proximity dimensions can affect each other in various ways. Substitutional effects of network proximity and geographical proximity on innovation are based on the notion that a firm may compensate for its disadvantages in one proximity dimension through an advantageous position in another proximity dimension. Or to put it another way, firms located in remote geographical regions may compensate for their location disadvantages by fostering cooperation. In contrast, complementary effects of network proximity and geographical proximity on innovation enhancement are regarded as mutually reinforcing.

We argue that network proximity and geographical proximity are not independent but rather complementary in nature (cf. Fig. 12.1, Arrow 3a and 3b). Accordingly, being well-embedded in both proximity dimensions implies that firms can gain extra-additive innovation effects due to their advantageous positioning in both the geographical space and the network space. To exemplify this point, firms benefiting from local knowledge spillovers, in terms of improved accessibility to professionals and graduates on the local labor market, may have better qualified employees. This would allow these firms to generate greater value from their interorganizational partnerships. Against the backdrop of these considerations we formulate the following two hypotheses:

H3a Combined proximity effects of a firm's network proximity and its geographical proximity to other laser source manufactures are complementary in nature.

H3b Combined proximity effects of a firm's network proximity and its geographical proximity to other laser-related public research organizations are complementary in nature.

12.4 Data Sources, Methodological Issues and Variable Specification

We employed a unique panel dataset for the full population of 233 German laser source manufactures between 1990 and 2010 to answer the questions raised above. Four main data sources were used to conduct this study: patent data, industry data, geographical data and network data.

Patent data[4] was used to measure innovative performance at the firm level. A lot has been written about the empirical challenges of measuring innovation processes. Despite the methodological constraints related to the use of patents to measure innovation performance (Patel and Pavitt 1995), patent indicators are commonly used for analyzing innovation processes (Jaffe 1989; Jaffe et al. 1993). The use of patent data as a proxy for firm innovativeness must be viewed critically for several reasons (cf. Fritsch and Slavtschev 2007, p. 204). Nonetheless, there are good reasons that advocate the use of patent data, especially in longitudinal settings (cf. Brenner and Broekel 2011, p. 13). In accordance with previous network studies (Ahuja 2000; Whittington et al. 2009; Stuart 1999), we used annual patent counts as a proxy for innovation output. Our database (cf. Sect. 6.1.2) includes patent applications as well as patents granted by the German Patent Office and by the European Patent Office. DEPATISnet (the German Patent and Trade Mark Office's online database) and *ESPACEnet* (the European Patent Office database) were used to check results for integrity and consistency. We ended up with patent data-based innovation indicators: annual count of patent applications *[pacnt]* and annual count of patent grants *[pgcnt]*.

Industry data[5] came from a proprietary dataset containing the entire population of German laser source manufacturers between 1969 and 2005 (Buenstorf 2007). Based on this initial dataset we used additional data sources to gather information about firm entries and exits after 2005. We ended up with an industry dataset encompassing 233 laser source manufacturers throughout the period under observation.

In addition, we used two methods to identify 145 universities and public research organizations that carried out laser-related activities. We started with the "expanding selection method" according to Doreian and Woodard (1992). Using this approach, we identified 138 laser-related public research organizations. This method, however, is limited insofar as it completely ignores non-cooperating laser-related PROs. Thus, we applied a second methodological approach to solve this problem and supplement our sample. Based on a bibliometric analysis we identified all German public research organizations which published laser papers, conference proceedings or articles in academic journals over the past two decades. Several raw data sources were tapped to conduct this analysis. We ended up with a final list of

---

[4] For a description of patent data sources and data gathering procedures, see Sect. 4.2.
[5] For a detailed description of industry data, see Sect. 4.2.1.

145 laser-related PROs for the time spanning 1990 and 2010. Then, entry and exit dates were retrieved for all of the PROs in the dataset.

Industry data was used for two reasons. On the one hand, we had to specify the boundaries of the network. On the other hand, two basic firm-level control variables were recorded and included in our panel dataset: a linear firm age variable *[firmage]* as well as a squared firm age variable *[firmage_sq]*.

Geographical data for all LSMs and PROs in the sample was reconstructed over the entire observation period from 1990 to 2010 (cf. Sect. 4.2). Data from Germany's official company register ("Bundesanzeiger") was used to reconstruct the firms' current addresses and address changes for the entire observation period. We employed the ESRI ArcMap 10.0 Software package and a freely accessible geo-coding application to gather GPS coordinates (latitudes and longitudes) on an annual basis for each firm in the sample. Based on this data we set up two types of localized density measures (LD).[6] The shortest distance on a curved surface can be calculated by using a simple geographical distance formula (cf. Sect. 5.3.1, Eq. 5.12). Unlike Sorenson and Audia (2000, p. 435) we calculated the distances in kilometers by using the natural earth radius constant ($c = 6{,}378$ km) and we split the overall population into two sub-populations, LSMs and PROs. Inspired by Whittington et al. (2009) we calculated the shortest distance on a curved surface not only for each LSM to all other LSMs but also for each LSM to all PROs in our sample. After these preparatory steps, both geographical co-location measures were calculated using the localized density formula (cf. Sect. 5.3.1, Eq. 5.13). Thus, we ended up with two types of localized density measures for each firm in the sample, each of which was calculated on an annual basis i.e. *[coloclsm]* and *[colocpro]*.

We also used a simple Herfindahl-Hirschman Index (Acar and Sankaran 1999) in order to set up two geographical concentration indices at the industry level. To do so, we proceeded as follows: First, we used the planning region scheme ("Raumordnungsregionen"), commonly used in Germany to classify of territorial units for statistical purposes. This divides the territory into 97 geographical areas. Next, we generated a count variable for both types of organizations – LSMs and PROs – that represented the number of organizations per planning region and year. Then we calculated for each planning region i (with $i = 1 \ldots 97$) the relative proportion of organizations on an annual basis. Finally, two concentration indices were established by applying an HHI formula (cf. Sect. 5.3.2, Eq. 5.14). The indices moved in the direction of zero if the organizations under observation are equally dispersed throughout the geographical space; the HHI had comparably large values if some organizations were widely dispersed whereas others showed a pronounced tendency of crowding together. We ended up with two normalized indicators that allowed us to quantify the intensity of LSM crowding *[hhi_lsm]* and PRO crowding *[hhi_pro]* in the geographical space.

---

[6] This measure was originally proposed by Sorenson and Audia (2000) and applied by Whittington et al. (2009) in order to quantify distinct and combined geographical proximity measures.

Network data[7] was gathered from two official databases on publicly funded R&D collaboration projects. The first source was the *Foerderkatalog* database provided by the German federal government which encompasses information on a total of more than 110,000 completed or ongoing subsidized research projects and provides detailed information on the starting point, duration, funding and characteristic features of the project partners involved. In total, we were able to identify, for the entire population of 233 German laser source manufacturers, 416 R&D projects with up to 33 project partners from various industry sectors, non-profit research organizations and universities. The second raw data source was an extract from the *CORDIS* project database which includes a complete collection of R&D projects for all German companies which were funded by the European Commission between 1990 and 2010. In total, this database extract consisted of a project dataset with over 31,000 project files and an organization dataset with over 57,100 German organizations and roughly 194,000 international project partners. Based on this raw data, we identified 154 R&D projects with up to 53 project partners for the entire sample of German laser source manufacturers.

Finally, both cooperation data sources were used to construct interorganizational innovation networks and to calculate network indicators on a yearly basis. We decided in favor of the degree centrality concept in order to quantify a firm's network position (cf. Sect. 5.2.1, Eq. 5.1). The degree centrality measure ranges from 0 to 1 and can be compared across networks of different sizes (Wasserman and Faust 1994, p. 179). We applied the data described above to calculate several network measures. Firstly, we calculated degree centrality measures on an annual basis for each actor in the sample *[ctr_degree]*. Then, two network level variables were calculated and included in the dataset to control for the structural network characteristics at the overall network level: overall network size *[nw_size]* and overall network density *[nw_density]*. Standard algorithms implemented in UCI-Net 6.2 were used to calculate the network measures (Borgatti et al. 2002).

Finally, we take a brief look at the variable description and basic summary statistics (cf. Table 12.1). We have a total of 2,645 firm-year observations in the time span between 1990 and 2010. The average number of observations per firm amounts to 11.35. Table 12.2 shows the correlation coefficients for all variables in our empirical models.

---

[7] For a detailed description of cooperation and network data see Sect. 4.2.3.

Table 12.1 Descriptive statistics – distinct and combined proximity effects

Variable	Variable definition	Summary statistics				
		Obs.	Mean	Std. dev.	Min	Max
Endogenous variables						
papcount	Patent applications (annual count)	2,645	2.662004	17.43323	0	366
pgrcount	Patent grants (annual count)	2,645	0.339130	1.635554	0	28
Control variables						
firmage	Age of the firm	2,645	8.055955	6.800477	0	43
firmage_sq	Age of the firm, squared	2,645	111.1274	177.8146	0	1,849
hhi_lsm	Spatial LSM concentration (industry level)	2,645	0.669407	0.078613	0.587833	1
hhi_pro	Spatial PRO concentration (industry level)	2,645	0.967246	0.020086	0.893634	1
nw_size	Network size (overall network level)	2,645	0.381853	0.060200	0.240506	0.472393
nw_density	Network density (overall network level)	2,645	0.088119	0.069955	0.037300	0.440500
Geograpical proximity						
coloclsm	Geographical proximity$_{(LSM-LSM)}$	2,645	1.870039	1.846850	0.140897	10.34948
colocpro	Geographical proximity$_{(LSM-PRO)}$	2,645	1.483881	1.290756	0.422707	6.757208
Network proximity						
ctr_degree	Network centrality$_{(degree)}$	2,645	0.023768	0.060601	0	0.645000
Combined proximity						
int_clsm_deg	NW-centrality$_{(degree)}$ × Geo_proximity$_{(LSM-LSM)}$	2,645	0.042164	0.124617	0	1.665512
int_cpro_deg	NW-centrality$_{(degree)}$ × Geo_proximity$_{(LSM-PRO)}$	2,645	0.035405	0.099654	0	1.476241

Source: Author's own calculations

Table 12.2 Correlation matrix – distinct and combined proximity effects

	pgr count	pap count	firm age	firm age_sq	ghhi_lsm	ghhi_pro	nw_size	nw_density	coloc_lsm	coloc_pro	ctr_degree	int_clsm_deg	int_cpro_deg
pgrcount	1.0000												
papcount	0.6506	1.0000											
firmage	0.0105	−0.0566	1.0000										
firmage_sq	−0.0047	−0.0455	0.9276	1.0000									
hhi_lsm	−0.0930	−0.0608	−0.2262	−0.1643	1.0000								
hhi_pro	0.0657	0.0453	0.0853	0.0579	−0.5250	1.0000							
nw_size	0.0670	0.0448	0.2131	0.1603	−0.5314	0.0911	1.0000						
nw_density	−0.0832	−0.0529	−0.2006	−0.1450	0.8395	−0.6617	−0.6576	1.0000					
coloclsm	0.0696	0.0061	−0.0572	−0.0661	−0.1867	0.2381	0.1135	−0.2080	1.0000				
colocpro	−0.0113	−0.0309	−0.1038	−0.1015	−0.0644	0.0899	0.0400	−0.0817	0.8194	1.0000			
ctr_degree	0.2006	0.1769	0.0709	0.0897	0.1486	−0.1377	−0.1079	0.1705	−0.0204	0.0017	1.0000		
int_clsm_deg	0.2283	0.1418	0.0391	0.0475	0.0276	−0.0180	0.0088	0.0092	0.2602	0.1770	0.6835	1.0000	
int_cpro_deg	0.1556	0.1149	−0.0042	0.0072	0.1029	−0.1087	−0.0671	0.1102	0.1349	0.2004	0.8548	0.8201	1.0000

Source: Author's own calculations

12.5   Model Specification, Estimation Strategy and Findings

Patents take non-negative integer values. We estimated a count model in line with Ahuja (2000), Stuart (2000) and Whittington et al. (2009).[8] In doing so, we employed panel count data techniques (cf. Sect. 6.1.2) and adopted the following estimation strategy to test our hypotheses. First we estimated a Poisson model (Hausman et al. 1984) in order to obtain an initial idea of the relationship between distinct proximity effects, combined proximity effects and a firm's patenting activity. We tested the significance of overdispersion using the procedure proposed by Cameron and Trivedi (1990) and rejected the null hypothesis of no overdispersion with a p-value of 0.000. As our endogenous variables exhibited strong overdispersion, we then turned to a negative binomial model specification. Like Whittington et al. (2009) we estimated all models using a negative binomial specification. In the next step we estimated both fixed effects and random effects models. We used the Standard Hausman Test (Hausman 1978) to decide which results to interpret. The basic idea was to test the null hypothesis that the unobserved effect is uncorrelated with the explanatory variables (Greene 2003, p. 301). If the null hypothesis cannot be rejected, both fixed effects estimates as well as random effects estimates are consistent and the model of choice is the random effects model due to its higher explanatory power. Under the alternative, random effects and fixed effects estimators diverge and it is argued that the latter model is the appropriate choice (Cameron and Trivedi 2009, p. 260). Finally, we ran several consistency checks to ensure robustness of the reported results. We set up several empirical settings with different time lags and we used data on patent grants as an additional innovation measure to ensure robustness of our results.

The presentation of our empirical findings was organized as follows. We specified a total of three empirical settings to test our hypotheses. The presentation and discussion of our estimation results was centered on a panel data count model for annual patent application counts with a time lag of 2 years, estimated by using both types of estimation techniques: fixed effects (cf. Table 12.3) and random effects (cf. Table 12.4). Next, we set up an additional empirical setting with a 3-year time lag structure to ensure the robustness of our results (cf. Table 12.5, fixed effects; Table 12.6, random effects). Finally, we employed an alternative innovation proxy to cross-check results and substantiate our initial findings. More precisely, we used data on annually granted patents, again with a time lag of 2 years (cf. Table 12.7; fixed effects: Table 12.8; random effects).

Each of the three empirical settings outlined above comprise eight models. In addition to a baseline model (BL Model), there were two models addressing distinct geographical proximity effects (Model I and Model II), one model addressing network proximity effects (Model III), and a fully specified model that incorporated

---

[8] We used STATA 10.1 (Stata 2007), a standard software package for statistical data analysis.

Table 12.3 Estimation results – distinct and combined proximity effects; negative-binomial panel data count model, patent applications, time lag (t − 2), fixed effects

Variables	Estimation results							
	BL model	Model I	Model II	Model III	Model IV	Model V	Model VI	Model VII
firmage	−0.0070568	−0.0010912	−0.0070276	−0.0006389	0.0047610	0.0062211	−0.0027705	0.0086218
firmage_sq	−0.0003997	−0.0006391	−0.0004155	−0.0006083	−0.0007814	−0.0008455	−0.0005790	−0.0008974
hhi_lsm	−1.6142121	−1.7800368[*]	−1.6116064	−1.5394755	−1.7431281[*]	−1.9012164[*]	−1.5757978	−2.1537482[**]
hhi_pro	6.9178791[**]	8.4401382[**]	6.9496824[**]	6.8495703[**]	8.5342712[**]	8.1344259[**]	5.9641374[*]	7.3994988[**]
nw_size	−2.1718702[*]	−1.9266211[*]	−2.1565523[*]	−2.1466816[*]	−1.8956025	−2.0601420[*]	−2.4277106[**]	−2.5453305[**]
nw_density	−0.7358115	−0.6177375	−0.7570526	−0.9822504	−0.7300402	−0.5730219	−1.3347972	−0.8415988
coloclsm		−0.0882829[*]			−0.1002982	−0.1025061[**]		−0.1870468[***]
colocpro			−0.0285081		0.0927344		0.0155661	0.2523224[*]
ctr_degree				1.2594943[**]	1.0343977[*]	0.1957331	3.1314868[**]	3.0766937[**]
int_clsm_deg						0.7410551[*]		1.2610741[***]
int_cpro_deg							−1.1181708	−2.2410614[**]
_cons	−4.8972226	−6.1193516[*]	−4.891807	−4.9790925	−6.5260827[*]	−5.876269	−3.9917104	−5.0343896
chi²	27.068238	30.14416	27.162722	32.169524	34.62112	37.727674	33.989394	44.544114
ll	−1,469.1716	−1,467.4622	−1,469.1285	−1,466.6479	−1,465.2776	−1,463.9246	−1,465.6147	−1,460.3665
aic	2,952.3432	2,950.9243	2,954.2571	2,949.2959	2,950.5551	2,947.8493	2,951.2295	2,944.7329
bic	2,988.5983	2,992.3588	2,995.6915	2,990.7304	3,002.3482	2,999.6423	3,003.0225	3,006.8846
N	1,312	1,312	1,312	1,312	1,312	1,312	1,312	1,312

Legend: [*] $p < .1$; [**] $p < .05$; [***] $p < .01$
Source: Author's own calculations

Table 12.4 Estimation results – distinct and combined proximity effects; negative-binomial panel data count model, patent applications, time lag $(t-2)$, random effects

Variables	Estimation results							
	BL model	Model I	Model II	Model III	Model IV	Model V	Model VI	Model VII
firmage	-0.0060462	-0.0028274	-0.0060985	0.0015188	0.0033155	0.0052971	-0.0016229	0.0051639
firmage_sq	-0.0002028	-0.0003496	-0.0002189	-0.0004725	-0.0005445	-0.0006026	-0.0004338	-0.0006401
hhi_lsm	-1.3658383	-1.4867359	-1.3658948	-1.2690390	-1.3576646	-1.6114669	-1.3422872	-1.8729508*
hhi_pro	6.6206425*	7.6562576**	6.6600820*	6.7147863**	7.4410194**	7.3833056**	5.5305042	6.1446808*
nw_size	-2.3674935**	-2.1877193*	-2.3524237**	-2.2588650**	-2.1336106*	-2.2869888**	-2.6533677**	-2.8568777**
nw_density	-0.7579469	-0.6660674	-0.7804461	-1.1329812	-1.0308966	-0.7268719	-1.6242155	-1.1824804
coloclsm		-0.0560484			-0.0407761	-0.0610071		-0.1286616**
colocpro			-0.0284377		0.0210489		0.0176933	0.1735485
ctr_degree				1.7573278***	1.6670734***	0.8198683	4.2397756***	4.2724021***
int_clsm_deg						0.7329593**		1.3067901***
int_cpro_deg							-1.4877841*	-2.6170781***
_cons	-4.7403179	-5.6278273	-4.7387944	-5.0527945	-5.6989654	-5.4002199	-3.6989475	-3.9321708
ln_r_cons	-0.1480217	-0.1520983	-0.147666	-0.1316727	-0.1356771	-0.1321681	-0.1226020	-0.1228937
ln_s_cons	-1.0284482***	-1.0379394***	-1.0276295***	-.97208921***	-.9814638***	-.97225467***	-.95659519***	-.95526803***
chi^2	26.645868	28.187945	26.799183	38.511101	39.054631	43.242461	42.308847	52.529661
ll	-2,094.5567	-2,093.6895	-2,094.4866	-2,089.0624	-2,088.7525	-2,086.8531	-2,086.9663	-2,082.1808
aic	4,207.1135	4,207.379	4,208.9732	4,198.1247	4,201.505	4,197.7062	4,197.9326	4,192.3616
bic	4,258.2931	4,264.2453	4,265.8394	4,254.9909	4,269.7445	4,265.9457	4,266.1721	4,271.9743
N	2,179	2,179	2,179	2,179	2,179	2,179	2,179	2,179

Legend: * $p<.1$; ** $p<.05$; *** $p<.01$
Source: Author's own calculations

Table 12.5 First robustness check – distinct and combined proximity effects; innovation proxy: patent applications, time lag (t − 3), fixed effects

Variables	Estimation results							
	BL model	Model I	Model II	Model III	Model IV	Model V	Model VI	Model VII
firmage	0.0099346	0.0123228	0.0100101	0.0179553	0.0186406	0.0195035	0.0173298	0.0204734
firmage_sq	−0.0008897	−0.0009866	−0.0009018	−0.0011638	−0.0011914	−0.0012183	−0.0011648	−0.0012589
hhi_lsm	1.2528634	1.1730014	1.2519674	1.3922195	1.3603151	1.0469149	1.4171171	0.9809708
hhi_pro	2.9088245	3.4804375	2.91113459	2.7592849	2.9743416	3.3649746	2.2844051	2.8998909
nw_size	−3.2414013**	−3.1233780**	−3.2236681**	−3.2678446**	−3.2221055**	−3.3634556**	−3.3490208***	−3.7701574***
nw_density	−3.8847853**	−3.8083308**	−3.8983664**	−4.2428895**	−4.2065788**	−3.8399364**	−4.4741649**	−4.1200171**
coloclsm		−0.0340206			−0.0127917	−0.0518489		−0.0978202
colocpro			−0.0188283		0.0018418		0.0075737	0.1346163
ctr_degree				1.4308354***	1.4039893**	0.4528551	2.3359341	2.2551724
int_clsm_deg						0.7240435**		0.9960981**
int_cpro_deg							−0.5331851	−1.3387819
_cons	−2.3622551	−2.850539	−2.3419099	−2.3910235	−2.5754673	−2.6644985	−1.8768692	−2.1240553
chi2	20.859551	21.255724	20.894274	26.848139	26.906763	31.378543	26.980168	33.187435
ll	−1,284.094	−1,283.8716	−1,284.0777	−1,281.141	−1,281.1132	−1,279.3498	−1,280.9204	−1,278.2686
aic	2,582.1881	2,583.7433	2,584.1554	2,578.2821	2,582.2264	2,578.6996	2,581.8409	2,580.5373
bic	2,617.2855	2,623.8546	2,624.2668	2,618.3934	2,632.3655	2,628.8387	2,631.98	2,640.7043
N	1,112	1,112	1,112	1,112	1,112	1,112	1,112	1,112

Legend: * $p < .1$; ** $p < .05$; *** $p < .01$
Source: Author's own calculations

Table 12.6 First robustness check – distinct and combined proximity effects; innovation proxy: patent applications, time lag (t − 3), random effects

Estimation results

Variables	BL model	Model I	Model II	Model III	Model IV	Model V	Model VI	Model VII
firmage	0.0088446	0.0096496	0.0088378	0.0178801	0.0156518	0.0173579	0.0165819	0.0161277
firmage_sq	−0.0006474	−0.0006841	−0.0006631	−0.0009872	−0.0009321	−0.0009581	−0.0009925	−0.0009844
hhi_lsm	1.4394205	1.4053019	1.4354348	1.6121212	1.7090741*	1.3086723	1.6452221	1.3050747
hhi_pro	2.7152379	2.9583686	2.7369444	2.7570393	1.9956988	2.8992505	1.9503382	1.6905532
nw_size	−3.3837269**	−3.3330255***	−3.3586336***	−3.3306055***	−3.4584151**	−3.5659041***	−3.6475968**	−4.161918***
nw_density	−3.9464491**	−3.9128700**	−3.9628332**	−4.4722968**	−4.6647148**	−4.1098795**	−4.8903974**	−4.6817768***
coloclsm		−0.0133028			0.0428654	−0.0208439		−0.0422407
colocpro			−0.0254499		−0.0762345		0.0074107	0.0592933
ctr_degree				1.9329724***	2.0313313***	1.0857883	3.4759637**	3.4839807**
int_clsm_deg						0.6929293**		0.9911912**
int_cpro_deg							−0.9135734	−1.6942147*
_cons	−2.2647666	−2.473552	−2.2523303	−2.5539052	−1.7898202	−2.4112379	−1.6584264	−1.0518407
ln_r_cons	−0.2415459**	−0.2423008**	−0.2411783*	−0.2211384*	−0.2161127*	−0.2181578*	−0.2158697*	−0.2103641*
ln_s_cons	−1.1590729***	−1.1608289***	−1.1580681***	−1.0982638***	−1.086712***	−1.089466***	−1.0890087***	−1.0735487**
chi²	21.074244	21.142033	21.173342	33.993848	34.309026	39.318376	34.811753	42.317453
ll	−1,863.761	−1,863.7183	−1,863.7136	−1,857.6986	−1,857.4306	−1,855.7657	−1,856.936	−1,853.9104
aic	3,745.522	3,747.4366	3,747.4273	3,735.3971	3,738.8613	3,735.5315	3,737.872	3,735.8207
bic	3,795.7022	3,803.1924	3,803.1831	3,791.153	3,805.7683	3,802.4385	3,804.779	3,813.8789
N	1,950	1,950	1,950	1,950	1,950	1,950	1,950	1,950

Legend: * $p < .1$; ** $p < .05$; *** $p < .01$
Source: Author's own calculations

Table 12.7 Second robustness check – distinct and combined proximity effects; innovation proxy: patent grants, time lag (t − 2), fixed effects

Variables	Estimation results							
	BL model	Model I	Model II	Model III	Model IV	Model V	Model VI	Model VII
firmage	0.0017402	.01374851	.0105495	0.0080566	0.0137267	0.0212033	0.0104301	.01114231
firmage_sq	−0.0011355	−0.0016033	−0.0016373	−0.0012725	−0.0016699	−0.0014955	−0.0015663	−0.0012378
hhi_lsm	−6.3318973***	−6.7522736***	−6.4555646***	−6.2968146***	−6.4439803***	−7.1400424***	−6.4596354***	−7.1546447***
hhi_pro	14.280471**	16.167111***	14.586829**	17.522724***	17.377065***	18.425687***	17.110529***	16.868151***
nw_size	1.9527517	2.1854892	1.9501869	2.3756571	2.2660454	2.1690657	2.1912249	1.7702146
nw_density	1.2784947	1.0769908	0.8693110	−0.1368867	−0.5696712	1.2599345	−0.8144882	.21062888
coloclsm		−0.1217267			0.0031799	−0.1115862		−0.0926783
colocpro			−0.4021544**		−0.3936555*		−0.3658763*	−0.1881526
ctr_degree				2.3458837***	2.4192056***	0.7741624	3.1700176	3.5442506*
int_clsm_deg						0.8377529**		
int_cpro_deg							−0.4040874	−1.5768709
_cons	−9.6822296*	−11.045597*	−9.2736236	−13.136683**	−12.249173**	−13.228773**	−11.960887**	−11.230924*
chi²	45.274615	46.534796	48.127590	50.391641	54.484330	55.907693	54.638471	59.639189
ll	−645.40637	−644.24851	−643.86910	−641.06797	−639.56633	−638.58815	−639.48815	−637.04097
aic	1,304.8127	1,304.4970	1,303.7382	1,298.1359	1,299.1327	1,297.1763	1,298.9763	1,298.0819
bic	1,338.8521	1,343.3991	1,342.6403	1,337.0380	1,347.7602	1,345.8039	1,347.6039	1,356.4350
N	956	956	956	956	956	956	956	956

Legend: * $p < .1$; ** $p < .05$; *** $p < .01$
Source: Author's own calculations

Table 12.8 Second robustness check – distinct and combined proximity effects; innovation proxy: patent grants, time lag $(t-2)$, random effects

Variables	BL model	Model I	Model II	Model III	Model IV	Model V	Model VI	Model VII
	Estimation results							
firmage	0.0012542	0.0040712	0.0071708	0.0082993	0.0060897	0.0089101	0.0032641	-0.0058626
firmage_sq	-0.0004622	-0.0005844	-0.0007884	-0.0006488	-0.0006458	-0.0005007	-0.0006391	-0.0001279
hhi_lsm	-6.0517034***	-6.1761479***	-6.1083006***	-5.8637916***	-5.4603195***	-6.5874285***	-5.9054915***	-6.4322381***
hhi_pro	14.344978***	14.887806***	14.471204***	18.899995***	16.596570***	18.987999***	17.565979***	16.003521***
nw_size	1.8486780	1.9308576	1.8270384	2.5740547	2.0115644	2.2706017	2.1530184	1.5281860
nw_density	1.9997066	1.9726547	1.7697568	-0.1548906	-0.6056288	1.3701019	-1.3023800	-0.1518409
coloclsm	-0.0327531	-0.0327531			0.1446847*	-0.0100873		0.0549705
colocpro			-0.2347595*		-0.3914245***		-0.1715719	-0.2244178
ctr_degree				3.2170032***	3.6699310***	2.0887043**	5.5692254***	6.0640612***
int_clsm_deg						0.7068339**		0.9877219**
int_cpro_deg							-1.2329184	-2.2731938*
_cons	-10.121501*	-10.518807*	-9.8349866*	-15.080252***	-12.658460**	-14.656144**	-13.242833**	11.276891*
ln_r_cons	0.7635622***	0.7633333***	0.7711876***	0.7882858***	0.8175586***	0.8153931***	0.8161902***	0.8770310***
ln_s_cons	-1.4801702***	-1.4904590***	-1.4811424***	-1.3727419***	-1.3036646***	-1.3732074***	-1.3522164***	-1.3023265***
chi²	47.504340	47.460668	50.274780	58.898058	67.191318	66.156507	64.936301	76.214753
ll	-1,040.4013	-1,040.2799	-1,038.8633	-1,031.3588	-1,028.0665	-1,029.2067	-1,028.8835	-1,024.6853
aic	2,098.8026	2,100.5599	2,097.7265	2,082.7176	2,080.1331	2,082.4133	2,081.7670	2,077.3706
bic	2,149.9822	2,157.4261	2,154.5927	2,139.5838	2,148.3725	2,150.6528	2,150.0065	2,156.9833
N	2,179	2,179	2,179	2,179	2,179	2,179	2,179	2,179

Legend: $^{*}p<.1$; $^{**}p<.05$; $^{***}p<.01$
Source: Author's own calculations

both combined proximity effects (Model VII). In addition, we specified three additional models to check whether the results remained stable when estimating distinct proximity effects together (Model IV, Model V) and when estimating combined proximity effects separately (Model VI). The results were reported in accordance with Standard Hausman Test results and interpreted on the basis of the fully specified models (Model VII).

We start the discussion with Tables 12.3 and 12.4. The baseline model consists of firm level, network level and industry level variables. At the firm level we included two very basic variables i.e. "firm age" and "firm age squared" in the model. Network size and network density variables were incorporated to control for the structural network topology at the overall network level. Finally, two geographical concentration measures were considered to account for geographical concentration patterns of LSMs and PROs at the overall industry level. Three findings stand out:

Firstly, the firm-level variables have no significant effect on a firm's patenting activity in terms of patent application counts in $t-2$. Or to put it another way, a young firm's patenting activity does not significantly differ from the patenting behavior of a mature firm. This result is robust over all empirical settings (cf. Tables 12.3, 12.4, 12.5, and 12.6) and fully confirmed by the patent grant model (cf. Tables 12.7 and 12.8).

Secondly, spatial concentration patterns of LSMs and PROs at the industry level turned out to have a significant impact on firm-level innovation outcomes. Estimation results from both fixed effects and random effects models (cf. Tables 12.3 and 12.4) indicate that a high level of geographical clustering of LSMs at the industry level is negatively related to a firm's patenting activity at later points in time. As above, the patent grant model fully confirms this finding (cf. Tables 12.7 and 12.8). These results suggest that being part of an industry with a high level of geographical clustering among firms at the same stage of the value chain negatively impacts firm innovativeness at later points in time.

In contrast, geographical concentration of PROs at the industry level reveals exactly the opposite. Again, estimation results for both patent applications (cf. Tables 12.3 and 12.4) and patent grants (cf. Tables 12.7 and 12.8) are significant and robust over almost all model specifications. In other words, being part of an industry that is characterized by a high level of PRO clustering is positively related to a firm's innovative performance measured by its patenting activities at later points in time.

Thirdly, it turns out that network size, measured by the number of participating laser source manufacturing firms, is negatively related to firm innovativeness at the 0.05 significance level (cf. Tables 12.3 and 12.4). This overall network size effect is fully consistent with the results reported by fixed effects and random effects models for grants with a t-3 time lag. Hence, firm-level innovativeness is negatively related to the increasing size of the industry's innovation network. A look at the network density measure, however, reveals a somewhat ambiguous picture. On the one hand, none of the patent application models with a 2-year time lag structure show statistically significant estimates (cf. Tables 12.3 and 12.4). The same is true for the patent grant models with a comparable time lag structure (cf. Tables 12.7 and 12.8). On the other hand, patent application models with a 3-year time lag structure indicate a negative relationship between overall network density and firm level patenting performance at the 0.01 significance level.

In summary, our findings suggest that geographical concentration patterns at the industry level, as well as the structural network topology itself, turn out to affect the innovativeness of the firms involved. In addition, the spatial concentration patterns at the overall industry level, especially PRO clustering, seem to have an earlier impact on firm innovativeness than structural network characteristics, especially network density effects.[9]

Now we look at geographical proximity effects. To start with, we examine a firm's geographical co-location to other firms at the same stage of the industry value chain. Estimation results from patent application models with a 2-year time lag indicate a negative relatedness between geographical proximity and firm innovativeness at later points in time (cf. Tables 12.3 and 12.4). Coefficient estimates from the fixed effects model are highly significant at a 0.01 significance level and results from the random effects model confirm this relationship at a 0.05 significance level. Robustness checks do not contradict these findings but they also fail to provide additional empirical support. As a consequence, we have, in the very least, modest empirical evidence for a negative co-location effect of an LSM's geographical proximity to other LSMs. Or to put it another way, being near to other LSMs is not beneficial per se; instead, geographical proximity can also hamper a firm's innovativeness in terms of its patenting activity. This result supports the theoretical argument stated by Boschma (2005b, p. 70) according to which spatial lock-in effects and a lack of openness to the outside world can result in a situation in which negative agglomeration effects prevail.

Next, we place our attention on a firm's geographical co-location to other laser-related research organizations. The fixed effects patent application model with a time lag of 2 years (cf. Table 12.3, Model VII) reports a positive and significant coefficient estimate for the co-location variable at the 0.1 significance level. The positive relationship between a firm's patenting performance and its geographical closeness to other laser-related public research organizations implies the presence of purely regional knowledge spillovers. However, empirical evidence for this relationship is fairly weak. Similar to what was mentioned above, robustness checks do not contradict this finding but they also do not reveal additional empirical evidence. In a nutshell, the potential emergence or existence of pure knowledge spillover effects due to a firm's geographical proximity to laser-related PROs is, in the very least, doubtful and the positive effect on firm innovativeness should not be overestimated.

---

[9] To substantiate this finding we repeated the estimations with a 1-year time lag. It turned out that coefficient estimates for PRO clustering were highly significant; all other coefficient estimates, including LSM clustering, showed no significant effects. Additional results are available upon request.

Estimation results for network proximity effects reveal a much clearer picture. Similar to what was mentioned above, we start the discussion by focusing on the patent application models with a 2-year time lag (cf. Tables 12.3 and 12.4). Results provide strong empirical evidence for a positive and highly significant relationship between a firm's nodal degree and its subsequent innovation output. Coefficient estimates are comparably high and turn out to be significant at the 0.05 level when using fixed effect estimation techniques. Random effects estimation techniques reveal even stronger empirical evidence at the 0.01 significance level. This result is fully confirmed by almost all model specifications, except for the fixed effects setting (cf. Table 12.5). The implications are straightforward: the more direct partners a firm has, the higher its innovative performance at later points in time. To recap, we found strong empirical support for a pronounced and highly significant relationship between distinct network proximity effects and firm innovativeness in the German laser industry.

Last but not least, we address the relationship between combined proximity effects and firm innovativeness. Coefficient estimates for combined proximity variables have to be interpreted as follows according to Whittington et al. (2009, p. 98): (a) insignificant estimates: the effects of geographical proximity and network proximity are independent; (b) positive significant estimates: geographical proximity effects and network proximity effects are complementary in nature; (c) negative significant estimates: geographical proximity effects can be substituted by network proximity effects and vice versa.

Estimation results for combined proximity effects of a firm's geographical co-location to other LSMs and its network centrality provide sound empirical evidence for a complementary proximity effect. Coefficient estimates are positive and highly significant at the 0.01 level (cf. Tables 12.3 and 12.4). Patent application models with a 3-year time lag (cf. Tables 12.5 and 12.6) as well as the patent grant specification (cf. Tables 12.7 and 12.8) fully confirm this result at the 0.05 significance level. This finding reveals some interesting implications. The negative relatedness outlined above between a firm's geographical proximity to other LSMs and its innovativeness only persists as long as these firms do not cooperate. Combined proximity effects – composed of a firm's co-location to other LSMs and a firm's network centrality measured by its nodal degree – are suggestive of complementary effects. Combined proximity effects of a firm's co-location to other PROs and its network proximity turn out to be substitutional in nature. Again we have sound empirical support for this finding (cf. Tables 12.3, 12.4, 12.6, and 12.8).

12.6 Discussion and Implications

What do our empirical findings tell us in relation to our previously formulated hypotheses? Our first hypothesis (H1) suggests a positive relationship between a firm's number of direct linkages and its patenting output at later points in time. We found strong empirical evidence for the relevance of distinct network proximity effects on the innovative performance of German laser source manufacturers. In other words, degree centrality, which measures the number of direct ties, turns out to be highly relevant for a firm's innovative performance at later points in time. This finding is in line with the results reported by Whittington and colleagues (2009) for the US biotech industry. They found a highly positive relatedness between a firm's eigenvector centrality and patenting performance.

With regard to distinct geographical proximity, Whittington et al. (2009) reported significant positive effects of co-location between US biotech firms and non-significant effects of geographical proximity to PROs. These results imply that the co-location to other biotech firms, rather than to research organizations, is what drives firm innovativeness. Our results for the German laser source industry paint quite a different picture. We have argued that geographical proximity – more precisely, co-location between a firm and other LSMs (H2a), or co-location between a firm and other PROs (H2b) – is positively related to firm innovativeness. Against our initial expectations, estimation results for co-location between laser source manufacturers turned out to have a significant negative effect on firm-level innovation outcomes. In other words, co-location between laser source manufactures reduces the innovative performance which leads to the rejection of Hypothesis H2a. This unexpected result is highly relevant for several reasons.

Firstly, it contradicts the empirical findings of Whittington et al. (2009). This indicates that distinct geographical proximity effects seem to be industry specific and follow a completely different logic in the German laser industry. Secondly, we found strong empirical evidence for the "dark side" of geographical proximity supporting the spatial lock-in arguments proposed by Boschma (2005b). Obviously, the positive knowledge spillover mechanisms do not unleash their effects. In contrast, the explanation for the negative co-location effect is straightforward. A firm's geographical proximity to competitors may create an atmosphere characterized by reticence, aversion and preconceived notions. As a consequence, a firm can become inward looking (Boschma 2005b) and may choose secrecy strategies (Liebeskind 1996) instead of opening itself up to other LSMs. Especially non-cooperative firms may face a situation in which their knowledge base erodes and becomes outdated (Boschma 2005b). In summary, non-cooperating firms trapped in an inward looking geographical surrounding are likely to be adversely affected in their efforts to innovate.

Our empirical findings for a firm's co-location to laser-related PROs are in line with our initial expectations. In contrast to Whittington and colleagues (2009) we found at least modest empirical support for Hypothesis H1b. Our result suggests that German laser source manufacturers may benefit from being located near laser-related public research organizations. Positive externalities in terms of scientific knowledge spillovers can be explained as follows. Firstly, it is commonly accepted that universities are an important source of new technological knowledge (Agrawal 2001, p. 285). It has been argued that especially the transfer of highly codified technological knowledge is facilitated by geographical proximity (Audretsch et al. 2004, p. 195). Others have put forward the argument that knowledge flows out of PROs, especially universities, much more readily than it does from privately owned firms (Jaffe et al. 1993; Owen-Smith and Powell 2004, p. 7). Secondly, firms located close to PROs may have a higher chance of attracting and hiring highly qualified graduates. It is plausible to assume that these employees bring new and creative ideas with them which affects a firm's ability to innovate. Our results suggest that at least one of these two transmission channels seems to create what we would call "pure" scientific knowledge spillovers.

Finally, our findings on combined geographical proximity and network proximity confirm our expectations that combined proximity effects are not independent. However, the story turns out to be more complex than initially expected. To start with, we take a look at Hypothesis H3a. We have assumed that combined proximity effects of a firm's network proximity and its geographical proximity to other laser source manufactures are complementary. Indeed, we found sound empirical support for Hypotheses H3a. The interpretation is straightforward. Being close to firms at the same stage in the value chain in an inward looking and non-cooperative environment seems to hinder non-cooperative actors in their efforts to innovate. In contrast, the complementary nature of combined proximity effects suggests that highly cooperative actors benefit from their geographical closeness to other LSMs. The results clearly show that being well-positioned in both types of proximity dimensions can lead to mutually reinforcing effects which, in turn, are positively related to a firm's innovative performance at later points in time.

Surprisingly, combined proximity effects of a firm's network proximity and its geographical proximity to other laser-related public research organizations turned out to be substitutional in nature. As a consequence we have to reject Hypothesis H3b. Nonetheless, this finding has some important implications. Firstly, scientific knowledge seems to be, at least to some extent, accessible via alternative transmission channels. In other words, firms can tap scientific knowledge by means of geographical spillovers or through cooperative linkages. Secondly, the two proximity dimensions seem to be exchangeable within certain limits. To illustrate this point, a peripheral firm located far away from laser-related research facilities can compensate for this geographical disadvantage by intensifying its cooperation activities. Following the same logic, firms located near laser-related PRO agglomerations are less dependent on having a high number of formal R&D partnerships. However, against the backdrop of the modest empirical support for the existence of "pure" scientific knowledge spillovers and the comparably strong empirical evidence for the existence of network effects, the latter implication should not be overstated.

To conclude with, the study provides us with some interesting insights and opens up at the same time several interesting research questions. Both, the limitations of this analysis (cf. Sect. 13.2) and fruitful avenues for further research on proximity and innovation (cf. Sect. 14.2) are the subject of discussion in the following chapters.

References

Acar W, Sankaran K (1999) The myth of unique decomposability: specializing the Herfindahl and entropy measures. Strateg Manag J 20(1):969–975

Acs ZJ, Audretsch DB, Deldman MP (1992) Real effects of academic research: comment. Am Econ Rev 82(1):363–367

Agrawal A (2001) University-to-industry knowledge transfer: literature review and unanswered questions. Int J Manag Rev 3(4):285–302

Ahuja G (2000) Collaboration networks, structural hole, and innovation: a longitudinal study. Adm Sci Q 45(3):425–455

Al-Laham A, Kudic M (2008) Strategische Allianzen. In: Corsten H, Goessinger R (eds) Lexikon der Betriebswirtschaftslehre, 5th edn. Oldenbourg Verlag, München, pp 39–41

Almeida P, Kogut B (1999) Localization of knowledge and the mobility of engineers in regional networks. Manag Sci 45(7):905–917

Amburgey TL, Dacin T, Singh JV (1996) Learning races, patent races, and capital races: strategic interaction and embeddedness within organizational fields. In: Baum JA (ed) Advances in strategic management. Elsevier, New York, pp 303–322

Amin A, Wilkinson F (1999) Learning, proximity and industrial performance: an introduction. Camb J Econ 23(2):121–125

Arrow KJ (1962) The economic implications of learning by doing. Rev Econ Stud 29(3):155–173

Audretsch DB (1998) Agglomeration and the location of innovative activity. Oxf Rev Econ Policy 14(2):18–29

Audretsch DB, Dohse D (2007) Location: a neglected determinant of firm growth. Rev World Econ 143(1):79–107

Audretsch DB, Feldman MP (1996) R&D spillovers and the geography of innovation and production. Am Econ Rev 86(3):630–640

Audretsch DB, Feldman MP (2003) Small-firm strategic research partnerships: the case of biotechnology. Tech Anal Strat Manag 15(2):273–288

Audretsch DB, Lehmann E, Warning S (2004) University spillovers: does the kind of science matter? Ind Innov 11(3):193–205

Baum JA, Calabrese T, Silverman BS (2000) Don't go it alone: alliance network composition and startup's performance in Canadian biotechnology. Strateg Manag J 21(3):267–294

Borgatti SP, Everett MG, Freeman LC (2002) Ucinet for windows: software for social network analysis. Analytic Technologies, Harvard

Boschma R (2005a) Role of proximity in interaction and performance: conceptual and empirical challenges. Reg Stud 39(1):41–45

Boschma R (2005b) Proximity and innovation: a critical assessment. Reg Stud 39(1):61–74

Boschma R, Frenken K (2010) The spatial evolution of innovation networks: a proximity perspective. In: Boschma R, Martin R (eds) The handbook of evolutionary economic geography. Edward Elgar, Cheltenham, pp 120–135

Brenner T, Broekel T (2011) Methodological issues in measuring innovation performance of spatial units. Ind Innov 18(1):7–37

Breschi S, Lissoni F (2001) Knowledge spillovers and local innovation systems: a critical survey. Ind Corp Chang 10(4):975–1005

Breschi S, Lissoni F (2009) Mobility of skilled workers and co-invention networks: an anatomy of localized knowledge flows. J Econ Geogr 9(4):439–468

Broekel T, Graf H (2011) Public research intensity and the structure of German R&D networks: a comparison of ten technologies. Econ Innov New Technol 21(4):345–372

Buckley PJ, Glaister KW, Klijn E, Tan H (2009) Knowledge accession and knowledge acquisition in strategic alliances: the impact of supplementary and complementary dimensions. Br J Manag 20(4):598–609

Buenstorf G (2007) Evolution on the shoulders of giants: entrepreneurship and firm survival in the German laser industry. Rev Ind Organ 30(3):179–202

Burt RS (1992) Structural holes: the social structure of competition. Harvard University Press, Cambridge

Burt RS (2000) The network structure of social capital. In: Staw BM, Sutton RI (eds) Research in organizational behavior, vol 22. JAI Press, Greenwich, pp 345–424

Burt RS (2005) Brokerage & closure – an introduction to social capital. Oxford University Press, New York

Cameron CA, Trivedi PK (1990) Regression based tests for overdispersion in the Poisson model. J Econ 46(3):347–364

Cameron CA, Trivedi PK (2009) Microeconometrics using Stata. Stata Press, College Station

Cassi L, Corrocher N, Malerba F, Vonortas N (2008) Research networks as infrastructure for knowledge diffusion in European regions. Econ Innov New Technol 17(7):665–678

Coenen L, Moodysson J, Asheim BT (2004) Nodes, networks and proximities: on the knowledge dynamics of the Medicon Valley Biotech Cluster. Eur Plan Stud 12(7):1003–1018

Coff RW (2003) The emergent knowledge-based theory of competitive advantage: an evolutionary approach to integrating economics and management. Manag Decis Econ 24(4):245–251

Cohen WM, Levinthal DA (1990) Absorptive capacity: a new perspective on learning and innovation. Adm Sci Q 35(3):128–152

Coleman JS (1988) Social capital in the creation of human capital. Am J Sociol 94:95–120

Doreian P, Woodard KL (1992) Fixed list versus snowball selection of social networks. Soc Networks 21(2):216–233

Fagerberg J (2005) Innovation: a guide to the literature. In: Fagerberg J, Mowery DC, Nelson RR (eds) The Oxford handbook of innovation. Oxford University Press, New York, pp 1–28

Feldman MP (1993) An examination of the geography of innovation. Ind Corp Chang 2 (3):451–470

Fornahl D, Broeckel T, Boschma R (2011) What drives patent performance of German biotech firms? The impact of R&D subsidies, knowledge networks and their location. Pap Reg Sci 90 (2):395–418

Fritsch M, Slavtschev V (2007) Universities and innovation in space. Ind Innov 14(2):201–218

Gilsing V, Nooteboom B, Vanhaverbeke W, Duysters G, van den Oord A (2008) Network embeddedness and the exploration of novel technologies: technological distance, betweenness centrality and density. Res Policy 37(10):1717–1731

Goerzen A (2005) Managing alliance networks: emerging practices of multinational corporations. Acad Manag Exec 19(2):94–107

Granovetter MS (1985) Economic action and social structure: the problem of embeddedness. Am J Sociol 91(3):481–510

Granovetter MS (2005) The impact of social structure on economic outcomes. J Econ Perspect 19 (1):33–50

Grant RM (1996) Towards a knowledge based theory of the firm. Strateg Manag J 17(2):109–122

Grant RM, Baden-Fuller C (2004) A knowledge accessing theory of strategic alliances. J Manag Stud 41(1):61–84

Greene WH (2003) Econometric analysis, 5th edn. Prentice Hall, Upper Saddle River

Grupp H (2000) Learning in a science driven market: the case of lasers. Ind Corp Chang 9 (1):143–172

Gulati R (2007) Managing network resources – alliances, affiliations and other relational assets. Oxford University Press, New York

Gulati R, Singh H (1998) The architecture of cooperation: managing coordination costs and appropriation concerns in strategic alliances. Adm Sci Q 43(4):781–814

Gulati R, Nohria N, Zaheer A (2000) Strategic networks. Strateg Manag J 21(3):203–215

Hamel G (1991) Competition for competence and inter-partner learning within international strategic alliances. Strateg Manag J 12(1):83–103

Hannan MT, Freeman J (1984) Structural inertia and organizational change. Am Sociol Rev 49 (2):149–164

Hanneman RA, Riddle M (2005) Introduction to social network methods. University of California, Riverside

Hanusch H, Pyka A (2007a) Principles of neo-Schumpeterian economics. Camb J Econ 31 (2):275–289

Hanusch H, Pyka A (2007b) Elgar companion to neo-Schumpeterian economics. Edward Elgar, Cheltenham

Hausman JA (1978) Specification tests in econometrics. Econometrica 46(6):1251–1271

Hausman JA, Hall BH, Griliches Z (1984) Econometric models for count data with an application to the patents – R&D relationship. Econometrica 52(4):909–938

Jacobs J (1969) The economy of cities. Random House, New York

Jaffe AB (1989) Real effects of academic research. Am Econ Rev 79(5):957–970

Jaffe AB, Trajtenberg M, Henderson R (1993) Geographic localization of knowledge spillovers as evidenced by patent citations. Q J Econ 108(3):577–598

Kale P, Singh H, Perlmutter H (2000) Learning and protection of proprietary assets in strategic alliances: building relational capital. Strateg Manag J 21(3):217–237

Kim T-Y, Oh H, Swaminathan A (2006) Framing interorganizational network change: a network inertia perspective. Acad Manag Rev 31(3):704–720

Knoben J, Oerlemans LA (2006) Proximity and inter-organizational collaboration: a literature review. Int J Manag Rev 8(2):71–89

Kogut B, Zander U (1992) Knowledge of the firm, combinative capabilities, and the replication of technology. Organ Sci 3(3):383–397

Kudic M, Boenisch P, Dominguez Lacasa I (2010) Network embeddedness, geographical collocation effects or both? The impact of distinct and combined proximity effects on firm-level innovation output in the German laser industry. In: Conference proceedings. The 36th EIBA annual conference, Porto, pp 1–29

Lamoreaux NR, Sokoloff KL (1999) The geography of the market for technology in the late nineteenth and early twentieth century United States. In: Libecap G (ed) Advances in the study of entrepreneurship, innovation, and economic growth. JAI Press, Stanford, pp 67–121

Laumann EO, Galaskiewicz J, Marsden PV (1978) Community structure as interorganizational linkages. Annu Rev Sociol 4:455–484

Liebeskind JP (1996) Knowledge, strategy and the theory of the firm. Strateg Manag J 17 (2):93–108

Lundvall B-A (1992) National systems of innovation – towards a theory of innovation and interactive learning. Pinter, London

Marschall A (1890) Principles of economics – an introductory (1920, 8th edn. Macmillan, London

Maskell P, Malmberg A (1999) Localized learning and industrial competitiveness. Camb J Econ 23(2):167–185

Nelson RR (1992) National innovation systems: a retrospective on a study. Ind Corp Chang 1 (2):347–374

Nonaka I (1991) The knowledge-creating company. Harv Bus Rev 69(6):96–104

Nooteboom B (2008) Learning and innovation in inter-organizational relationships. In: Cropper S, Ebers M, Huxham C, Ring PS (eds) The Oxford handbook of interorganizational relations. Oxford University Press, New York, pp 607–634

OECD (2005) Oslo manual: guidelines for collecting and interpreting innovation data, 3rd edn. OECD Publishing, Paris

Oerlemans LA, Meeus MT (2005) Do organizational and spatial proximity impact on firm performance? Reg Stud 39(1):89–104

Oerlemans LA, Meeus MT, Boekema FW (2001) Firm clustering and innovation: determinants and effects. Pap Reg Sci 80(3):337–356

Owen-Smith J, Powell WW (2004) Knowledge networks as channels and conduits: the effects of spillovers in the Boston biotechnology community. Organ Sci 15(1):5–21

Patel P, Pavitt K (1995) Patterns of technological activity: their measurement and interpretation. In: Stoneman P (ed) Handbook of the economics of innovation and technological change. Blackwell, Oxford, UK, pp 14–51

Polanyi M (1958) Personal knowledge: towards a post-critical philosophy. University of Chicago Press, Chicago

Polanyi M (1967) The tacit dimension. Doubleday, New York

Powell WW, Koput KW, Smith-Doerr L (1996) Interorganizational collaboration and the locus of innovation – networks of learning in biotechnology. Adm Sci Q 41(1):116–145

Pyka A (1997) Informal networking. Technovation 17(4):207–220

Pyka A (2002) Innovation networks in economics: from the incentive-based to the knowledge based approaches. Eur J Innov Manag 5(3):152–163

Romer P (1986) Increasing returns and long-run growth. J Polit Econ 94(5):1002–1037

Rowley TJ, Behrens D, Krackhardt D (2000) Redundant governance structures: an analysis of structural and relational embeddedness in the steel and semiconductor industries. Strateg Manag J 21(3):369–386

Saxanian A (1990) Regional networks and the resurgence of Silicon Valley. Calif Manag Rev 33 (1):89–112

Scherngell T, Barber MJ (2009) Spatial interaction modeling of cross-region R&D collaborations: empirical evidence from the 5th EU framework programme. Pap Reg Sci 88(3):531–546

Scherngell T, Barber MJ (2011) Distinct spatial characteristics of industrial and public research collaborations: evidence from the fifth EU framework programme. Ann Reg Sci 46 (2):247–266

Schilke O, Goerzen A (2010) Alliance management capability: an investigation of the construct and its measurement. J Manag 36(5):1192–1219

Schoenmakers W, Duysters G (2006) Learning in strategic technology alliances. Tech Anal Strat Manag 18(2):245–264

Shan W, Walker G, Kogut B (1994) Interfirm cooperation and startup innovation in the biotechnology industry. Strateg Manag J 15(5):387–394

Simonin BL (1999) Ambiguity and the process of knowledge transfer in strategic alliances. Strateg Manag J 20(1):595–623

Sorenson O, Audia PG (2000) The social structure of entrepreneurial activity: geographic concentration of footwear production in the Unites States, 1940–1989. Am J Sociol 106(2):424–462

Spence M (1976) Informational aspects of market structure: an introduction. Q J Econ 90 (4):591–597

Spence M (2002) Signaling in retrospect and the informational structure of markets. Am Econ Rev 92(3):434–459

Stata (2007) Stata statistical software: release 10. StataCorp LP, College Station

Stuart TE (1999) A structural perspective on organizational performance. Ind Corp Chang 8 (4):745–775

Stuart TE (2000) Interorganizational alliances and the performance of firms: a study of growth and innovational rates in a high-technology industry. Strateg Manag J 21(8):791–811

Stuart TE, Hoang H, Hybles RC (1999) Interorganizational endorsements and the performance of entrepreneurial ventures. Adm Sci Q 44(2):315–349

Torre A, Rallet A (2005) Proximity and localization. Reg Stud 39(1):47–59

Uzzi B (1996) The sources and consequences of embeddedness for the economic performance of organizations: the network effect. Am Sociol Rev 61(4):674–698

Uzzi B (1997) Social structure and competition in interfirm networks : the paradox of embeddedness. Adm Sci Q 42(1):35–67

Visser E-J (2009) The complementary dynamic effects of clusters and networks. Ind Innov 16 (2):167–195

Wasserman S, Faust K (1994) Social network analysis: methods and applications. Cambridge University Press, Cambridge

Whittington KB, Owen-Smith J, Powell WW (2009) Networks, propinquity, and innovation in knowledge-intensive industries. Adm Sci Q 54(1):90–122

Zaheer A, Bell GG (2005) Benefiting from network position: firm capabilities, structural holes, and performance. Strateg Manag J 26(9):809–825

Zollo M, Reuer JJ, Singh H (2002) Interorganizational routines and performance in strategic alliances. Organ Sci 13(6):701–713

Zucker LG, Darby MR, Armstrong JS (1998) Geographically localized knowledge: spillovers or markets? Econ Inq 36(1):65–86

Part V
Summary, Conclusion and Outlook

Chapter 13
Findings and Limitations

This neglect of other aspects of the system has been made easier by another feature of modern economic theory – the growing abstraction of the analysis, which does not seem to call for a detailed knowledge of the actual economic system or, at any rate, has managed to proceed without it.

(Ronald H. Coase 1991)

Abstract The question that naturally arises at this point is what have we learned so far and, maybe even more importantly, what have we yet to learn? This chapter addresses precisely these questions. In Sect. 13.1 we start with a brief discussion of some general issues and limitations. In doing so we focus primarily on general limitations related to the scope of data sources employed for the purpose of this study. In Sect. 13.2 we provide a summary of key findings for the analyses conducted in Parts III and IV. In the same breath, we also address theoretical and empirical issues that deserve particular attention in future studies.

13.1 General Issues and Limitations

Each empirical research project bears some inherent risks that cannot be excluded even by a conscious selection of the empirical setting. After careful and critical reflection we chose the German laser industry to put our theoretical considerations to the test. This proved to be a good decision for at least two reasons. Firstly, even though the majority of laser source manufacturing firms are micro and small enterprises, they demonstrate high cooperation and innovation activities. Secondly, the underlying technological developments that fuel technological change processes in the industry have, by no means, reached the end of the road. This study concentrates on laser source manufacturers (LSMs) that are at the very heart of the laser industry's value chain. The underlying assumption is straightforward; these

© Springer International Publishing Switzerland 2015

M. Kudic, *Innovation Networks in the German Laser Industry,*
Economic Complexity and Evolution, DOI 10.1007/978-3-319-07935-6_13

firms are considered to be heavily involved in the development of laser sources. Cooperation activities between LSMs and laser-related public research organizations (PROs) were explicitly considered in this study. However, R&D cooperation between LSMs and up-stream firms (e.g. component suppliers) or down-stream firms (e.g. laser system producers) were beyond the scope of this study.

In general, the scope of this study is limited in several ways. Firstly, we chose a window of time between 1990 and 2010 to conduct our analyses. The reason for this is simple. There are considerable gaps in information on firm characteristics and cooperation activities in the period before 1990. In contrast, data availability for the time period after 1990 is much higher and firm histories can be more easily reconstructed based on these historical raw data sources. Because the laser industry database was still under construction in 2011, our window of observation was restricted to the time before 2011.

A second point relates to the R&D cooperation data sources used in this study. Data on publicly funded R&D cooperation projects was employed to construct annual innovation network layers between 1990 and 2010. As already highlighted throughout the study, the use of data on publicly funded research projects can cause selectivity problems. These problems are usually caused by some unobserved actor-specific characteristics which can lead to a systematic pre-selection of a sub-set of actors in a given population. Against this backdrop one could argue that the empirical findings in this study that higher innovativeness is related to cooperation-related determinants might simply be caused by the inherent superiority of those actors who were preselected because they were awarded more grants. These concerns seem to be of limited salience because the optical industry is considered to be one of the key technologies affecting the innovativeness and prosperity of the German economy as a whole (BMBF 2010). The very aim of German policy-makers was to increase the international competiveness of the industry as a whole (Fabian 2011). Since the early 1980s, German technology policy has strongly supported not only large but also small and micro-sized firms in the optical industry (ibid). In other words, funding decisions were primarily motivated by the aim to make German actors more competitive than their international rivals; spurring on domestic competition through highly selective merit-based funding decisions appears to have been of secondary importance. Basically the same arguments hold true with regard to European funding decisions. Scherngell and Barber (2009, p. 534) point out that one of the main EU Framework program objectives is to strengthen the scientific and technological bases of European industries and foster international competitiveness. Nonetheless, we believe that our current R&D cooperation database can and should be supplemented in several ways (cf. Chap. 14).

Thirdly, the use of patent data in constructing innovation indicators was frequently criticized in the literature. Other indicators such as survey-based innovation indicators are simply not available retrospectively over a period of more than two decades. Nonetheless, we agree that more appropriate proxies for measuring innovation output could be applied, especially for industries characterized by a high number of micro and small-sized firms. A promising way to gather additional information on innovation activities of LSMs and PROs may be the use of market-based innovation outcome measures. These and many other challenges constitute the next steps in our research agenda (cf. Chap. 14).

Finally, this study is restricted to the national level. An interesting observation throughout the data compilation procedure was the high involvement of international partners in CORDIS cooperation projects. In addition, we know from other studies that German laser source manufacturers have a strong position in international markets and they export their products to a large extent. Both observations substantiate the assumption that a national analysis provides some highly interesting results but may not go far enough.

Each of the areas addressed above provides interesting starting points for enriching our database and solidifying the empirical findings made so far.

13.2 Summary of Our Main Findings and Open Questions

In this section we briefly summarize the most salient descriptive findings (Part III) and address the insights of our four explanatory analyses (Part IV).

The descriptive exploration of industry data has revealed some interesting patterns. The initial descriptive analyses provide a comparison of industry dynamics and spatial distribution patterns for laser source manufacturers (LSMs), laser system providers (LSPs) and laser-related public research organizations (PROs). We start with a brief look at the overall industry dynamics and find that LSPs dominate in terms of numbers over the entire observation period. All in all, the period between 1990 and 2005 is characterized by a more-or-less stable growth trend with only some minor fluctuations in all three types of organizations. The last 5 years are characterized by a slight decrease in the number of LSMs and LSPs whereas the number of PROs continues to grow at a moderate rate. Next, we employed geographical Herfindahl-Hirschman Indices to track the geographical concentration patterns at the overall industry level. Concentration indices for both LSMs and LSPs start at a high level in 1990 and, after some minor fluctuations, level off at around 0.06 index points in 2010. Analyzing spatial patterns at the regional level reveals a concentration of laser-related organizations in four geographical regions: Munich, Thuringia, Berlin and Stuttgart. These geographical areas still constitute the centers of the German laser industry.

Our study concentrates on the full population of German laser source manufacturers between 1990 and 2010. A closer look at the entry and exit dynamics and the size distribution of LSMs provides some interesting insights. Data indicates the highest number of firm entries in 1995, 1999 and 2001. In contrast, the total number of firm exits peaked in 2000 with 11 LSMs leaving the industry. The overall trend indicates a 3.4-fold increase in firms over the course of just 15 years with a peak in 2005 followed by an overall decrease by 2010 with some minor fluctuations. The descriptive analysis of the firm size distribution over the entire observation period reveals interesting insights. At the beginning of our observation period, more than half of all firms are micro firms. Even though micro firms lose ground over time, the total number of micro firms in the sample remains comparably high. Small firms show the highest average growth rates, followed by medium, large and very large firms. A closer look at the size distribution of LSMs at the regional level shows a notable number of LSMs in 10 out of the 16 federal states. The majority of large and very large firms are located in Bavaria and Baden-Wurttemberg. In contrast, the situation in Berlin and North Rhine-Westphalia is characterized by a comparably high number of either micro firms or small firms. In Bremen, Saxony-Anhalt and Mecklenburg-Western Pomerania we found a very low presence of LSMs over the entire observation period.

Not only LSMs but also laser-related public research organizations (PROs) play an important role in this study. PROs are considered to be important sources of applied and basic scientific knowledge. The descriptive analysis of the composition of laser-related PROs in Germany reveals some interesting insights. Fraunhofer institutes make up the largest percentage of non-university research organizations in our sample at about 22 %. The proportion of technical universities, universities and universities of applied science was 10 %, 34 %, and 4 % respectively. About 20 % of the laser-related PROs were members of the Leibniz, Helmholtz or Max-Planck societies. Finally, about 10 % of the overall population of all laser-related PROs in Germany do not belong any of the four major German research societies. Accordingly applied research facilities seem to play a key role in the Germany laser industry.

Cooperation data for this study came from two sources: *Foerderkatalog* data and *CORDIS* data. From the first source, we identified 416 laser-related R&D cooperation projects and the second source produced R&D projects. The findings show that *CORDIS* projects are considerably larger than *Foerderkatalog* projects. The average size of *CORDIS* projects, measured by the number of partners involved, was 10.44 with a standard deviation of 8.02. In the case of *Foerderkatalog* projects, we found an involvement of 6.38 organizations per project with a standard deviation of 3.96. Both data sources were used to construct innovation networks on an annual basis. In general, our analysis of cooperation project involvement of LSMs and PROs at the national level shows an increasing proportion of organizations participating in publicly funded research projects over time. The average percentage of PROs participating in either *CORDIS* or *Foerderkatalog* R&D cooperation projects was 42.74 % and the maximum percentage of cooperation reached nearly 60 % in 2008. A look at the overall participation of LSMs in both types of publicly funded cooperation projects reveals a minimum participation of 24.05 % in 1990, a maximum participation of 47.24 % in 2008 and an average participation of 36.92 %.

In Chap. 8 we focused on an exploratory analysis of structural evolution of the industry's innovation network. We applied two strategies to gain a comprehensive picture of evolutionary network change processes in the German laser industry. On the one hand, we made use of exploratory social network analysis methods (De Nooy et al. 2005) to explore structural change patterns over time. We also conducted an in-depth analysis of large-scale properties by using more sophisticated network models (Barabasi and Albert 1999; Watts and Strogatz 1998; Borgatti and Everett 1999).

We start by reporting the findings of the scale-free analysis. The German industry innovation network displays no perfect power law behavior. Nonetheless, the log-log plot for the degree distribution over the entire observation period reveals systematic differences between the real world and the random network. This indicates at least a pronounced tendency towards the emergence of scale-free properties. Our results are in line with the findings of Powell et al. (2005) for the US biotech industry. Next, we looked at the small-world properties of the innovation network. We employed graph theoretical concepts and measures, i.e. the "clustering coefficient ratio", the "path length ratio" and the "small-world Q" to test for the existence of the network's small-world nature. Data clearly reveals the emergence of small-world properties in the German laser industry innovation network. Robustness checks substantiate this finding. In addition, we found an increasing tendency towards a solidification of small-world properties over time. Last but not least, we checked for the existence of core-periphery patterns by using complementary indicators. Our findings give us good reason to assume that the German laser industry innovation network exhibited a pronounced core-periphery structure during three time periods – (I) 1994–1997, (II) 1999–2002 and (III) 2004–2008. At least two out of four indicators substantiate these findings in all three time periods.

In Part IV we conducted four empirical investigations. Each of the empirical studies addresses a quite narrowly defined problem and provides new empirical evidence for innovation networks in the German laser industry, a still widely unexplored topic in the literature.

The overall objective of the first empirical part (Chap. 9) was to contribute to an in-depth understanding of the causes and consequences of evolutionary network change processes at the micro-level. A natural starting point to throw some light on the evolution of networks is to look at a firm's initial cooperation event and the determinants that affect the timing of network entry processes. In particular, we included three types of determinants in our analysis: firm size, cooperation type, and geographical location. Estimation results from a non-parametric event history model indicate that micro firms enter the network later than small and large firms. An in-depth analysis of the size effects for medium-sized firms provides some unexpected yet quite interesting findings. These findings show that the choice of cooperation type makes no significant difference to a firm's timing in entering the network. Finally, the analysis of contextual determinants reveals that cluster membership can, but does not necessarily, affect a firm's timing to cooperate. It appears that firms in some regions (e.g. Thuringia) tend to cooperate earlier and to have a significantly higher propensity to cooperate than those in other regions (e.g. Bavaria).

From a theoretical perspective, a lot remains to be done. For instance, our conceptual framework still requires further refinement. Organizational, relational and contextual determinates have to be concretized and interdependencies between these three dimensions have to be addressed more explicitly. An interesting theoretical study presented by Hagedoorn (2006) moves in this direction. The proximity concept (cf. Sect. 2.3.3.2) provides another promising starting point for addressing the role of interdependencies between these dimensions for micro-level network change processes. Our theoretical framework raises awareness of the importance of network paths. We included a very specific type of network path dependency to account for a network entrant's cooperation behavior in the subsequent cooperation rounds. We refer to this idea as "cooperation imprinting". We believe that the sequential analysis of cooperation processes, against the backdrop of new cooperation options and revised strategies, is crucial in understanding structural network change. A refinement of this idea constitutes one of the next steps in our research agenda.

From an empirical point of view we are still at an early stage. This study concentrates exclusively on a firm's first cooperation event. Cooperation events between incumbents were not addressed. Consequently, the next steps in our research agenda are straightforward. Firstly, repeated cooperation events have to be included in our empirical analysis. An initial step in this direction has already been made (cf. Kudic et al. 2013). Regression results of a parametric event history model reveal that a firm's knowledge endowment (and cooperation experience) shortens the duration to first (and consecutive) cooperation events. The study conducted by Kudic et al. (2013) also shows that previous occupation of strategic network positions is closely related to the swift establishment of further R&D cooperation at later points in time. Secondly, we have to find a way to analyze the structural consequences of micro-level network change processes empirically. Not only the formation entry processes and the network formation phase, but also dissolution processes and network fragmentation tendencies, have to be explored more in detail.

The focus of the second empirical part (Chap. 10) is quite different but closely related to the issues addressed in the first study. The key objective was to analyze how firm innovativeness is related to individual cooperation events and the structure and dynamics of a firm's ego network. We applied panel data count models to accomplish this task. Estimation results, from a fixed effects model, are suggestive of direct innovation effects due to individual cooperation events, but only as long as structural ego network characteristics are ignored. These effects, however, partially diminish when individual cooperation events and ego network characteristics are looked at simultaneously. Innovation effects of ego network size, as well as ego network brokerage, remain stable whereas ego network density reveals some surprising results. It is also interesting to note that, because we include firm-level funding as a control variable in all models, our findings relativize the argument that a firm's innovative performance is affected more by public funding than by the cooperation activities themselves.

However, we still face some theoretical and methodological challenges. The structural configuration of an ego network can be analyzed from various theoretical perspectives. Not only the size, brokerage and density of the ego network but also additional structural features have to be explicitly considered in future research. For instance, various dimensions of node-level structural heterogeneity of ego networks (i.e. nationality, financial power, organizational form etc.) have to be integrated into the analysis. Additionally, a fine-grained differentiation between different types of collaboration (i.e. funded vs. non-funded collaborations, various types of strategic alliances etc.) can significantly improve our understanding in this research area. There are also some methodological limitations. For instance, the use of more sophisticated indicators of a firm's ego network structure is needed to account for additional ego network characteristics that go beyond the scope of this analysis. To accomplish these tasks, our laser industry database has to be refined and completed in several ways.

The third empirical investigation (Chap. 11) addresses the relationship between large-scale network properties and innovation outcomes at the micro-level. More precisely, we analyzed how small-world properties affect firm innovativeness in a longitudinal empirical setting. The estimation results for the network's average path lengths are as expected. Thus, a short average path length at the overall network level is positively related to a firm-level innovative performance at later points in time. Our results for the clustering coefficient are in line with our theoretical expectations. We found a positive relatedness between clustering at the overall network level and firm innovativeness. Finally, estimation results provide empirical evidence for a positive relatedness between a network's small-world nature and a firm's subsequent innovativeness. This result is in sharp contrast to the findings of Fleming et al. (2007, p. 949) but in line with previous findings by Schilling and Phelps (2007).

Both theoretical and methodological limitations are closely related to graph theoretical concepts. Firstly, concerns were expressed that bipartite networks significantly exaggerate the network's true level of clustering and understate the true path length (Uzzi and Spiro 2005, p. 453). We checked for this issue by conducting several consistency checks. Consequently, we have to address the bipartite nature of the networks more explicitly. Not only a network's small world nature but also an in-depth analysis of other types of large-scale network characteristics, such as core-periphery patterns, provide promising opportunities for further research.

Secondly, we did not specify an empirical model that incorporates path-length, clustering and small-world indicators simultaneously. The reason for this is straightforward. In this study we were particularly interested in investigating the relatedness between three distinct structural patterns at the overall network level and firm-level innovativeness. A more integrated estimation approach would be the next logical step towards an in-depth understanding of how systemic parameters affect the innovativeness of the actors involved.

The last of four analytical parts (Chap. 12) explicitly addresses the proximity concept and analyzes the extent to which firm innovativeness is positively or negatively related to various proximity dimensions. More precisely, we investigated the relatedness between firm innovativeness and distinct and/or combined network positioning effects, and geographical co-location effects. Firstly, we found strong empirical evidence for the relevance of distinct network proximity effects on the innovative performance of German laser source manufacturers. In other words, a firm's degree centrality turned out to be positively related to its innovative performance at later points in time. Against our initial expectations, estimation results for co-location between laser source manufacturers turned out to be negatively correlated with firm-level innovation outcomes. Findings on combined geographical proximity and network proximity confirm our theoretical expectations that combined proximity effects are not independent.

From a theoretical point of view, we are at the onset. The proximity concept proposed by Boschma (2005) opens up rich opportunities to study the relatedness between network proximity and other proximity dimensions. For instance, Nooteboom (2008) heightened our awareness for the importance of cognitive proximity in this context. Another interesting theoretical perspective could be the integration of the isolation concept (cf. Ehrenfeld et al. 2014). Hall and Wylie (2014, p. 358) argue that isolation only rarely appears as a stringent analytical concept in the literature on economics and innovation. It is usually used in a descriptive or metaphoric way without being clearly defined (ibid). They make the point that isolation is a pervasive element of all kinds of social and economic systems which can be exogenous but also self-imposed (Hall and Wylie 2014, p. 373). The consequences of isolation for technological innovation are not yet fully understood. However, it is important to note that isolation in a geographical, social or cognitive sense is not necessarily negatively related to innovativeness (ibid). Instead isolation can provide a unique environment and induces innovation processes that otherwise would not have happened (Hall and Wylie 2014, p. 374).

Like any empirical investigation, this analysis also has its methodological limitations. For instance, we used the localized density measure according to Sorenson and Audia (2000) to quantify two types of geographical proximity dimensions: geographical proximity between an LSM and other LSMs and geographical proximity between an LSM and other PROs. This approach is limited in several ways. It ignores, for instance, the effects of geographical proximity in the exploitation of inter-industry knowledge spillovers. Further research could include indicators capturing the effects of a firm's geographical embeddedness in diversified industrial agglomerations and in urban areas.

References

Barabasi A-L, Albert R (1999) Emergence of scaling in random networks. Science 286 (15):509–512

BMBF (2010) Ideas, innovation, prosperity – high-tech strategy 2020 for Germany. Federal Ministry of Education and Research, Bonn

Borgatti SP, Everett MG (1999) Models of core/periphery structures. Soc Networks 21 (4):375–395

Boschma R (2005) Proximity and innovation: a critical assessment. Reg Stud 39(1):61–74

De Nooy W, Mrvar A, Batagelj V (2005) Exploratory social network analysis wit PAJEK. Cambridge University Press, Cambridge

Ehrenfeld W, Kudic M, Pusch T (2014) On the trail of core-periphery patterns – measurement and new empirical findings from the German laser industry. Conference proceedings, 17th Uddevalla symposium, Udevalla, pp 1–22

Fabian C (2011) Technologieentwicklung im Spannungsfeld von Industrie, Wissenschaft und Staat: Zu den Anfängen des Innovationssystems der Materialbearbeitungslaser in der Bundesrepublik Deutschland 1960 bis 1997. Dissertation, TU Bergakademie Freiberg

Fleming L, King C, Juda AI (2007) Small worlds and regional innovation. Organ Sci 18 (6):938–954

Hagedoorn J (2006) Understanding the cross-level embeddedness of interfirm partnership formation. Acad Manag Rev 31(3):670–680

Hall P, Wylie R (2014) Isolation and technological innovation. J Evol Econ 24(2):357–376

Kudic M, Pyka A, Sunder M (2013) Network formation: R&D cooperation propensity and timing among German laser source manufacturers. IWH discussion papers, 01/2013, pp 1–25

Nooteboom B (2008) Learning and innovation in inter-organizational relationships. In: Cropper S, Ebers M, Huxham C, Ring PS (eds) The Oxford handbook of interorganizational relations. Oxford University Press, New York, pp 607–634

Powell WW, White DR, Koput KW, Owen-Smith J (2005) Network dynamics and field evolution: the growth of the interorganizational collaboration in the life sciences. Am J Sociol 110 (4):1132–1205

Scherngell T, Barber MJ (2009) Spatial interaction modeling of cross-region R&D collaborations: empirical evidence from the 5th EU framework programme. Pap Reg Sci 88(3):531–546

Schilling MA, Phelps CC (2007) Interfirm collaboration networks: the impact of large-scale network structure on firm innovation. Manag Sci 53(7):1113–1126

Sorenson O, Audia PG (2000) The social structure of entrepreneurial activity: geographic concentration of footwear production in the Unites States, 1940–1989. Am J Sociol 106 (2):424–462

Uzzi B, Spiro J (2005) Collaboration and creativity: the small world problem. Am J Sociol 111 (2):447–504

Watts DJ, Strogatz SH (1998) Collective dynamics of 'small-world' networks. Nature 393 (6684):440–442

Chapter 14
Further Research and Conclusions

Believing in progress does not mean believing that any progress has yet been made.

(Franz Kafka)

Abstract The paradox underlying each scientific research project is that once it comes to an end we face more questions than at the beginning of the process. Of course, not all of the interesting new questions can be addressed here in detail. Nonetheless, we believe that a comprehensive understanding of network dynamics is essential for nearly all other fields of cooperation and network research. The complexity of network change processes calls for the application of unconventional methods. In my point of view this opens up a most promising field of research and constitutes, at the same time, the core of the outlook that follows in Sect. 14.1. Finally, we conclude with some final remarks in Sect. 14.2.

14.1 Fruitful Avenues for Further Research

The preceding discussion shows that our database has to be extended in several ways. Even though data and methods used in this study provide a good starting point for the analysis of network change processes, they are limited in several ways and the German laser industry still has many interesting secrets to divulge. Widely unexplored archival raw data sources contain valuable information on firm characteristics and cooperation activities that are waiting to be explored. We have recently started to extend the database in all four of the following areas: industry data, firm data, network data and innovation data. Our efforts encompass not only data gathering but also the construction of more sophisticated indicators. For instance, a promising way to gather additional information on R&D cooperation activities between LSMs and PROs is the exploration and utilization of bibliometric data.

© Springer International Publishing Switzerland 2015

M. Kudic, *Innovation Networks in the German Laser Industry*,
Economic Complexity and Evolution, DOI 10.1007/978-3-319-07935-6_14

Bibliographical sources can also be employed to gather more comprehensive information on innovation activities at the firm level. Information on product placement and advertisement can be used to gather and construct market-based innovation indicators. Quite recently we started systematically exploring data on new product launches based on several archival raw data sources in order to gain a more appropriate picture of innovation processes at the firm level. One of our next steps will be to focus on the inclusion of international linkages in our database to lay the groundwork for studying networks in an international context.

Secondly, more sophisticated empirical estimation methods are needed to address some of the empirical limitations. Both parametric and semi-parametric estimation approaches (Blossfeld et al. 2007) provide a broad range of empirical models that can be used for an in-depth analysis of tie formation and tie termination processes at the firm level. Moreover, we used standard panel data count models for our estimation in Chaps. 10, 11, and 12. These methods are limited in at least two ways. Firstly, the conditional fixed effects estimation approach, which is usually implemented in standard software packages, has been criticized (Allison and Waterman 2002). Secondly, more sophisticated methods have recently been proposed in the literature to handle selection biases in panel data (Imbens and Wooldridge 2009). These empirical challenges need to be addressed in future.

In addition to the issues addressed above, other powerful methods are now available such as agent-based simulation approaches. We are convinced that the use of different methodological approaches adds value in understanding a specific phenomenon. Two classes of agent-based models seem to have the potential to break new ground in the field of interorganizational network research.

The first class of models, so-called stochastic agent-based models (Snijders 2004; Snijders et al. 2010; Huisman and Snijders 2003; Huisman and Steglich 2008), can be applied to explore the mechanism that fuels the structural change of networks between two or more discrete points in time. The main focus of stochastic actor-based simulation models is the analysis of network evolution processes and co-evolutionary processes between social networks and changeable actor attributes (Snijders 2004). At their core, stochastic agent-based models combine a random utility model, continuous-time Markov process, and Monte Carlo simulation (Buchmann et al. 2014, p. 27). One processing avenue is to apply these models to gain a more profound understanding of how and why interorganizational innovation networks change over time.[1]

Stochastic actor-based models possess several distinctive features, including flexibility and accessibility of procedures to estimate as well as to test parameters which support the description of mechanisms or tendencies (Snijders et al. 2010, p. 2). Therefore, they reflect "network dynamics as being driven by many different tendencies" (Snijders et al. 2010, p. 1). These tendencies may be, for example, reciprocity, transitivity or homophily (ibid). Stochastic actor-based models are

---

[1] The following discussion is guided by Snijders et al. (2010). See also, Huisman and Snijders (2003).

based on some basic assumptions (cf. Snijders et al. 2010 pp. 1–3). Firstly, the time parameter t is continuous. This postulation enables the representation of dependencies between ties which are the consequence of processes where one tie occurs due to the existence of others. Secondly, the modifications of the network are the result of a Markov process, i.e. that "for any point in time, the current state of the network determines probabilistically its further evolution, and there are no additional effects of the earlier past" (Snijders et al. 2010, p. 2). The third assumption is that the actors are in control of their outgoing ties. Therefore, the changes of ties occur as a result of the actions of the actors instigating the tie which is influenced by their and other actors' attributes, their location in the network as well as their awareness of the rest of the network. Fourthly, at a certain point in time one probabilistically chosen actor ('ego') may have the occasion to change one outgoing tie. This postulation ends by decomposing the process of change into its minimum of possible components and consequently in the implication that alterations are not implemented coordinately, but merely depends on each other sequentially (Snijders et al. 2010, p. 3).

In the application of stochastic actor-based models, the focal actor – the one who can make a change – has to be selected with equal probabilities or with probabilities that depend on features like network position or other attributes. His reaction possibilities include the opportunity to change one outgoing tie or to do nothing. Hence, the set of permissible actions includes n elements ($n-1$ changes and one non-change). "The probabilities for a choice depend on the so-called objective function" (Snijders et al. 2010, p. 3) which is the heart of this model. The objective function ultimately determines the probabilities of modification in the network. The occurring effects can be divided into two groups: (a) endogenous effects, such as basic effects, transitivity and other triadic effects and degree-related effects that solely depend on the network itself, (b) exogenous effects (covariates) and interactions that, in contrast, are external in nature.[2] Moreover, it needs to be emphasized that issues regarding statistical modeling may arise. This means, among other things, that certain data requirements have to be met e.g. number of actors, number of observation moments, and the total number of observations (Snijders et al. 2010, p. 6).

By now, there are some excellent studies using stochastic agent-based methods in an economic context (Van de Bunt and Groenewegen 2007; Balland et al. 2012; Ter Wal and Boschma 2011; Giuliani 2010). One of our current research projects also moves in this direction. The study conducted by Buchmann et al. (2014) explores evolutionary network change processes in the German laser and automotive industry by using a stochastic actor-based simulation approach. The results provide empirical evidence for the explanatory power of network-related determinants in both industries.

Another class of models, the so-called KENE approach (Gilbert et al. 2001, 2007; Pyka et al. 2007) allows a firm's knowledge base, learning processes and knowledge transfer in complex network structures to be modeled. These types of

---

[2] For further explanation, see Snijders et al. (2010, pp. 4–6).

agent-based models can be applied to simulate micro-level firm behavior which shapes the macro-level network patterns.

Work has already started in this research area. Mueller and colleagues (2014) draw upon the KENE approach to analyze the evolution of interfirm innovation networks. In this study we focus on the evolutionary change of innovation networks which are composed of and driven by individual strategies and goals of heterogeneous actors. These actors follow a number of well-defined cooperation partner selection strategies. The agent-based simulation model (ABSM) that was implemented allows the causal relationships between firm strategies and the emerging network structures to be analyzed.

Mueller and colleagues (2014) applied the model to test the following well-known mechanisms that are assumed to affect a firm's cooperation activities and affect the evolution of the overall network over time: homophily, reputation and cohesion mechanisms. An initial, simplified version of the model was extended by adding a market mechanism which linked the knowledge base of a firm with the rewards a firm receives and with its incentives to cooperate. The results of our study show that a transitive closure mechanism, combined with a tendency for preferential attachment, produces networks that exhibit both small-world characteristics and a power-law degree distribution. Moreover our simulation results suggest that diversity in the selection of cooperation partners is important when we consider an evolving network.

14.2   Some Concluding Considerations

An in-depth understanding of collective innovation processes and technological change patterns is a necessary prerequisite for creating appropriate conditions for economic growth and prosperity. Indeed, there are still a lot of open questions to be addressed in order to provide a more comprehensive understanding of the evolutionary nature of innovation networks.

This study demonstrates that the neo-Schumpeterian approach in economics provides an appropriate theoretical framework for studying firm innovativeness in evolving networks. We chose this theoretical framework and decided in favor of a longitudinal empirical setting because we were convinced that factors influencing the creation of novelty are best understood from a dynamic perspective. Similarly, methods used for the purpose of this study were selected on the basis of two criteria. One the one hand, they must allow for an exact measurement of industry, firm, network and innovation characteristics at multiple analytical levels. On the other hand, they must be able to account for change processes over time. In principle, all applied indicators and methods, i.e. basic descriptive indicators, social network analysis methods and empirical estimation techniques, meet these requirements.

All in all our results show that R&D cooperation and innovation network involvement affects the innovativeness of science-driven firms in multiple ways. We believe that this book makes a valuable contribution to innovation network literature by exploring how and why firm-specific R&D cooperation activities and network positions, large-scale network patterns and evolutionary network change processes affect the innovative performance of laser source manufacturers in Germany. Nonetheless results should always be accessed and interpreted carefully in light of the limitations raised above. Current follow-up studies, using alternative methodological approaches, have already confirmed some of our findings and contributed towards a better understanding of network entry processes (Kudic et al. 2013; Kudic et. al. 2015) and network evolution processes (Mueller et al. 2014; Buchmann et al. 2014; Kudic and Guenther 2014). In a similar vein, recently started research projects on core-periphery patterns in Large-scale networks (Ehrenfeld et al. 2014) aim to complement and enhance our current picture of collective innovation processes in the German laser industry.

While this book is certainly a good starting point, there is yet much to be done to fully understand evolutionary network change, strategic positioning, and firm innovativeness in the German laser industry.

References

Allison PD, Waterman R (2002) Fixed-effects negative binomial regression models. Sociol Methodol 32(1):247–265

Balland PA, De Vaan M, Boschma R (2012) The dynamics of interfirm networks along the industry life cycle: the case of the global video game industry, 1987–2007. J Econ Geogr 13 (5):1–25

Blossfeld H-P, Golsch K, Rohwer G (2007) Event history analysis with Stata. Lawrence Erlbaum, London

Buchmann T, Hain D, Kudic M, Mueller M (2014) Exploring the evolution of innovation networks in science-driven and scale-intensive industries – new evidence from a stochastic actor-based approach. IWH discussion papers 01/2014, pp 1–42

Ehrenfeld W, Kudic M, Pusch T (2014) On the trail of core-periphery patterns – measurement and new empirical findings from the German laser industry. Conference proceedings, 17th Uddevalla symposium, Udevalla, pp 1–22

Gilbert N, Pyka A, Ahrweiler P (2001) Innovation networks – a simulation approach. J Artif Soc Soc Simul 4(3):1–13

Gilbert N, Ahrweiler P, Pyka A (2007) Learning in innovation networks: some simulation experiments. Phys A Stat Mech Appl 378(1):100–109

Giuliani E (2010) Network dynamics in regional clusters: the perspective of an emerging economy. Pap Evolut Econ Geogr (PEEG) (Online) 10(14)

Huisman M, Snijders TA (2003) Statistical analysis of longitudinal network data with changing composition. Sociol Methods Res 32:253–287

Huisman M, Steglich CE (2008) Treatment of non-response in longitudinal network data. Soc Networks 30:297–308

Imbens GW, Wooldridge JM (2009) Recent developments in the econometrics of program evaluation. J Econ Lit 47:5–86

Kudic M, Guenther J (2014) Towards an in-depth understanding of structural network change processes in innovation networks. In: International Schumpeter society conference proceedings, Jena, 27.30 July

Kudic M, Pyka A, Sunder M (2013) Network formation: R&D cooperation propensity and timing among German laser source manufacturers. IWH discussion papers, 01/2013, pp 1–25

Kudic M, Pyka A, Guenther J (2015) Taking the first step – what determines German laser source manufacturers' entry into innovation networks? Int J Innov Manag IJIM (forthcoming)

Mueller M, Buchmann T, Kudic M (2014) Micro strategies and macro patterns in the evolution of innovation networks: an agent-based simulation approach. In: Gilbert N, Ahrweiler P, Pyka A (eds) Simulating knowledge dynamics in innovation networks. Springer, Heidelberg/New York, pp. 73–95

Pyka A, Gilbert N, Ahrweiler P (2007) Simulating knowledge-generation and distribution processes in innovation collaborations and networks. Cybern Syst 38(7):667–693

Snijders TA (2004) Explained variation in dynamic network models. Math Soc Sci 42(168):5–15

Snijders TA, Van De Bunt GG, Steglich CE (2010) Introduction to actor-based models for network dynamics. Soc Networks 32(1):44–60

Ter Wal AL, Boschma R (2011) Co-evolution of firms, industries and networks in space. Reg Stud 45(7):919–933

Van de Bunt GG, Groenewegen P (2007) An actor-oriented dynamic network approach. Organ Res Methods 10(3):463–482

Appendix

Appendix 1: Bibliometric Analysis

Bibliometric Analysis: Data Sources, Search Terms and List of Explored Journals

We used several online databases to conduct our search. These are among others: ISI Web of knowledge (http://wokinfo.com), EBSCO (http://web.ebscohost.com), JSTORE (http://www.jstor.org). In addition we explored an IWH literature database to conduct our analysis. The differentiation of journals by scientific field was conducted on the basis of the journals' main focus.

The following search terms were used to identify publications on cooperation in general: "coop", "collab", "co-op", "partnership", "dyad", "dyadic", "alliance", "inter-org", "interorg", "inter-firm", "interfirm", "network", "linkage", "link", "relationship", "relation", "hybrid", "joint", "franchising", "licensing".

The following search terms were used to identify publications on alliance portfolios, ego-networks and multi-partner alliances: "portfolio", "ego", "constell", "multi-partner".

The following search terms were used to identify publications on networks in general and large-scale network features: "network", "net", "small-world", "small world", "large-scale", "core-periphery", "scaling", "scale-free", "fat-tailed".

The following search terms were used to identify alliance and network-related publications that deal with dynamic issues: "change", "growth", "dyn", "evol", "process", "transition", "formation", "emerge", "fragmentation".

We found relevant articles in 242 academic journals. The full list of journals is provided below.

© Springer International Publishing Switzerland 2015   329
M. Kudic, *Innovation Networks in the German Laser Industry*,
Economic Complexity and Evolution, DOI 10.1007/978-3-319-07935-6

Academy of Management Executive
Academy of Management Journal
Academy of Management Review
Acta Sociologica
Administrative Science Quarterly
Advances in Complex Systems
Advances in Strategic Management
American Economic Review
American Journal of Sociology
American Physical Society
American Sociological Review
Annals of Regional Science
Annual Review of Sociology
Artificial Intelligence and Society
B.E. Journal of Theoretical Economics
Bell Journal of Economics
Biol Philos
British Journal of Management
Brokering Digest
Brooking Institution
Bulletin of Economic Research
Bulletin of Mathematical Biophysics
Business and Economic History
Business Digest
Business Quarterly
California Management Review
Cambridge Journal of Economics
Chain and Network Science
Comparative Economic Studies
Competition & Change
Competitiveness Review
Complexity
Complexity International
Computational & Mathematical Organization Theory
Computation and Economics
Connections
Creativity and Innovation Management
Current Politics and Economics of Europe
Das Wirtschaftsstudium
Die Unternehmung

Econometrica
Economics and Business Review
Economica New Series
Economic Geography
Economic Journal
Economics of Innovation and New Technology
Economic Theory
Economy and Society
Entrepreneurship & Regional Development
Entrepreneurship Theory and Practice
European Journal of Innovation Management
European Journal of Marketing
European Management Journal
European Management Review
European Planning Studies
European Sociological Review
Europhysics Letters
Experimental Economics
Games and Economic Behavior
Growth and Change
Harvard Business Manager
Harvard Business Review
History of Political Economy
Human Relations
Industrial and Corporate Change
Industrial Marketing Management
Industry & Innovation
Infor
Information Storage Industry Center
Institute of Global Management Studies
International Business Review
International Journal of Behavioral Development
International Journal of Economics of Business
International Journal of Entrepreneurship and Innovation Management
International Journal of Game Theory
International Journal of Healthcare Technology and Management
International Journal of Industrial Organization
International Journal of Innovation Management
International Journal of Management Reviews
International Journal of Organizational Analysis

International Journal of Physical Distribution & Logistics
International Journal of Research in Marketing
International Journal of Sociology and Social Policy
International Journal of Technology Management
International Journal of Technology Management and Sustainable Development
International Journal of the Economics of Business
International Marketing Review
International Sociology
International Studies of Management and Organization
Ivey Business Journal
Journal of American Academy of Business
Journal of Applied Econometrics
Journal of Business & Industrial Marketing
Journal of Business and Enterprise Development
Journal of Business Chemistry
Journal of Business Research
Journal of Business Review
Journal of Business Strategies
Journal of Business Venturing
Journal of Commercial Biotechnology
Journal of Computer-Mediated Communication
Journal of Econometrics
Journal of Economic Behavior and Organization
Journal of Economic Geography
Journal of Economic Issues
Journal of Economic Literature
Journal of Economic Perspectives
Journal of Economics of Business
Journal of Economic Theory
Journal of Enterprising Culture
Journal of Evolutionary Economics
Journal of Financial Economics
Journal of High Technology Management Research
Journal of Industrial Economics
Journal of Industry Competition and Trade
Journal of Industry Studies
Journal of Institutional and Theoretical Economics
Journal of International Business Studies
Journal of International Entrepreneurship
Journal of International Marketing

Journal of Labor Research
Journal of Law Economics & Organization
Journal of Management
Journal of Management and Governance
Journal of Management Research
Journal of Management Science
Journal of Management Studies
Journal of Marketing
Journal of Marketing Management
Journal of Marketing Research
Journal of New Business Ideas and Trends
Journal of Organizational Behavior
Journal of Organizational Change Management
Journal of Political Economy
Journal of Product Innovation Management
Journal of Public Administration Research and Theory
Journal of Regional Science
Journal of Small Business and Enterprise Development
Journal of Small Business Management
Journal of Statistical Mechanics
Journal of Statistical Planning and Inference
Journal of Strategic Marketing
Journal of Technology Transfer
Journal of the Academy of Marketing Science
Journal of the Economics of Business
Journal of the European Economic Association
Journal of the Korean Physical Society
Journal of the Operational Research Society
Journal of World Business
Knowledge and Process Management
Kyklos
Laser Technik Journal
Local Economy
Long Range Planning
Management Decision
Management International Review
Management Revue
Management Science
Managerial and Decision Economics
Marketing Review

Mathematical Social Sciences
Minerva
MIT Sloan Management Review
Nanotechnology Law and Business
Nature
Nonlinear Dynamics and Evolutionary Economics
Organizational Dynamics
Organization Science
Organization Studies
Papers in Regional Science
Philosophical Transansitions in the Royal Society B
Physica A
Physical Review Letters
PNAS
Problems and Perspectives in Management
Progress in Human Geography
Psychometrika
Public Choice
Public Money & Management
Public Policy and Administration
Quarterly Journal of Economics
Quarterly Journal of Political Science
R&D Management
Rand Journal of Economics
Regional Science
Regional Studies
Research in Management
Research in Organizational Behavior
Research Policy
Research Technology Management
Review of Austrian Economics
Review of Industrial Organization
Review of International Political Economy
Review of Social Economy
Reviews of Modern Physics
Revue D'Economic Industrielle
RIKEN Review
SAM Advanced Management Journal
Scandinavian Journal of Economics
Scandinavian Journal of Management

Schmalenbach Business Review
Science
Science Technology and Human Values
Scientific American
Scientometrics
SIGKDD Explorations
Simulating Social Phenomena
Singapore Management Review
Small Business Economics
Social Forces
Social Methodology
Social Networks
Social Science Journal
Social Science Research
Social Studies of Science
Sociological Methods and Research
Sociometry
Stochastic Models
Strategic Management Journal
Strategic Organization
Strategies in Global Competition
Strategy & Leadership
Structural Change and Economic Dynamics
Technological Forecasting & Social Change
Technology Analysis & Strategic Management
Technovation
The Information Society
The Journal of Operational Research Society
The Journal of Strategic Management
The Leadership Quarterly
The Policy Studies Journal
The Royal Society
Tijdschrift voor Economische en Sociale Geografie
Trends in Ecology and Evolution
Urban Studies
Wirtschaftsinformatik
Work Employment and Society
World Development
Zeitschrift für Betriebswirtschaft
Zeitschrift für betriebswirtschaftliche Forschung
Zeitschrift für Soziologie
Zeitschrift für Wirtschaftsgeographie

Appendix 2: Complete List of the 97 Planning
Regions as Applied in this Analysis

ZIP code	Number of planning region	Name of planning region	State
24901	1	Schleswig-Holstein Nord	SH
25534	2	Schleswig-Holstein Süd-West	SH
24011	3	Schleswig-Holstein Mitte	SH
23400	4	Schleswig-Holstein Ost	SH
21018	5	Schleswig-Holstein Süd	SH
20001	6	Hamburg	HH
19001	7	Westmecklenburg	MV
18001	8	Mittleres Mecklenburg/Rostock	MV
17461	9	Vorpommern	MV
17013	10	Mecklenburgische Seeplatte	MV
28001	11	Bremen	HB
26691	12	Ost-Friesland	NI
27501	13	Bremerhaven	HB
21202	14	Hamburg-Umland-Süd	NI
27201	15	Bremen-Umland	NI
26001	16	Oldenburg	NI
26851	17	Emsland	NI
49001	18	Osnabrück	NI
30001	19	Hannover	NI
29200	20	Südheide	NI
29431	21	Lüneburg	NI
38001	22	Braunschweig	NI
31013	23	Hildesheim	NI
34331	24	Göttingen	NI
16501	25	Prignitz-Oberhavel	BB
16201	26	Uckermark-Barnim	BB
15201	27	Oderland-Spree	BB
3001	28	Lausitz-Spreewald	BB
14731	29	Havelland-Fläming	BB
10001	30	Berlin	BE

(continued)

ZIP code	Number of planning region	Name of planning region	State
39511	31	Altmark	LSA
39001	32	Magdeburg	LSA
6811	33	Dessau	LSA
6002	34	Halle/Saale	LSA
48001	35	Münster	NW
33501	36	Bielefeld	NW
32833	37	Paderborn	NW
34418	38	Arnsberg	NW
44001	39	Dortmund	NW
46201	40	Emscher-Lippe	NW
47001	41	Duisburg/Essen	NW
40001	42	Düsseldorf	NW
44701	43	Bochum/Hagen	NW
50400	44	Köln	NW
52001	45	Aachen	NW
53001	46	Bonn	NW
57341	47	Siegen	NW
34001	48	Nordhessen	HE
35301	49	Mittelhessen	HE
36001	50	Osthessen	HE
60001	51	Rhein-Main	HE
64100	52	Starkenburg	HE
37301	53	Nordthüringen	TH
99001	54	Mittelthüringen	TH
98490	55	Südthüringen	TH
7490	56	Ostthüringen	TH
4001	57	Westsachsen	SN
1001	58	Oberes Elbtal/Osterzgebirge	SN
2806	59	Oberlausitz-Niederschlesien	SN
9001	60	Chemnitz-Erzgebirge	SN
8501	61	Südwestsachsen	SN
56001	62	Mittelrhein-Westerwald	RP
54181	63	Trier	RP
55438	64	Rheinhessen-Nahe	RP
67601	65	Westpfalz	RP
67201	66	Rheinpfalz	RP
66001	67	Saar	SL
69001	68	Unterer Neckar	BW
74001	69	Franken	BW
76481	70	Mittlerer Oberrhein	BW

(continued)

ZIP code	Number of planning region	Name of planning region	State
75090	71	Nordschwarzwald	BW
70001	72	Stuttgart	BW
89166	73	Ostwürtemberg	BW
89001	74	Donau-Iller (BW)	BW
72120	75	Neckar-Alb	BW
72168	76	Schwarzwald-Baar-Heuberg	BW
79001	77	Südlicher Oberrhein	BW
78201	78	Hochrhein-Bodensee	BW
78352	79	Bodensee-Oberschwaben	BW
63701	80	Bayerischer Untermain	BY
97001	81	Würzburg	BY
97401	82	Main-Rhön	BY
96001	83	Oberfranken-West	BY
95400	84	Oberfranken-Ost	BY
92200	85	Oberpfalz-Nord	BY
91001	86	Industrieregion Mittelfranken	BY
91501	87	Westmittelfranken	BY
86001	88	Augsburg	BY
85001	89	Ingolstadt	BY
84042	90	Regensburg	BY
94001	91	Donau-Wald	BY
84001	92	Landshut	BY
80001	93	München	BY
87681	94	Donau-Iller (BY)	BY
87571	95	Allgäu	BY
82055	96	Oberland	BY
83001	97	Südostoberbayern	BY

Appendix 3: Consistency Check: Small World Properties in the German Laser Industry Innovation Network

Consistency Check: Small World Properties

Procedural steps:

1. Use of multi-partner R&D project data (cf. Sect 4.2.3).
2. Assumption: Project partners are not fully connected. A focal actor is assumed to have only direct linkages to other project partners; these partners are not interconnected ("star networks") (cf. Sect. 4.2.3).
3. Annual network layers are compiled based on the R&D project data.
4. Calculation of small world indicators (cf. Sects. 5.2.3 and 8.3.2).

Clustering coefficient ratio

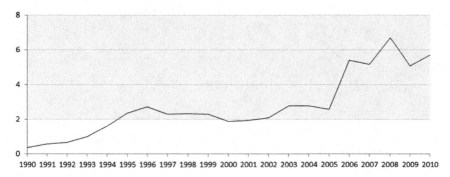

Path length ratio

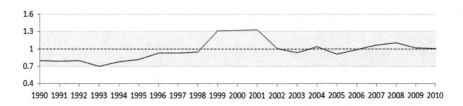

Small world (Q)

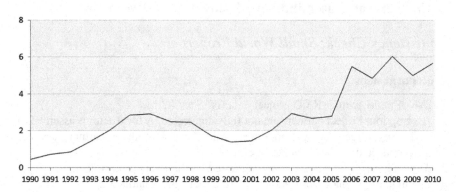

Source: Author's own calculation

Bibliography

Acar W, Sankaran K (1999) The myth of unique decomposability: specializing the Herfindahl and entropy measures. Strateg Manag J 20(1):969–975

Ackerlof GA (1970) The market for "lemons". Quality uncertainty and the market mechanism. Q J Econ 84(3):488–500

Acs ZJ, Audretsch DB, Deldman MP (1992) Real effects of academic research: comment. Am Econ Rev 82(1):363–367

Agrawal A (2001) University-to-industry knowledge transfer: literature review and unanswered questions. Int J Manag Rev 3(4):285–302

Agyris C, Schön DA (1978) Organizational learning: a theory of action perspective. Addison-Wesley, Reading

Ahuja G (2000) Collaboration networks, structural hole, and innovation: a longitudinal study. Adm Sci Q 45(3):425–455

Ahuja G, Soda G, Zaheer A (2012) The genesis and dynamics of organizational networks. Organ Sci 23(2):434–448

Albert R, Barabasi A-L (2000) Topology of evolving networks: local events and universality. Phys Rev Lett 85(24):5234–5237

Albert R, Barabasi A-L (2002) Statistical mechanics of complex networks. Rev Mod Phys 74(1):47–97

Albrecht H (1997) Eine vergleichende Studie zur Frühgeschichte von Laserforschung und Lasertechnik in der Bundesrepublik und der Deutschen Demokratischen Republik. Habilitation: Universität Stuttgart, Stuttgart

Albrecht H (2010a) The German Research Foundation and the early days of laser research at West German universities during the 1960s. In: Trischler H, Walker M (eds) Physics and politics – research and research support in twentieth century Germany in international perspective. Franz Steiner Verlag, Stuttgart, pp 161–196

Albrecht H (2010b) Innovationen im Zeichen von Planwirtschaft und SED-Diktatur – Die Anfänge der Entwicklung der Laser-Technologie in Jena in den 1969er Jahren. In: Dicke K, Cantner U, Ruffert M (eds) Die Rolle der Universität in Wirtschaft und Gesellschaft. IKS Garamond, Jena, pp 171–201

Albrecht H, Buenstorf G, Fritsch M (2011) System? What system? The (co-) evolution of laser research and laser innovation in Germany since 1960. Working paper, pp 1–38

Aldrich HE, Ruef M (2006) Organizations evolving, 2nd edn. Sage, London

Alic JA (1990) Cooperation in R&D. Technovation 10(5):319–332

Al-Laham A, Kudic M (2008) Strategische Allianzen. In: Corsten H, Goessinger R (eds) Lexikon der Betriebswirtschaftslehre, 5th edn. Oldenbourg Verlag, München, pp 39–41

Allison PD (1984) Event history analysis – regression for longitudinal event data. Sage, London

Allison PD, Waterman R (2002) Fixed-effects negative binomial regression models. Sociol Methodol 32(1):247–265

Almeida P, Kogut B (1999) Localization of knowledge and the mobility of engineers in regional networks. Manag Sci 45(7):905–917

Alter M (1982) Carl Menger and homo oeconomicus: some thoughts on Austrian theory and methodology. J Econ Issue XVI(1):149–160

Amburgey T, Al-Laham A (2005) Islands in the net. Conference paper: 22nd EGOS colloquium, Bergen, pp 1–42

Amburgey TL, Rao H (1996) Organizational ecology: past, present, and future directions. Acad Manag J 39(5):1265–1286

Amburgey TL, Singh JV (2005) Organizational evolution. In: Baum JA (ed) The Blackwell companion to organizations. Blackwell, Malden, pp 327–343

Amburgey TL, Dacin T, Singh JV (1996) Learning races, patent races, and capital races: strategic interaction and embeddedness within organizational fields. In: Baum JA (ed) Advances in strategic management. Elsevier, New York, pp 303–322

Amburgey TL, Al-Laham A, Tzabbar D, Aharonson BS (2008) The structural evolution of multiplex organizational networks: research and commerce in biotechnology. In: Baum JA, Rowley TJ (eds) Advances in strategic management – network strategy, vol 25. Emerald Publishing, Bingley, pp 171–212

Amburgey T, Aharonson BS, Tzabbar D (2009) Heterophily in inter-organizational network ties. Conference paper: 25th EGOS colloquium, Barcelona, pp 1–40

Amin A, Wilkinson F (1999) Learning, proximity and industrial performance: an introduction. Camb J Econ 23(2):121–125

Anand BN, Khanna T (2000) Do firms learn to create value? The case of alliances. Strateg Manag J 21(3):295–315

Anthonisse JM (1971) The rush in the directed graph – technical report BN 9/71. Stichting Mathematisch Centrum, Amsterdam

Antonelli C (2009) The economics of innovation: from the classical legacies to the economics of complexity. Econ Innov New Technol 18(7):611–646

Antonelli C (2011) Handbook on the economic complexity of technological change. Edward Elgar, Cheltenham

Araujo L, Harrison D (2002) Path dependence, agency and technological evolution. Tech Anal Strat Manag 14(1):5–19

Arend RJ (2009) Reputation for cooperation: contingent benefits in alliance activity. Strateg Manag J 30(4):371–385

Arino A, De La Torre J (1998) Learning from failure: towards an evolutionary model of collaborative ventures. Organ Sci 9(3):306–325

Arrow KJ (1962) The economic implications of learning by doing. Rev Econ Stud 29(3):155–173

Arrow KJ (1994) Methodological individualism and social knowledge. Am Econ Rev 84(2):1–9

Arthur BW (1989) Competing technologies, increasing returns, and lock-in by historical events. Econ J 99(394):116–131

Arthur BW (2007) Complexity and the economy. In: Hanusch H, Pyka A (eds) Elgar companion to neo-Schumpeterian economics. Edward Elgar, Cheltenham, pp 1102–1110

Audretsch DB (1998) Agglomeration and the location of innovative activity. Oxf Rev Econ Policy 14(2):18–29

Audretsch DB, Dohse D (2007) Location: a neglected determinant of firm growth. Rev World Econ 143(1):79–107

Audretsch DB, Feldman MP (1996) R&D spillovers and the geography of innovation and production. Am Econ Rev 86(3):630–640

Audretsch DB, Feldman MP (2003) Small-firm strategic research partnerships: the case of biotechnology. Tech Anal Strat Manag 15(2):273–288

Audretsch DB, Lehmann E, Warning S (2004) University spillovers: does the kind of science matter? Ind Innov 11(3):193–205

Bain JS (1950) Workable competition in oligopoly: theoretical consideration and some empirical economics. Am Econ Rev 40(2):35–47

Bain JS (1951) Relation of profit rate to industry concentration: American manufacturing, 1936–1940. Q J Econ 65:293–324

Balland PA, De Vaan M, Boschma R (2012) The dynamics of interfirm networks along the industry life cycle: the case of the global video game industry, 1987–2007. J Econ Geogr 13 (5):1–25

Barabasi A-L, Albert R (1999) Emergence of scaling in random networks. Science 286(15): 509–512

Barabasi A-L, Bonabeau E (2003) Scale-free networks. Sci Am 288(5):50–59

Barkema HG, Bell JH, Pennings JM (1996) Foreign entry, cultural barriers, and learning. Strateg Manag J 17(2):151–166

Barney JB (1991) Firm resources and sustained competitive advantage. J Manag 17(1):99–120

Barney JB (2001) Is the resource-based "view" a useful perspective for strategic management research? Yes. Acad Manag Rev 26(1):41–56

Barney JB (2002) Gaining and sustaining competitive advantage. Prentice Hall, Upper Saddle River

Barron DN, West E, Hannan MT (1994) A time to grow and a time to die: growth and mortality of Credit Unions in New York City, 1914–1990. Am J Sociol 100(2):381–421

Basov NG, Prokhorov AM (1950) About possible methods for obtaining active molecules (English translation). Zh Eksp Teor Fiz 28(2):249–250

Basov NG, Danilychev VA, Popov YM, Khodkevich DD (1970) Laser operating in the vacuum region of the spectrum by excitation of liquid xenon with an electron beam (english translation). ZhETF Pis Red 12(10):473–474

Baum JA, Calabrese T, Silverman BS (2000) Don't go it alone: alliance network composition and startup's performance in Canadian biotechnology. Strateg Manag J 21(3):267–294

Baum JA, Shipilov AW, Rowley TJ (2003) Where do small worlds come from? Ind Corp Chang 12(4):697–725

Bavelas A (1948) A mathematical model for group structure. Hum Organ 7(3):16–30

Beauchamp MA (1965) An improved index of centrality. Behav Sci 10(2):161–163

Becker MC (2004) Organizational routines: a review of the literature. Ind Corp Chang 13(4): 643–678

Benjamin BA, Podolny JM (1999) Status, quality, and social order in the California wine industry. Adm Sci Q 44(3):563–589

Bergenholtz C, Waldstrom C (2011) Inter-organizational network studies – a literature review. Ind Innov 18(6):539–562

Bertalanffy LV (1951) Problems of general system theory. Hum Biol 23(4):302–312

Bertalanffy LV (1968) General system theory: foundations, development, applications. George Braziller, New York

Bertolotti M (2005) The history of the laser. Institute of Physics Publishing, Bristol

Bessant J, Alexander A, Tsekouras G, Rush H, Lamming R (2012) Developing innovation capability through learning networks. J Econ Geogr 12(5):1087–1112

Betton J, Dess GG (1985) The application of population ecology models to the study of organizations. Acad Manag Rev 10(4):750–757

Bidault F, Cummings T (1994) Innovating through alliances: expectations and limitations. R&D Manag 24(1):33–45

Bierly PE, Kessler EH, Christensen EW (2000) Organizational learning, knowledge and wisdom. J Organ Chang Manag 13(6):595–618

Bleeke J, Ernst D (1991) The way to win in cross-border alliances. Harv Bus Rev 69(6):127–135

Bloch I, Haensch T, Esslinger T (1999) Atom laser with a cw output coupler. Phys Rev Lett 82(15):3008–3011

Blossfeld H-P, Rohwer G (2002) Techniques of event history analysis – new approaches to causal analysis. Lawrence Erlbaum, London

Blossfeld H-P, Golsch K, Rohwer G (2007) Event history analysis with Stata. Lawrence Erlbaum, London

BMBF (2010) Ideas, innovation, prosperity – high-tech strategy 2020 for Germany. Federal Ministry of Education and Research, Bonn

Bonacich P (1972) Factoring and weighting approaches to status scores and clique identification. J Math Sociol 2(1):113–120

Bonacich P (1987) Power and centrality: a family of measures. Am J Sociol 92(5):1170–1182

Borgatti SP (2002) NetDraw: graph visualization software. Analytic Technologies, Harvard

Borgatti SP (2005) Centrality and network flow. Soc Networks 27(1):55–71

Borgatti SP, Everett MG (1999) Models of core/periphery structures. Soc Networks 21 (4):375–395

Borgatti SP, Foster PC (2003) The network paradigm in organizational research: a review and typology. J Manag 29(6):991–1013

Borgatti SP, Halgin DS (2011) On network theory. Organ Sci 22(5):1168–1182

Borgatti SP, Everett MG, Freeman LC (2002) Ucinet for windows: software for social network analysis. Analytic Technologies, Harvard

Borgatti SP, Everett MG, Johnson JC (2013) Analyzing social networks. Sage, London

Borys B, Jemison DB (1989) Hybrid arrangements as strategic alliances: theoretical issues in organizational combinations. Acad Manag Rev 14(2):234–249

Boschma R (2005a) Role of proximity in interaction and performance: conceptual and empirical challenges. Reg Stud 39(1):41–45

Boschma R (2005b) Proximity and innovation: a critical assessment. Reg Stud 39(1):61–74

Boschma RA, Frenken K (2006) Why is economic geography not an evolutionary science? Towards an evolutionary economic geography. J Econ Geogr 6(3):273–302

Boschma R, Frenken K (2010) The spatial evolution of innovation networks: a proximity perspective. In: Boschma R, Martin R (eds) The handbook of evolutionary economic geography. Edward Elgar, Cheltenham, pp 120–135

Boschma R, Martin R (2010) The aims and scope of evolutionary economic geography. In: Boschma R, Martin R (eds) The handbook of evolutionary economics geography. Edward Elgar, Cheltenham, pp 3–43

Boulding KE (1956) General system theory – the skeleton of science. Manag Sci 2(3):197–208

Boulding KE (1966) The economics of knowledge and the knowledge of economics. Am Econ Rev 56(1):1–13

Bourdieu P (1986) The forms of capital. In: Richardson J (ed) Handbook of theory and research for the sociology of education. Greenwood, New York, pp 241–258

Box GEP (1979) Robustness in the strategy of scientific model building. In: Launer RL, Wilkinson GN (eds) Robustness in statistics. Academic, New York, pp 201–236

Brass DJ, Galaskiewicz J, Greve HR, Tsai W (2004) Taking stock of networks and organizations: a multilevel perspective. Acad Manag J 47(6):795–817

Brenner T, Broekel T (2011) Methodological issues in measuring innovation performance of spatial units. Ind Innov 18(1):7–37

Brenner T, Cantner U, Graf H (2011) Innovation networks: measurement, performance and regional dimensions. Ind Innov 18(1):1–5

Breschi S, Lissoni F (2001) Knowledge spillovers and local innovation systems: a critical survey. Ind Corp Chang 10(4):975–1005

Breschi S, Lissoni F (2009) Mobility of skilled workers and co-invention networks: an anatomy of localized knowledge flows. J Econ Geogr 9(4):439–468

Brewer A (2010) The making of the classical theory of economic growth. Routledge, New York

Bridges WB (1964) Laser oscillation in single ionized argon in the visible spectrum. Appl Phys Lett 4(7):128–130

Broekel T, Graf H (2011) Public research intensity and the structure of German R&D networks: a comparison of ten technologies. Econ Innov New Technol 21(4):345–372

Broekel T, Hartog M (2013) Explaining the structure of inter-organizational networks using Exponential Random Graph models. Ind Innov 20(3):277–295

Bruderer E, Singh JV (1996) Organizational evolution, learning, and selection: a genetic-algorithm-based model. Acad Manag J 39(5):1322–1349

Brüderl J, Preisendörfer P (1998) Network support and the success of newly founded businesses. Small Bus Econ 10(3):213–225

Buchmann T, Hain D, Kudic M, Mueller M (2014) Exploring the evolution of innovation networks in science-driven and scale-intensive industries – new evidence from a stochastic actor-based approach. IWH discussion papers 01/2014, pp 1–42

Buckley PJ, Glaister KW, Klijn E, Tan H (2009) Knowledge accession and knowledge acquisition in strategic alliances: the impact of supplementary and complementary dimensions. Br J Manag 20(4):598–609

Buenstorf G (2007) Evolution on the shoulders of giants: entrepreneurship and firm survival in the German laser industry. Rev Ind Organ 30(3):179–202

Burt RS (1992) Structural holes: the social structure of competition. Harvard University Press, Cambridge

Burt RS (2000) The network structure of social capital. In: Staw BM, Sutton RI (eds) Research in organizational behavior, vol 22. JAI Press, Greenwich, pp 345–424

Burt RS (2005) Brokerage & closure – an introduction to social capital. Oxford University Press, New York

Bush V (1945) Science: the endless frontier – a report to the President on a program for postwar scientific research. US Government, Washington, DC

Camagni R (1993) Inter-firm industrial networks: the costs and benefits of cooperative behaviour. J Ind Stud 1(1):1–15

Cameron CA, Trivedi PK (1986) Econometric models based on count data: comparisons and applications of some estimators and tests. J Appl Econ 1:29–53

Cameron CA, Trivedi PK (1990) Regression based tests for overdispersion in the Poisson model. J Econ 46(3):347–364

Cameron CA, Trivedi PK (2009) Microeconometrics using Stata. Stata Press, College Station

Campbell DT (1969) Variation and selective retention in sociocultural evolution. Gen Syst 14:69–85

Cantner U, Graf H (2011) Innovation networks: formation, performance and dynamics. In: Antonelli C (ed) Handbook on the economic complexity of technological change. Edward Elgar, Cheltenham, pp 366–394

Capaldo A (2007) Network structure and innovation: the leveraging of a dual network as a distinctive relational capability. Strateg Manag J 28(6):585–608

Carlsson B, Jacobsson S, Holmen M, Rickne A (2002) Innovation systems: analytical and methodological issues. Res Policy 31(2):233–245

Carrington PJ, Scott J, Wasserman S (2005) Models and methods in social network analysis. Cambridge University Press, Cambridge

Carroll GR (1984) Organizational ecology. Annu Rev Sociol 10:71–93

Carson C (2000) The origins of the quantum theory. Beam Line (Stanf Linear Accel Center) 30(2):6–19

Cassi L, Zirulia L (2008) The opportunity cost of social relations: on the effectiveness of small worlds. J Evol Econ 18(1):77–101

Cassi L, Corrocher N, Malerba F, Vonortas N (2008) Research networks as infrastructure for knowledge diffusion in European regions. Econ Innov New Technol 17(7):665–678

Ciesa V, Toletti G (2004) Network of collaborations for innovation: the case of biotechnology. Tech Anal Strat Manag 16(1):73–96

Cincera M (1997) Patents, R&D, and technological spillovers at the firm level: some evidence from econometric count models for panel data. J Appl Econ 12(3):265–280

Cleves MA, Gould WW, Gutierrez RG, Marchenko YU (2008) An introduction to survival analysis using Stata, 2nd edn. Stata Press, College Station

Coase RH (1937) The nature of the firm. Economica (New Series) 4(16):386–405

Coenen L, Moodysson J, Asheim BT (2004) Nodes, networks and proximities: on the knowledge dynamics of the Medicon Valley Biotech Cluster. Eur Plan Stud 12(7):1003–1018

Coff RW (2003) The emergent knowledge-based theory of competitive advantage: an evolutionary approach to integrating economics and management. Manag Decis Econ 24(4):245–251

Cohen WM, Levinthal DA (1989) Innovation and learning: the two faces of R&D. Econ J 99(397):569–596

Cohen WM, Levinthal DA (1990) Absorptive capacity: a new perspective on learning and innovation. Adm Sci Q 35(3):128–152

Coleman JS (1988) Social capital in the creation of human capital. Am J Sociol 94:95–120

Conner KR (1991) A historical comparison of resource-based theory and five schools of thought within industrial organization economics: do we have a new theory of the firm? J Manag 17(1):121–154

Contractor FJ, Lorange P (2002) The growth of alliances in the knowledge-base economy. Int Bus Rev 11(4):485–502

Cooke P (2001) Regional innovation systems, clusters, and the knowledge economy. Ind Corp Chang 10(4):945–974

Corolleur F, Courlet C (2003) The Marshallian Industrial District, an organizational and institutional answer to uncertainty. Entrepren Reg Dev 15:299–307

Corrado R, Zollo M (2006) Small worlds evolving: governance reforms, privatizations, and ownership networks in Italy. Ind Corp Chang 15(2):319–352

Cowan R, David PA, Foray D (2000) The explicit economics of knowledge codification and tacitness. Ind Corp Chang 9(2):211–253

Cowan R, Jonard N, Zimmermann J-B (2006) Evolving networks of inventors. J Evol Econ 16(1):155–174

Cropper S, Ebers M, Huxham C, Ring PS (2008) Introducing inter-organizational relations. In: Cropper S, Ebers M, Huxham C, Ring PS (eds) Interorganizational relations. Oxford University Press, New York, pp 3–25

Cyert R, March JG (1963) Behavioral theory of the firm. Prentice-Hall, Englewood Cliffs

Czepiel JA (1974) Word of mouth processes in diffusion of a major technological innovation. J Mark Res 11:172–180

Dacin TM, Hitt MA, Levitas E (1997) Selecting partners for successful international alliances: examination of U.S. and Korean firms. J World Bus 32(1):3–16

Das TK, Teng B-S (2000) Instabilities of strategic alliances: an internal tensions perspective. Organ Sci 11(1):77–101

Das TK, Teng B-S (2002) Alliance constellations: a social exchange perspective. Acad Manag J 27(3):445–456

David PA (1985) Clio and the economics of QWERTY. Am Econ Rev 75(2):332–337

De Nooy W, Mrvar A, Batagelj V (2005) Exploratory social network analysis wit PAJEK. Cambridge University Press, Cambridge

De Propris L (2000) Innovation and inter-firm co-operation: the case of the West Midlands. Econ Innov New Technol 9(5):421–446

De Rond M, Bouchiki H (2004) On the dialectics of strategic alliances. Organ Sci 15(1):56–69

De Vaus D (2001) Research design in social science research. Sage, London

Decarolis DM, Deeds DL (1999) The impact of stocks and flows of organizational knowledge on firm performance: an empirical investigation of the biotechnology industry. Strateg Manag J 20(10):953–968

Degenne A, Forse M (1999) Introducing social networks. Sage, London

Dierickx I, Cool K (1989) Asset stock accumulation and sustainability of competitive advantage. Manag Sci 35(12):1504–1511

Dingle R, Wiegemann W, Henry CH (1974) Quantum states of confined carriers in very thin AlxGa1-xAs-GaAs-AlxGa1-x as heterostructures. Phys Rev Lett 33(14):827–830

Dittrich K, Duysters G, De Man A-P (2007) Strategic repositioning by means of alliance networks: the case of IBM. Res Policy 36(10):1496–1511

Dixit A, Skeath S (2004) Games of strategy, 2nd edn. W. W. Norton, New York

Dodgson M (1993) Organizational learning – a review of some literature. Organ Stud 14(3):375–394

Dodgson M (2011) Exploring new combinations in innovation and entrepreneurship: social networks, Schumpeter, and the case of Josiah Wedgwood (1730–1795). Ind Corp Chang 20(4):1119–1151

Domar ED (1948) The problem of capital accumulation. Am Econ Rev 38:777–794

Dopfer K (2005) The evolutionary foundation of economics. Cambridge University Press, Cambridge

Dopfer K (2011) Economics in a cultural key: complexity and evolution revisited. In: Davis JB, Hand WD (eds) The Elgar companion to recent economic methodology. Edward Elgar, Cheltenham, pp 319–341

Dopfer K, Foster J, Potts J (2004) Micro–meso–macro. J Evol Econ 14(3):263–279

Doreian P (2008) Actor utilities, strategic action and network evolution. In: Baum JA, Rowley TJ (eds) Advances in strategic management – network strategy, vol 25. Emerald Publishing, Bingley, pp 247–271

Doreian P, Stokman FN (2005) The dynamics and evolution of social networks. In: Doreian P, Stokman FN (eds) Evolution of social networks, 2nd edn. Gordon and Breach, New York, pp 1–17

Doreian P, Woodard KL (1992) Fixed list versus snowball selection of social networks. Soc Networks 21(2):216–233

Doreian P, Woodard KL (1994) Defining and locating cores and boundaries of social networks. Soc Networks 16(1994):267–293

Dosi G (1982) Technological paradigms and technological trajectories. Res Policy 11(3):147–162

Dosi G (1988) Sources, procedures, and microeconomic effects of innovation. J Econ Lit 26(3):1120–1171

Dosi G, Nelson RR (1994) An introduction to evolutionary theories in economics. J Evol Econ 4(3):153–172

Doz YL (1996) The evolution of cooperation in strategic alliances: initial conditions or learning processes? Strateg Manag J 17(1):55–83

Doz Y, Hamel G (1997) The use of alliances in implementing technology strategies. In: Tushman MT, Anderson P (eds) Managing strategic innovation and change. Oxford University Press, New York, pp 556–580

Duysters G, Lemmens C (2004) Alliance group formation – enabling and constraining effects of embeddedness and social capital in strategic technology alliance networks. Int Stud Manag Org 33(2):49–68

Duysters G, De Man A-P, Wildeman L (1999) A network approach to alliance management. Eur Manag J 17(2):182–187

Dwyer RF, Schurr PH, Oh S (1987) Developing buyer-seller relationships. J Mark 51(2):11–27

Dyer JH, Nobeoka K (2000) Creating and managing a high-performance knowledge-sharing network: the Toyota case. Strateg Manag J 21(3):345–367

Dyer JH, Singh H (1998) The relational view: cooperative strategy and sources of international competitive advantage. Acad Manag Rev 23(4):660–680

Eades P (1984) A heuristic for graph drawing. Congr Numer 42:149–160

Easterby-Smith M, Lyles MA, Peteraf MA (2009) Dynamic capabilities: current debates and future directions. Br J Manag 20:1–8

Ehrenfeld W, Kudic M, Pusch T (2014) On the trail of core-periphery patterns – measurement and new empirical findings from the German laser industry. Conference proceedings, 17th Uddevalla symposium, Udevalla, pp 1–22

Einstein A (1905) On a heuristic viewpoint concerning the production and transformation of light (english translation). Ann Phys 17:132–148

Einstein A (1917) On the quantum mechanics of radiation (english translation). Physikalische Zeitschrift 18:121–128

Ejermo O (2009) Regional innovation measured by patent data – does quality matter? Ind Innov 16(2):141–165

Elfring T, Hulsink W (2007) Networking by entrepreneurs: patterns of tie formation in emerging organizations. Organ Stud 28(12):1849–1872

Elias LR, Fairbank WM, Madey JM, Schwettman AH, Smith TI (1976) Observation of stimulated emission of radiation by relativistic electrons in a spatially periodic transverse magnetic field. Phys Rev Lett 36(13):717–720

Elliott JE (1978) Marx's "Grundrisse": vision of capitalism's creative destruction. J Post Keynes Econ 1(2):148–169

Eraydin A, Aematli-Köroglu B (2005) Innovation, networking and the new industrial clusters: the characteristics of networks and local innovation capabilities in the Turkish industrial clusters. Entrepren Reg Dev 17(4):237–266

European Commission (2005) The new SME definition – user guide and model declaration. Enterprise and Industry Publications, Brussels

Fabian C (2011) Technologieentwicklung im Spannungsfeld von Industrie, Wissenschaft und Staat: Zu den Anfängen des Innovationssystems der Materialbearbeitungslaser in der Bundesrepublik Deutschland 1960 bis 1997. Dissertation, TU Bergakademie Freiberg

Fagerberg J (2003) Schumpeter and the revival of evolutionary economics: an appraisal of the literature. J Evol Econ 13(2):125–159

Fagerberg J (2005) Innovation: a guide to the literature. In: Fagerberg J, Mowery DC, Nelson RR (eds) The Oxford handbook of innovation. Oxford University Press, New York, pp 1–28

Faist J, Capasso F, Sivco DL, Sirtori C, Hutchinson AL, Cho AY (1994) Quantum cascade laser. Science 264(5158):553–556

Fang AW, Park H, Jones R, Cohen O, Paniccia MJ, Bowers JE (2006) A continuous-wave hybrid AlGaInAs–silicon evanescent laser. IEEE Photon Technol Lett 18(10):1143–1145

Faulkner D (2006) Cooperative strategy – strategic alliances and networks. In: Campbell A, Faulkner DO (eds) The Oxford handbook of strategy – a strategic overview and competitive strategy. Oxford University Press, New York, pp 610–648

Feldman MP (1993) An examination of the geography of innovation. Ind Corp Chang 2(3): 451–470

Feldman MP (1999) The new economics of innovation, spillovers and agglomeration: a review of empirical studies. Econ Innov New Technol 8(1):5–25

Fischer EP (2010) Laser – Eine deutsche Erfolgsgeschichte von Einstein bis heute. Siedler Verlag, München

Fleming L, King C, Juda AI (2007) Small worlds and regional innovation. Organ Sci 18(6): 938–954

Fombrun CJ, Shanley M (1990) What's in a name? Reputation building and corporate strategy. Acad Manag J 33(2):233–258

Fornahl D, Broeckel T, Boschma R (2011) What drives patent performance of German biotech firms? The impact of R&D subsidies, knowledge networks and their location. Pap Reg Sci 90(2):395–418

Forrest JE, Martin MJ (1992) Strategic alliances between large and small research intensive organizations: experiences in the biotechnology industry. R&D Manag 22(1):41–53

Foss NJ, Ishikawa I (2007) Towards a dynamic resource-based view: insights from Austrian capital and entrepreneurship theory. Organ Stud 28(5):749–772

Foster J (1987) Evolutionary macroeconomics. Unwin Hyman, London

Foster J (2005) From simplistic to complex adaptive systems in economics. Camb J Econ 29(6):873–892

Frank O (2005) Network sampling and model fitting. In: Carrington PJ, Scott J, Wasserman S (eds) Models and methods in social network analysis. Cambridge University Press, Cambridge, pp 31–56

Freel MS, Harrison RT (2006) Innovation and cooperation in the small firm sector: evidence from 'Northern Britain'. Reg Stud 40(4):289–305

Freeman C (1974) The economics of industrial innovation. Penguin, Harmondsworth

Freeman LC (1977) A set of measures of centrality based on betweenness. Sociometry 40(1): 35–41

Freeman LC (1979) Centrality in social networks: I. conceptual clarification. Soc Networks 1(3): 215–239

Freeman C (1988) Japan: a new national system of innovation. In: Dosi G, Nelson RR, Silverberg G, Soete L (eds) Technical change and economic theory. Pinter, London, pp 330–348

Freeman C (1991) Networks of innovators: a synthesis of research issues. Res Policy 20(5): 499–514

Fritsch M, Medrano L (2010) The spatial diffusion of a knowledge base – laser technology research in West Germany, 1960–2005. Jena economic research papers, pp 1–52

Fritsch M, Medrano LF (2015) New technology in the region – agglomeration and absorptive capacity effects on laser technology research in West Germany, 1960–2005. Econ Innov New Technol 24 (forthcoming)

Fritsch M, Slavtschev V (2007) Universities and innovation in space. Ind Innov 14(2):201–218

Fruchterman T, Reingold E (1991) Graph drawing by force-directed placement. Softw Pract Exp 21(11):1129–1164

Gargiulo M, Benassi M (2000) Trapped in your own net? Network cohesion, structural holes, and the adaptation of social capital. Organ Sci 11(2):183–196

George G, Zahra SA, Wheatley KK, Khan R (2001) The effects of alliance portfolio characteristics and absorptive capacity on performance: a study of biotechnology firms. J High Technol Manag Res 12(2):205–226

Geusic JE, Marcos HM, Van Uitert LG (1964) Laser oscillations in Nd-doped yttrium aluminium, yttrium gallium, and gadolinium garnet. Appl Phys Lett 4(10):182–184

Giesekus J (2007) Die Industrie für Strahlquellen und optische Komponenten – Eine aktuelle Marktübersicht von SPECTARIS. Laser Technik J 4(5):11–13

Gilbert N, Pyka A, Ahrweiler P (2001) Innovation networks – a simulation approach. J Artif Soc Soc Simul 4(3):1–13

Gilbert N, Ahrweiler P, Pyka A (2007) Learning in innovation networks: some simulation experiments. Phys A Stat Mech Appl 378(1):100–109

Gilsing V, Nooteboom B, Vanhaverbeke W, Duysters G, van den Oord A (2008) Network embeddedness and the exploration of novel technologies: technological distance, betweenness centrality and density. Res Policy 37(10):1717–1731

Giuliani E (2010) Network dynamics in regional clusters: the perspective of an emerging economy. Pap Evolut Econ Geogr (PEEG) (Online) 10(14)

Glueckler J (2007) Economic geography and the evolution of networks. J Econ Geogr 7(5): 619–634

Goerzen A (2005) Managing alliance networks: emerging practices of multinational corporations. Acad Manag Exec 19(2):94–107

Golbeck J, Mutton P (2005) Spring-embedded graphs for semantic visualization. In: Geroimenko V, Chen C (eds) Visualizing the semantic web. Springer, Heidelberg/New York, pp 172–182

Gomes-Casseres B (2003) Competitive advantage in alliance constellations. Strateg Organ 1(3): 327–335

Gordon JP, Zeiger HJ, Townes CH (1954) Molecular microwave oscillator and new hyperfine structure in the microwave spectrum of NH3. Phys Rev 95(1):282–284

Gordon JP, Zeiger HJ, Townes CH (1955) The maser – new type of microwave amplifier, frequency standard, and spectrometer. Phys Rev 99(4):1264–1274

Goss D (2005) Schumpeter's legacy? Interaction and emotions in the sociology of entrepreneurship. Enterp Theory Pract 29(2):205–218

Gould GR (1959) The laser: light amplification by stimulated emission of radiation. In: Ann Arbor conference on optical pumping, conference proceeding, 15–18 June 1959, pp 128–130

Graf H (2006) Networks in the innovation process. Edward Elgar, Cheltenham

Graf H, Krueger JJ (2011) The performance of gatekeepers in innovator networks. Ind Innov 18(1):69–88

Granovetter MS (1973) The strength of weak ties. Am J Sociol 78(6):1360–1380

Granovetter MS (1985) Economic action and social structure: the problem of embeddedness. Am J Sociol 91(3):481–510

Granovetter MS (2005) The impact of social structure on economic outcomes. J Econ Perspect 19(1):33–50

Grant RM (1996) Towards a knowledge based theory of the firm. Strateg Manag J 17(2):109–122

Grant RM, Baden-Fuller C (2004) A knowledge accessing theory of strategic alliances. J Manag Stud 41(1):61–84

Greene WH (2003) Econometric analysis, 5th edn. Prentice Hall, Upper Saddle River

Grunwald R, Kieser A (2007) Learning to reduce interorganizational learning: an analysis of architectural product innovation in strategic alliances. J Prod Innov Manag 24(4):369–391

Grupp H (2000) Learning in a science driven market: the case of lasers. Ind Corp Chang 9(1):143–172

Guimera R, Uzzi B, Spiro J, Armaral LA (2005) Team assembly mechanisms determine collaboration network structure and team performance. Science 308(29):697–702

Gulati R (1995) Social structure and alliance formation pattern: a longitudinal analysis. Adm Sci Q 40(4):619–652

Gulati R (1998) Alliances and networks. Strateg Manag J 19(4):293–317

Gulati R (2007) Managing network resources – alliances, affiliations and other relational assets. Oxford University Press, New York

Gulati R, Gargiulo M (1999) Where do interorganizational networks come from? Am J Sociol 104(5):1439–1493

Gulati R, Singh H (1998) The architecture of cooperation: managing coordination costs and appropriation concerns in strategic alliances. Adm Sci Q 43(4):781–814

Gulati R, Nohria N, Zaheer A (2000) Strategic networks. Strateg Manag J 21(3):203–215

Gunasekaran A (1997) Essentials of international and joint R&D projects. Technovation 17(11):637–647

Hagedoorn J (1993) Understanding the rational of strategic technology partnering – organizational modes of cooperation and sectoral differences. Strateg Manag J 14(5):371–385

Hagedoorn J (2002) Inter-firm R&D partnership: an overview of major trends and patterns since 1960. Res Policy 31(4):477–492

Hagedoorn J (2006) Understanding the cross-level embeddedness of interfirm partnership formation. Acad Manag Rev 31(3):670–680

Hagedoorn J, Schakenraad J (1994) The effects of strategic technology alliances on company performance. Strateg Manag J 15(4):291–309

Hakansson H, Johanson J (1988) Formal and informal cooperation – strategies in international industrial networks. In: Contractor FJ, Lorange P (eds) Cooperative strategies in international business. Lexington Books, Lexington, pp 369–379

Hakansson H, Snehota I (1995) Stability and change in business networks. In: Hakansson H, Snetota I (eds) Developing relationships in business networks. Thomson, London, pp 24–49

Hall P, Wylie R (2014) Isolation and technological innovation. J Evol Econ 24(2):357–376

Halinen A, Salmi A, Havila V (1999) From dyadic change to changing business networks: an analytical framework. J Manag Stud 36(6):779–794

Hamel G (1991) Competition for competence and inter-partner learning within international strategic alliances. Strateg Manag J 12(1):83–103

Hamel G, Doz YL, Prahalad CK (1989) Collaborate with your competitors – and win. Harv Bus Rev 67(1):133–139

Hannan MT, Freeman J (1977) The population ecology of organizations. Am J Sociol 82(5):929–964

Hannan MT, Freeman J (1984) Structural inertia and organizational change. Am Sociol Rev 49(2): 149–164

Hanneman RA, Riddle M (2005) Introduction to social network methods. University of California, Riverside

Hanusch H, Pyka A (2007a) Principles of neo-Schumpeterian economics. Camb J Econ 31(2): 275–289

Hanusch H, Pyka A (2007b) Schumpeter, Joseph Alois (1883–1950). In: Hanusch H, Pyka A (eds) Elgar companion on neo-Schumpeterian economics. Edward Elgar, Cheltenham, pp 19–27

Hanusch H, Pyka A (2007c) Elgar companion to neo-Schumpeterian economics. Edward Elgar, Cheltenham

Harabi N (2002) The impact of vertical R&D cooperation on firm innovation: an empirical investigation. Econ Innov New Technol 11(2):93–108

Hargrove LE, Fork RL, Pollack MA (1964) Locking of He-Ne laser modes induced by synchronous intracavity modulation. Appl Phys Lett 5(4):4–5

Harrigan KR (1988) Joint ventures and competitive strategy. Strateg Manag J 9(2):141–158

Harrod RF (1948) Towards a dynamic economics. Macmillan, London

Hausman JA (1978) Specification tests in econometrics. Econometrica 46(6):1251–1271

Hausman JA, Hall BH, Griliches Z (1984) Econometric models for count data with an application to the patents – R&D relationship. Econometrica 52(4):909–938

Hecht J (2005) Beam – the race to make the laser. Oxford University Press, New York

Hecht J (2010) The first half-century of laser development – how a solution that once was looking for a problem has become part of everyday life. Laser Technik J 7(4):20–25

Heertje A (2004) Schumpeter and methodological individualism. J Evol Econ 14(2):153–156

Hellwarth RW, McClung FJ (1962) Giant pulsations from ruby. J Appl Phys 33(3):838–841

Hellwarth RW, McClung FF (1963) Characteristics of giant optical pulsations from ruby. Proc IEEE 51(1):46–53

Hite JM (2008) The role of dyadic multi-dimensionality in the evolution of strategic network ties. In: Baum JA, Rowley TJ (eds) Advances in strategic management – network strategy, vol 25. Emerald Publishing, Bingley, pp 133–170

Hite JM, Hesterly WS (2001) The evolution of firm networks: from emergence to early growth of the firm. Strateg Manag J 22(3):275–286

Hodgson GM (2006) Economics in the shadows of Darwin and Marx – essays on institutional and evolutionary economics. Edward Elgar, Cheltenham

Hoffmann WH (2005) How to manage a portfolio of alliances. Long Range Plan 38(2):121–143

Hoffmann WH (2007) Strategies for managing alliance portfolios. Strateg Manag J 28(8):827–856

Hofstede G (2001) Culture's consequences: comparing values, behaviors, institutions, and organizations across nations, 2nd edn. Sage, Thousand Oaks

Holland PW, Leinhardt S (1970) A method for detecting structure in sociometric data. Am J Sociol 76(3):492–513

Holland PW, Leinhardt S (1976) Local structure in social networks. Sociol Methodol 7:1–45

Huisman M, Snijders TA (2003) Statistical analysis of longitudinal network data with changing composition. Sociol Methods Res 32:253–287

Huisman M, Steglich CE (2008) Treatment of non-response in longitudinal network data. Soc Networks 30:297–308

Imbens GW, Wooldridge JM (2009) Recent developments in the econometrics of program evaluation. J Econ Lit 47:5–86

Inkpen A (2009) Strategic alliances. In: Rugman A (ed) The Oxford handbook of international business. Oxford University Press, New York, pp 389–414

Inkpen AC, Beamish PW (1997) Knowledge, bargaining power, and the instability of international joint ventures. Acad Manag Rev 22(1):177–202

Jackson MO (2008) Social and economic networks. Princeton University Press, Princeton

Jackson MO, Watts A (2002) The evolution of social and economic networks. J Econ Theory 106(2):265–295

Jacobs J (1969) The economy of cities. Random House, New York

Jaffe AB (1989) Real effects of academic research. Am Econ Rev 79(5):957–970

Jaffe AB, Trajtenberg M, Henderson R (1993) Geographic localization of knowledge spillovers as evidenced by patent citations. Q J Econ 108(3):577–598

Jarillo CJ (1988) On strategic networks. Strateg Manag J 9(1):31–41

Jarvenpaa SL, Majchrzak A (2008) Knowledge collaboration among professionals protecting national security: role of transactive memories in ego-centered knowledge networks. Organ Sci 19(2):260–276

Javan A, Bennett WR, Herriott DR (1961) Population inversion and continuous optical maser oscillation in a gas discharge containing a He-Ne mixture. Phys Rev Lett 6(3):106–110

Jeong H, Neda Z, Barabasi A-L (2003) Measuring preferential attachment in evolving networks. Europhys Lett 61(4):567–572

Jevons WS (1871) The theory of political economy (1888, 3rd edn. Macmillan, London

Johanson J, Mattson L-G (1988) Internationalization in industrial systems – a network approach. In: Hood N, Vahlen J-E (eds) Strategies in global competition. Croom Helm, New York, pp 287–331

Jones GR, Hill CW (1988) Transaction cost analysis of strategy-structure choice. Strateg Manag J 9(2):159–172

Joshi AM, Nerkar A (2011) When do strategic alliances inhibit innovation by firms? Evidence from patent pools in the global optical disc industry. Strateg Manag J 32(11):1139–1160

Jun T, Sethi R (2009) Reciprocity in evolving social networks. J Evol Econ 19(3):379–396

Kale P, Singh H, Perlmutter H (2000) Learning and protection of proprietary assets in strategic alliances: building relational capital. Strateg Manag J 21(3):217–237

Kao CK, Hockham GA (1966) Dielectric-fibre surface waveguides for optical frequencies. IEE Proc 113(7):1151–1159

Kaplan EL, Meier P (1958) Nonparametric estimation from incomplete observations. J Am Stat Assoc 53:457–481

Kasper JV, Pimentel GC (1965) HCl chemical laser. Phys Rev Lett 14(10):352–354

Katkalo VS, Pitelis CN, Teece DJ (2010) Introduction: on the nature and scope of the dynamic capabilities. Ind Corp Chang 19(4):1175–1186

Katz L (1953) A new status index derived from sociometric analysis. Psychometrika 18(1):39–43

Kazarinov RF, Suris RA (1971) Possibility of amplification of electromagnetic waves in a semiconductor with a superlattice (english translation). Sov Phys Semicond 5:707–709

Kenis P, Knoke D (2002) How organizational field networks shape interorganizational tie-formation rates. Acad Manag Rev 27(2):275–293

Kenis P, Oerlmans L (2008) The social network perspective – understanding the structure of cooperation. In: Cropper S, Ebers M, Huxham C, Ring PS (eds) The Oxford handbook of inter-organizational relations. Oxford University Press, New York, pp 289–312

Kennedy P (2003) A guide to econometrics. Blackwell, Oxford

Ketterle W, Misner H-J (1997) Coherence properties of Bose-Einstein condensates and atom lasers. Phys Rev A 56(4):3291–3293

Khanna T, Gulati R, Nohria N (1998) The dynamics of learning alliances: competition, cooperation, and relative scope. Strateg Manag J 19(3):193–210

Kim T-Y, Oh H, Swaminathan A (2006) Framing interorganizational network change: a network inertia perspective. Acad Manag Rev 31(3):704–720

Kirman A (1989) The intrinsic limits of modern economic theory: the emperor has no clothes. Econ J 99(395):126–139

Kirman A (1993) Ants, rationality, and recruitment. Q J Econ 108(1):137–156

Klein J, Kafka JD (2010) The Ti:Sapphire laser: the flexible research tool. Nat Photonics 4(5): 288–289

Klein B, Crawford RG, Alchian AA (1978) Vertical integration, appropriable rents, and the competitive contracting process. J Law Econ 21(2):297–326

Klepper S (1997) Industry life cycles. Ind Corp Chang 6(1):145–181

Klepper S, Malerba F (2010) Demand, innovation and industrial dynamics: an introduction. Ind Corp Chang 19(5):1515–1520

Kline SJ (1985) Innovation is not a linear process. Res Manag 28(4):36–45

Kline SJ, Rosenberg N (1986) An overview of innovation. In: Landau R, Rosenberg N (eds) The positive sum strategy: harnessing technology for economic growth. National Academy Press, Washington, DC, pp 275–304

Knoben J, Oerlemans LA (2006) Proximity and inter-organizational collaboration: a literature review. Int J Manag Rev 8(2):71–89

Knoke D, Yang S (2008) Social network analysis. Sage, London

Kogut B (1991) Joint ventures and the option to expand and acquire. Manag Sci 37(1):19–33

Kogut B, Zander U (1992) Knowledge of the firm, combinative capabilities, and the replication of technology. Organ Sci 3(3):383–397

Koka BR, Presscott JE (2008) Designing alliance networks: the influence of network position, environmental change, and strategy on firm performance. Strateg Manag J 29:639–661

Koka BR, Madhavan R, Prescott JE (2006) The evolution of interfirm networks: environmental effects on patterns of network change. Acad Manag Rev 31(3):721–737

Kraaijenbrink J, Spender JC, Groen AJ (2010) The resource-based view: a review and assessment of its critiques. J Manag 36(1):349–372

Krackhardt D (1992) The strength of strong ties – the importance of philos in organizations. In: Nohria N, Eccles RG (eds) Networks and organizations – structure, form, and action. Harvard Business Press, Boston, pp 216–239

Kudic M (2012) Innovation networks in the German laser industry – evolutionary change, strategic positioning and firm innovativeness. PhD thesis (unpublished manuscript), University of Hohenheim, Stuttgart

Kudic M, Banaszak M (2009) The economic optimality of sanction mechanisms in interorganizational ego networks – a game theoretical analysis. In: 35th European International Business Academy conference, Valencia, pp 1–40

Kudic M, Boenisch P, Dominguez Lacasa I (2010) Network embeddedness, geographical collocation effects or both? The impact of distinct and combined proximity effects on firm-level innovation output in the German laser industry. In: Conference proceedings. The 36th EIBA annual conference, Porto, pp 1–29

Kudic M, Buenstorf G, Guhr K (2011a) Analyzing the relationship between cooperation events, ego-networks and firm innovativeness – empirical evidence from the German laser industry. In: Conference proceedings. The 5th international EMNet conference, Limassol, pp 1–42

Kudic M, Guhr K, Bullmer I, Guenther J (2011b) Kooperationsintensität und Kooperationsförderung in der deutschen Laserindustrie. Wirtschaft im Wandel 17(3):121–129

Kudic M, Pyka A, Guenther J (2012) Determinants of evolutionary network change processes in innovation networks – empirical evidence from the German laser industry. In: Conference proceedings. The 14th international Schumpeter Society conference, Brisbane, pp 1–29

Kudic M, Pyka A, Sunder M (2013) Network formation: R&D cooperation propensity and timing among German laser source manufacturers. IWH discussion papers, 01/2013, pp 1–25

Kudic M, Guenther J (2014) Towards an in-depth understanding of structural network change processes in innovation networks. In: International Schumpeter society conference proceedings, Jena, 27.30 July

Kudic M, Pyka A, Guenther J (2015) Taking the first step – what determines German laser source manufacturers' entry into innovation networks? Int J Innov Manag IJIM (forthcoming)

Kumar R, Nti KO (1998) Differential learning and interaction in alliance dynamics: a process and outcome discrepancy model. Organ Sci 9(3):356–367

Lamoreaux NR, Sokoloff KL (1999) The geography of the market for technology in the late nineteenth and early twentieth century United States. In: Libecap G (ed) Advances in the study of entrepreneurship, innovation, and economic growth. JAI Press, Stanford, pp 67–121

Lane PJ, Lubatkin MH (1998) Relative absorptive capacity and interorganizational learning. Strateg Manag J 19(5):461–477

Lane PJ, Salk JE, Lyles MA (2001) Absorptive capacity, learning, and performance in international joint ventures. Strateg Manag J 22(12):1139–1161

Lane PJ, Koka BR, Pathak S (2006) The reification of absorptive capacity: a critical review and rejuvenation of the construct. Acad Manag Rev 31(4):833–863

Larson A (1992) Network dyads in entrepreneurial settings: a study of the governance of exchange relationships. Adm Sci Q 37(3):76–104

Laumann EO, Galaskiewicz J, Marsden PV (1978) Community structure as interorganizational linkages. Annu Rev Sociol 4:455–484

Laumann EO, Marsden PV, Prensky D (1989) The boundary specification problem in network analysis. In: Freeman LC, White DR, Romney KA (eds) Research methods in social network analysis. George Mason University Press, Fairfax, pp 61–87

Lavie D (2007) Alliance portfolios and firm performance: a study of value creation and appropriation in the U.S. software industry. Strateg Manag J 28(12):1187–1212

Lavie D, Miller SR (2008) Alliance portfolio internationalization and firm performance. Organ Sci 19(4):623–646

Lavie D, Rosenkopf L (2006) Balancing exploration and exploitation in alliance formation. Acad Manag J 49(4):497–818

Lavie D, Lechner C, Singh H (2007) The performance implications of timing of entry and involvement in multipartner alliances. Acad Manag J 50(3):578–604

Leavitt HJ (1951) Some effects of communication patterns on group performance. J Abnorm Soc Psychol 46(1):38–50

Lee V (1992) Organizational dynamics of market transition: hybrid forms, property rights, and mixed economy in China. Adm Sci Q 37(1):1–27

Lee JJ (2010) Heterogeneity, brokerage, and innovative performance: endogenous formation of collaborative inventor networks. Organ Sci 21(4):804–822

Lerch F (2009) Netzwerkdynamiken im Cluster: Optische Technologien in der Region Berlin-Brandenburg. Dissertation, Freien Universität Berlin, Berlin

Leven P, Holmström J, Mathiassen L (2014) Managing research and innovation networks: evidence from a government sponsored cross-industry program. Res Policy 43(1):156–168

Levin DZ, Cross R (2004) The strength of weak ties you can trust: the mediating role of trust in effective knowledge transfer. Manag Sci 50(11):1477–1490

Levinthal DA (1991) Organizational adaptation and environmental selection-interrelated processes of change. Organ Sci 2(1):140–145

Levitt T (1965) Exploit the product life cycle. Harv Bus Rev 43(6):81–94

Levitt B, March JG (1988) Organizational learning. Annu Rev Sociol 14:319–340

Liebeskind JP (1996) Knowledge, strategy and the theory of the firm. Strateg Manag J 17(2): 93–108

Lin N (2002) Social capital: a theory of social structure and action. Cambridge University Press, Cambridge

Lippman S, Rumelt R (1982) Uncertain imitability: an analysis of interfirm differences in efficiency and competition. Bell J Econ 13:418–438

Lomi A, Negro G, Fonti F (2008) Evolutionary perspectives on inter-organizational relations. In: Cropper S, Ebers M, Huxham C, Ring SP (eds) The Oxford handbook of interorganizational relations. Oxford University Press, New York, pp 313–339

Lorenzoni G, Ornati OA (1988) Constellations of firms and new ventures. J Bus Ventur 3(1):41–57

Lu JW, Beamish PW (2006) Partnering strategies and performance of SMEs' international joint ventures. J Bus Ventur 21(4):461–486

Lui SS (2009) Interorganizational learning the roles of competence trust, formal contract, and time horizon in interorganizational learning. Organ Stud 30(4):333–353

Lundvall B-A (1988) Innovation as an interactive process: from user-producer interaction to the national system of innovation. In: Dosi G, Freeman C, Nelson RR, Silverberg G, Soete L (eds) Technical change and economic theory. Pinter, London, pp 349–369

Lundvall B-A (1992) National systems of innovation – towards a theory of innovation and interactive learning. Pinter, London

Maiman TH (1960) Stimulated optical radiation in ruby. Nature 187(4736):493–494

Malerba F (1992) Learning by firms and incremental technical change. Econ J 102(413):845–859

Malerba F (2002) Sectoral systems of innovation and production. Res Policy 31(2):247–264

Malerba F (2007) Innovation and the evolution of industries. In: Cantner U, Malerba F (eds) Innovation, industrial dynamics and structural transformation. Springer, Heidelberg/New York, pp 7–29

Malthus RT (1798) An essay on the principle of population. Johnson, London

Malthus RT (1820) Principles of political economy. Murray, London

March JG (1991) Exploration and exploitation in organizational learning. Organ Sci 2(1):71–87

Markowitz H (1952) Portfolio selection. J Financ 7(1):77–91

Marschall A (1890) Principles of economics – an introductory (1920, 8th edn. Macmillan, London

Marsden PV (2002) Egocentric and sociocentric measures of network centrality. Soc Netw 24:407–422

Marsden PV (2005) Recent developments in network measurement. In: Carrington PJ, Scott J, Wasserman S (eds) Models and methods in social network analysis. Cambridge University Press, Cambridge, pp 8–30

Marx K (1857) Grundrisse der Kritik der politischen Ökonomie (The Grundrisse – introduction to the critique of political economy,1973). Vintage, New York

Marx K (1867) Capital: a critical analysis of capitalist production, vol 1 (1974). Lawrence & Wishart, London

Maskell P, Malmberg A (1999) Localized learning and industrial competitiveness. Camb J Econ 23(2):167–185

Mason E (1939) Price and production policies of large scale enterprise. Am Econ Rev 29:61–74

Mason E (1957) Economic concentration and the monopoly problem. Harvard University Press, Cambridge

Mayer A (2004) Laser in der Materialbearbeitung – Eine Marktübersicht. Laser Technik J 1(1): 9–12

Mayer A (2006) Laser materials processing systems in 2005 – the world market reaches record volume. Laser Technik J 3(1):10–11

McPherson M, Smith-Lovin L, Cook JM (2001) Birds of a feather: homophily in social networks. Annu Rev Sociol 27(1):415–444

Mears RJ, Reekie L, Poole SB, Payne DN (1986) Low-threshold tunable CW and Q-switched fibre laser operating at 1.55 μm. Electron Lett 22(3):159–160

Menger C (1871) Grundsätze der Volkswirthschaftslehre – Allgemeiner Teil (Principles of economics). Wilhelm Braumüller, Wien

Menzel M-P, Fornahl D (2009) Cluster life cycles – dimensions and rationales of cluster evolution. Ind Corp Chang 19(1):205–238

Metcalfe SJ (2010) The open, evolving economy: Alfred Marshall on knowledge, management and innovation. In: Gaffard J-L, Salies E (eds) Innovation, economic growth and the firm – theory and evidence of industrial dynamics. Edward Elgar, Cheltenham, pp 3–30

Milgram S (1967) The small-world problem. Psychol Today 1(1):60–67

Mill JS (1848) Principles of political economy. Longmans, London

Mill JS (1859) On liberty. Watts, London

Morroni M (2006) Knowledge, scale and transactions in the theory of the firm. Cambridge University Press, Cambridge

Mowery DC, Oxley JE, Silverman BS (1996) Strategic alliances and interfirm knowledge transfer. Strateg Manag J 17(2):77–92

Mueller A, Faist J (2010) The quantum cascade laser: ready for take-off. Nat Photonics 4(5): 290–291

Mueller M, Buchmann T, Kudic M (2014) Micro strategies and macro patterns in the evolution of innovation networks: an agent-based simulation approach. In: Gilbert N, Ahrweiler P, Pyka A (eds) Simulating knowledge dynamics in innovation networks. Springer, Heidelberg/New York

Muldur U, Corvers F, Delanghe H, Dratwa J, Heimberge D, Sloan B, Vanslembrouck S (2006) A new deal for an effective European research policy: the design and impacts of the 7th Framework Programme. Springer Netherlands, Dordrecht

Muniz AS, Raya AM, Carvajal CR (2010) Core periphery valued models in input–output field: a scope from network theory. Pap Reg Sci 90(1):111–121

Murray EA, Mahon JF (1993) Strategic alliances: gateway to new Europe. Long Range Plan 26(4):102–111

Nakamura S, Senoh M, Nagahama S-I, Iwasa N, Yamada T, Matsushita T et al (1996) InGaN-based multi-quantum-well-structure laser diodes. Jpn J Appl Phys 35(1b):74–76

Nakamura M, Vertinsky I, Zietsam C (1997) Does culture matter in inter-firm cooperation? Research consortia in Japan and the USA. Manag Decis Econ 18:153–175

Narula R, Hagedoorn J (1999) Innovating through strategic alliances: moving towards international partnerships and contractual agreements. Technovation 19(5):283–294

Nature (2010) Technology focus – laser anniversary. Nat Photonics 4(5):278–295

Nelson RR (1992) National innovation systems: a retrospective on a study. Ind Corp Chang 1(2): 347–374

Nelson RR (2007) Understanding economic growth as the central task of economic analysis. In: Hanusch H, Pyka A (eds) Elgar companion to neo-Schumpeterian economics. Edward Elgar, Cheltenham, pp 840–853

Nelson RR, Winter SG (1974) Neoclassical vs. evolutionary theories of economic growth: critique and prospectus. Econ J 84(336):886–905

Nelson RR, Winter SG (1982) An evolutionary theory of economic change. Harvard University Press, Cambridge

Nelson RR, Winter SG (2002) Evolutionary theorizing in economics. J Econ Perspect 16(2):23–46

Newman ME (2010) Networks – an introduction. Oxford University Press, New York

Newman ME, Strogatz S, Watts D (2001) Random graphs with arbitrary degree distributions and their applications. Phys Rev E 64:1–17

Nonaka I (1991) The knowledge-creating company. Harv Bus Rev 69(6):96–104

Nonaka I, Toyama R, Nagata A (2000) A firm as a knowledge-creating entity: a new perspective on the theory of the firm. Ind Corp Chang 9(1):1–20

Nooteboom B (2008) Learning and innovation in inter-organizational relationships. In: Cropper S, Ebers M, Huxham C, Ring PS (eds) The Oxford handbook of interorganizational relations. Oxford University Press, New York, pp 607–634

Nowak MA, Tarnita CE, Antal T (2010) Evolutionary dynamics in structured populations. Philos Trans R Soc B 365(1537):19–30

Noyons E, Raan VA, Grupp H, Schoch U (1994) Exploring the science and technology interface: inventor-author relations in laser medicine research. Res Policy 23(4):443–457

OECD (2005) Oslo manual: guidelines for collecting and interpreting innovation data, 3rd edn. OECD Publishing, Paris

OECD (2008) OECD science, technology and industry outlook. OECD, Paris

Oerlemans LA, Meeus MT (2005) Do organizational and spatial proximity impact on firm performance? Reg Stud 39(1):89–104

Oerlemans LA, Meeus MT, Boekema FW (2001) Firm clustering and innovation: determinants and effects. Pap Reg Sci 80(3):337–356

Ohmae K (1989) The global logic of strategic alliances. Harv Bus Rev 67(3/4):143–154

Oliver C (1990) Determinants of interorganizational relationships: integration and future directions. Acad Manag Rev 15(2):241–265

Osborn RN, Hagedoorn J (1997) The institutionalization and evolutionary dynamics of interorganizational alliances and networks. Acad Manag J 40(2):261–278

Ostrom E (2009) Beyond markets and states: Polycentric Governance of Complex Economic Systems. Prize Lecture, December 8

Ouchi WG (1980) Markets, bureaucracies, and clans. Adm Sci Q 25(1):129–141

Ouimet M, Landry R, Amara N (2007) Network position and efforts to innovate in small Canadian optics and photonics clusters. Int J Entrep Innov Manag 7:251–271

Owen-Smith J, Powell WW (2004) Knowledge networks as channels and conduits: the effects of spillovers in the Boston biotechnology community. Organ Sci 15(1):5–21

Owen-Smith J, Riccaboni M, Pammolli F, Powell WW (2002) A comparison of U.S. and European University-Industry relations in the Life Sciences. Manag Sci 48(1):24–43

Oxley JE, Sampson RC (2004) The scope and governance of international R&D alliances. Strateg Manag J 25(8):723–749

Ozman M (2009) Inter-firm networks and innovation: a survey of literature. Econ Innov New Technol 18(1):39–67

Ozman M (2013) Networks, irreversibility, and knowledge creation. J Evol Econ 23(2):431–453

Page SE (2006) Path dependence. Q J Polit Sci 1(1):87–115

Paniccia M, Krutul V, Jones R, Cohen O, Bowers J, Fang A et al. (2006) A hybrid silicon laser – silicon photonics technology for future tera-scale computing. Intel White Paper, pp 1–6

Parise S, Casher A (2003) Alliance portfolios: designing and managing your network of business-partner relationships. Acad Manag Exec 17(4):25–39

Park SH, Russo MV (1996) When competition eclipses cooperation: an event history analysis of joint venture failure. Manag Sci 42(6):875–890

Parkhe A (1993) Strategic alliance structuring: a game theoretic and transaction cost examination of interfirm cooperation. Acad Manag J 36(4):794–829

Parkhe A, Wasserman S, Ralston DA (2006) New frontiers in network theory development. Acad Manag Rev 31(3):560–568

Patel KC (1964) Selective excitation through vibrational energy transfer and optical maser action in N_2-CO_2. Phys Rev Lett 13(21):617–619

Patel P, Pavitt K (1995) Patterns of technological activity: their measurement and interpretation. In: Stoneman P (ed) Handbook of the economics of innovation and technological change. Blackwell, Oxford, UK, pp 14–51

Pavitt K (1984) Sectoral patterns of technical change: towards a taxonomy and a theory. Res Policy 13(6):343–373

Pavitt K (1998) Technologies, products and organization in the innovating firm: what Adam Smit tells us and Joseph Schumpeter doesn't. Ind Corp Chang 7(3):433–452

Penrose ET (1959) The theory of the growth of the firm. Wiley, New York

Perlmutter HV, Heenan DA (1986) Cooperate to compete globally. Harv Bus Rev 64(2):136–152

Peteraf MA (1993) The cornerstones of competitive advantage: a resource-based view. Strateg Manag J 14(3):179–191

Phelps C (2010) A longitudinal study of alliance network structure and composition firm exploratory innovation. Acad Manag J 53(4):890–913

Phillips AC (2003) Introduction to quantum mechanics. Wiley, Sussex

Pittaway L, Robertson M, Munir K, Denyer D, Neely A (2004) Networking and innovation: a systematic review of the evidence. Int J Manag Rev 5(6):137–168

Podolny JM (1993) A status-based model of market competition. Am J Sociol 98(4):829–872

Podolny JM (1994) Market uncertainty and the social character of economic exchange. Adm Sci Q 39(3):458–483

Podolny JM (2001) Networks as the pipes and prisms of the market. Am J Sociol 7(1):33–60

Podolny JM, Page KL (1998) Network forms of organization. Annu Rev Sociol 24(1):57–76

Podolny JM, Stuart TE (1995) A role-based ecology of technological change. Am J Sociol 100(5):1224–1260

Podolny JM, Stuart TE, Hannan MT (1996) Networks, knowledge, and niches: competition in the worldwide semiconductor industry, 1984–1991. Am J Sociol 102(3):659–689

Polanyi M (1958) Personal knowledge: towards a post-critical philosophy. University of Chicago Press, Chicago

Polanyi M (1967) The tacit dimension. Doubleday, New York

Poprawe R (2010) Part 1 – coherent light: an invention that once was searching for its applications became a versatile enabling tool. Laser Technik J 7(2):31–36

Porter ME (1981) The contributions of industrial organization to strategic management. Strateg Manag J 6(4):609–620

Porter ME (1985) Competitive advantage. Free Press, New York

Powell WW (1987) Hybrid organizational arrangements: new form of transitional development? Calif Manag Rev 30(1):67–87

Powell WW (1990) Neither market nor hierarchy: networks forms of organization. Res Organ Behav 12(1):295–336

Powell WW, Koput KW, Smith-Doerr L (1996) Interorganizational collaboration and the locus of innovation – networks of learning in biotechnology. Adm Sci Q 41(1):116–145

Powell WW, White DR, Koput KW, Owen-Smith J (2005) Network dynamics and field evolution: the growth of the interorganizational collaboration in the life sciences. Am J Sociol 110(4): 1132–1205

Priem RL, Buttler JE (2001) Is the resource-based "view" a useful perspective for strategic management research? Acad Manag Rev 26(1):22–40

Prokhorov AM (1964) Quantum electronics. Nobel Lecture, Elsevier Publishing Company, Amsterdam, pp. 110–116. http://www.nobelprize.org/nobel_prizes/physics/laureates/1964/prokhorov-lecture.html

Protogerou A, Caloghirou Y, Siokas E (2010) Policy-driven collaborative research networks in Europe. Econ Innov New Technol 19(4):349–372

Provan KG, Kenis P (2007) Modes of network governance: structure, management, and effectiveness. Econ Innov New Technol 18(2):229–252

Provan KG, Fish A, Sydow J (2007) Interorganizational networks at the network level: a review of the empirical literature on whole networks. J Manag 33(3):479–516

Pyka A (1997) Informal networking. Technovation 17(4):207–220

Pyka A (2002) Innovation networks in economics: from the incentive-based to the knowledge based approaches. Eur J Innov Manag 5(3):152–163

Pyka A (2007) Innovation networks. In: Hanusch H, Pyka A (eds) Elgar companion to neo-Schumpeterian economics. Edward Elgar, Cheltenham, pp 360–377

Pyka A, Scharnhorst A (2009) Network perspectives on innovations: innovative networks – network innovation. In: Pyka A, Scharnhorst A (eds) Innovation networks. Springer, Heidelberg/New York, pp 1–16

Pyka A, Gilbert N, Ahrweiler P (2007) Simulating knowledge-generation and distribution processes in innovation collaborations and networks. Cybern Syst 38(7):667–693

Rank C, Rank O, Wald A (2006) Integrated versus core-periphery structures in regional biotechnology networks. Eur Manag J 24(1):73–85

Rank ON, Robins GL, Pattison PE (2010) Structural logic of intraorganizational networks. Organ Sci 21(3):745–764

Ricardo D (1817) On the principles of political economy and taxation. Dent & Sons, London

Ring PS, Van De Ven AH (1994) Developmental processes of cooperative interorganizational relationships. Acad Manag Rev 19(1):90–118

Robertson DT, Stuart TE (2007) Network effects in the governance of strategic alliances. J Law Econ Org 23(1):242–273

Rodan S, Galunic C (2004) More than network structure: how knowledge heterogeneity influences managerial performance and innovativeness. Strateg Manag J 25(6):541–562

Romer P (1986) Increasing returns and long-run growth. J Polit Econ 94(5):1002–1037

Rosenberg N (1973) Innovative responses to materials shortages. Am Econ Rev 63(2):111–118

Rosenberg N (1974) Science, invention and economic growth. Econ J 84(333):90–108

Rosenberg N (2011) Was Schumpeter a Marxist? Ind Corp Chang 20(4):1215–1222

Rosenkopf L, Tushman ML (1998) The coevolution of community networks and technology: lessons from the flight simulation industry. Ind Corp Chang 7(2):311–346

Rothaermel FT (2001) Incumbent's advantage through exploiting complementary assets via interfirm cooperation. Strateg Manag J 22(6):687–699

Rothaermel FT, Deeds DL (2004) Exploration and exploitation alliances in biotechnology: a system of new product development. Strateg Manag J 25(3):201–221

Rothaermel FT, Deeds DL (2006) Alliance type, alliance experience and alliance management capability in high-technology ventures. J Bus Ventur 21(4):429–460

Rothwell R (1994) Towards the fifth-generation innovation process. Int Mark Rev 11(1):7–31

Rowley TJ, Behrens D, Krackhardt D (2000) Redundant governance structures: an analysis of structural and relational embeddedness in the steel and semiconductor industries. Strateg Manag J 21(3):369–386

Rutherford D (2007) Economics – the key concepts. Routledge, New York

Sabidussi G (1966) The centrality index of graph. Psychmetrika 31(4):581–603

Saloner G (1991) Modelling, game theory and strategic management. Strateg Manag J 12(2):119–136

Sato R (1964) The Harrod-Domar model vs the Neo-Classical Growth model. Econ J 74(294):380–387

Saviotti PP (2011) Knowledge, complexity and networks. In: Antonelli C (ed) Handbook on the economic complexity of technological change. Edward Elgar, Cheltenham, pp 120–141

Saxanian A (1990) Regional networks and the resurgence of Silicon Valley. Calif Manag Rev 33(1):89–112

Schawlow AL, Townes CH (1958) Infrared and optical masers. Phys Rev 112(6):1940–1949

Scherngell T, Barber MJ (2009) Spatial interaction modeling of cross-region R&D collaborations: empirical evidence from the 5th EU framework programme. Pap Reg Sci 88(3):531–546

Scherngell T, Barber MJ (2011) Distinct spatial characteristics of industrial and public research collaborations: evidence from the fifth EU framework programme. Ann Reg Sci 46(2):247–266

Schilke O, Goerzen A (2010) Alliance management capability: an investigation of the construct and its measurement. J Manag 36(5):1192–1219

Schilling MA (2009) Understanding the alliance data. Strateg Manag J 30(3):233–260

Schilling MA, Phelps CC (2007) Interfirm collaboration networks: the impact of large-scale network structure on firm innovation. Manag Sci 53(7):1113–1126

Schmalensee R (1988) Industrial economics: an overview. Econ J 98:643–681

Schoenmakers W, Duysters G (2006) Learning in strategic technology alliances. Tech Anal Strat Manag 18(2):245–264

Schramm M (2005) Präzision als Leitbild? Carl Zeiss und die deutsche Innovationskultur in Ost und West 1945–1990. Technikgeschichte 72(1):35–49

Schumpeter JA (1908) Das Wesen und der Hauptinhalt der theoretischen Nationalökonomie. Dunker & Humblot, Leipzig

Schumpeter JA (1912) Theorie der wirtschaftlichen Entwicklung (The theory of economic developmnt 1934). Duncker & Humblot, Berlin

Schumpeter JA (1939) Business cycles – a theoretical, historical and statistical analysis of the capitalism process. McGraw-Hill, New York

Schumpeter JA (1942) Kapitalismus, Sozialismus und Demokratie (Capitalism, socialism and democracy, 1950). Harper & Bros, New York

Schwartz M (2010) A control group study of incubators' impact to promote firm survival. IWH discussion papers, 11, pp 1–36

Schwerk A (2000) Dynamik von Unternehmenskooperationen. Duncker & Humbolt, Berlin

Shan W, Walker G, Kogut B (1994) Interfirm cooperation and startup innovation in the biotechnology industry. Strateg Manag J 15(5):387–394

Shimbel A (1953) Structural parameters of communication networks. Bull Math Biophys 15(4):501–507

Shimizu H, Hirao T (2009) Inter-organizational collaborative research networks in semiconductor laser 1975–1994. Soc Sci J 46(2):233–251

Silverberg G, Verspagen B (2005) Evolutionary theorizing on economic growth. In: Dopfer K (ed) The evolutionary foundation of economics. Cambridge University Press, Cambridge, pp 506–539

Simon HA (1955) A behavioral model of rational choice. Q J Econ 69(1):99–118

Simon HA (1991) Bounded rationality and organizational learning. Organ Sci 2(1):125–134

Simonin BL (1999) Ambiguity and the process of knowledge transfer in strategic alliances. Strateg Manag J 20(1):595–623

Sivadas E, Dwyer RF (2000) An examination of organizational factors influencing new product success in internal and alliance-based processes. J Mark 64(1):31–49

Slusher RE (1999) Laser technology. Rev Modern Phys 71(2):471–479

Smith A (1776) An inquiry into the nature and causes of the wealth of nations [1904, Edwin Cannan ed.]. Methuen, London

Smith K (2005) Measuring innovation. In: Fagerberg J, Mowery DC, Nelson RR (eds) The Oxford handbook of innovation. Oxford University Press, New York, pp 148–177

Snijders TA (2004) Explained variation in dynamic network models. Math Soc Sci 42(1,68):5–15

Snijders TA, Van De Bunt GG, Steglich CE (2010) Introduction to actor-based models for network dynamics. Soc Networks 32(1):44–60

Snitzer E (1961) Optical maser action of Nd+3 in barium crown glass. Phys Rev Lett 7(3):444–446

Soda G, Zaheer A (2004) Network memory: the influence of past and current networks on performance. Acad Manag J 47(6):893–906

Solow RM (1956) A contribution to the theory of economic growth. Q J Econ 70(1):65–94

Solow RM (1957) Technical change and the aggregate production function. Rev Econ Stat 39 (3):312–320

Sorenson O, Audia PG (2000) The social structure of entrepreneurial activity: geographic concentration of footwear production in the Unites States, 1940–1989. Am J Sociol 106(2): 424–462

Sornn-Fries H (2000) Frontiers of research in industrial dynamics and national systems of innovation. Ind Innov 7(1):1–13

Sorokin PP, Lankard JR (1966) Stimulated emission observed from an organic dye, chloro-aluminum phtatocyanine. IBM J Res Dev 10(2):162–163

Sorokin PP, Stevenson MJ (1960) Stimulated infrared emission from trivalent uranium. Phys Rev Lett 5(12):557–559

Sorokin PP, Stevenson MJ (1961) Solid-state optical maser divalent samarium in calcium fluoride. IBM J Res Dev 5(1):56–58

Spence M (1976) Informational aspects of market structure: an introduction. Q J Econ 90(4): 591–597

Spence M (2002) Signaling in retrospect and the informational structure of markets. Am Econ Rev 92(3):434–459

Spender JC, Grant RM (1996) Knowledge and the firm: overview. Strateg Manag J 17(2):5–10

Staber U (1998) Inter-firm co-operation and competition in industrial districts. Organ Stud 19(4): 701–724

Stata (2007) Stata statistical software: release 10. StataCorp LP, College Station

Stokman FN, Doreian P (2005) Evolution of social networks: processes and principles. In: Doreian P, Stockman FN (eds) Evolution of social networks, 2nd edn. Gordan Breach, New York, pp 233–251

Stuart TE (1999) A structural perspective on organizational performance. Ind Corp Chang 8(4): 745–775

Stuart TE (2000) Interorganizational alliances and the performance of firms: a study of growth and innovational rates in a high-technology industry. Strateg Manag J 21(8):791–811

Stuart TE, Hoang H, Hybles RC (1999) Interorganizational endorsements and the performance of entrepreneurial ventures. Adm Sci Q 44(2):315–349

Suitor JJ, Wellman B, Morgan DL (1997) It's about time: how, why, and when networks change. Soc Networks 19(1):1–7

Swan T (1956) Economic growth and capital accumulation. Econ Record 32(63):334–361

Swann PG (2009) The economics of innovation – an introduction. Edward Elgar, Cheltenham

Swedberg R (2000) Entrepreurship. In: Swedberg R (ed) The social science view of entrepreneur-ship. Oxford University Press, New York, pp 7–44

Sydow J (2003) Dynamik von Netzwerkorganisationen – Entwicklung, Evolution, Strukturation. In: Hoffmann WH (ed) Die Gestaltung der Organisationsdynamik – Konfiguration und Evolution. Schäffer-Poeschel, Stuttgart, pp 327–357

Sydow J, Lerch F, Staber U (2010) Planning for path dependence? The case of a network in the Berlin-Brandenburg optics cluster. Econ Geogr 86(2):173–195

Teece DJ (2007) Explicating dynamic capabilities: the nature and microfoundations of (sustain-able) enterprise performance. Strateg Manag J 28(13):1319–1350

Teece DJ, Pisano GP, Shuen A (1997) Dynamic capabilities and strategic management. Strateg Manag J 18(7):509–533

Ter Wal AL, Boschma R (2011) Co-evolution of firms, industries and networks in space. Reg Stud 45(7):919–933

Thorelli HB (1986) Networks: between markets and hierarchies. Strateg Manag J 7(1):37–51

Tiberius V (2008) Prozesse und Dynamik des Netzwerkwandels. Gaber, Wiesbaden

Tidd J (2006) A review of innovation models. Imperial College discussion paper series, 06/1, pp 1–15

Tidd J, Bessant J, Pavitt K (2005) Managing innovation: integrating technological, market and organizational change, 3rd edn. Wiley, Chichester

Torre A, Rallet A (2005) Proximity and localization. Reg Stud 39(1):47–59

Townes CH (1964) Production of coherent radiation by atoms and molecules. Nobel Lecture 12(11):58–86

Townes CH (1999) How the laser happened – adventures of a scientist. Oxford University Press, New York

TSB (2010) Laser technology report – Berlin Brandenburg. TSB Innovationsagentur GmbH, Berlin

Tushman ML, Nelson RR (1990) Introduction: technology, organizations, and innovation. Adm Sci Q 35(1):1–8

Uzzi B (1996) The sources and consequences of embeddedness for the economic performance of organizations: the network effect. Am Sociol Rev 61(4):674–698

Uzzi B (1997) Social structure and competition in interfirm networks : the paradox of embeddedness. Adm Sci Q 42(1):35–67

Uzzi B, Spiro J (2005) Collaboration and creativity: the small'world problem. Am J Sociol 111(2):447–504

Uzzi B, Amaral LA, Reed-Tsochas F (2007) Small-world networks and management science research: a review. Eur Manag Rev 4(2):77–91

Van de Bunt GG, Groenewegen P (2007) An actor-oriented dynamic network approach. Organ Res Methods 10(3):463–482

Van De Ven AH, Poole MS (1995) Explaining development and change in organizations. Acad Manag Rev 20(3):510–540

Van Den Bosch FA, Volberda HW, De Boer M (1999) Coevolution of firm absorptive capacity and knowledge environment: organizational forms and combinative capabilities. Organ Sci 10(5):551–568

Van Witteloostuijn A (2000) Organizational ecology has a bright future. Organ Stud 21(2):X–XIV

Veblen T (1898) Why is economics not an evolutionary science? Q J Econ 12(3):373–397

Venkatraman S, Lee C-H (2004) Preferential linkage and network evolution: a conceptual model and empirical test in the U.S. video game sector. Acad Manag J 47(6):876–892

Visser E-J (2009) The complementary dynamic effects of clusters and networks. Ind Innov 16(2):167–195

Von Hippel E (1986) Lead users: a source of novel product concepts. Manag Sci 32(7):791–805

Von Hippel E (1988) The sources of innovation. Oxford University Press, New York von Mises L (1969) The historical setting of the Austrian school of economics. Arlington House, New York

Walker G, Kogut B, Shan W (1997) Social capital, structural holes and the formation of an industry network. Organ Sci 8(2):109–125

Walras L (1874) Elements of pure economics (1954). George Allen and Unwin, London

Wang C, Rodan S, Fruin M, Xu X (2014) Knowledge networks, collaboration networks, and exploratory innovation. Acad Manag J 57:484–514

Wasserman S, Faust K (1994) Social network analysis: methods and applications. Cambridge University Press, Cambridge

Wassmer U (2010) Alliance portfolios: a review and research agenda. J Manag 36(1):141–171

Watts DJ (1999) Small worlds – the dynamics of networks between order and randomness. Princeton University Press, Princeton

Watts DJ, Strogatz SH (1998) Collective dynamics of 'small-world' networks. Nature 393 (6684):440–442

Weick KE (1979) The social psychology of organization. Addison-Wesley, Reading

Weigel K, Camerer C (1988) Reputation and corporate strategy: a review of recent theory and applications. Strateg Manag J 9(5):443–454

Wernerfelt B (1984) A resource based view of the firm. Strateg Manag J 5(2):171–180

White S (2005) Cooperation costs, governance choice and alliance evolution. J Manag Stud 42 (7):1383–1413

Whittington KB, Owen-Smith J, Powell WW (2009) Networks, propinquity, and innovation in knowledge-intensive industries. Adm Sci Q 54(1):90–122

Williamson OE (1973) Organizational forms and internal efficiency – markets and hierarchies: some elementary considerations. Am Econ Rev 63(2):316–325

Williamson OE (1975) Markets and hierarchies: analysis and antitrust implications. Free Press, New York

Williamson OE (1985) The economic institutions of capitalism. Free Press, New York

Williamson OE (1991) Comparative economic organization: the analysis of discrete structural alternatives. Adm Sci Q 36(2):269–296

Winkelmann R (2003) Econometric analysis of count data, 4th edn. Springer, Heidelberg/New York

Winter S (2000) The satisficing principle in capability learning. Strateg Manag J 21:981–996

Winter SG (2003) Understanding dynamic capabilities. Strateg Manag J 24(10):991–995

Winter SG (2006) Toward a neo-Schumpeterian theory of the firm. Ind Corp Chang 15(1):125–141

Witt U (2003) Evolutionary economics and the extension of evolution to the economy. In: Witt U (ed) The evolving economy. Edward Elgar, Cheltenham, pp 3–37

Witt U (2008a) What is specific about evolutionary economics? J Evol Econ 18(5):547–575

Witt U (2008b) Recent developments in evolutionary economics. Edward Elgar, Cheltenham

Witt U, Broekel T, Brenner T (2012) Knowledge and its economic characteristics: a conceptual clarification. In: Arena R, Festre A, Lazaric N (eds) Handbook of economics and knowledge. Edward Elgar, Cheltenham

Wooldridge JM (2002) Econometric analysis of cross sectional and panel data. MIT Press, Cambridge, MA

Wuyts S, Dutta S, Stremersch S (2004) Portfolios of interfirm agreements in technology-intensive markets: consequences for innovation and profitability. J Mark 68(2):88–100

Yamakawa Y, Yang H, Lin Z (2011) Exploration versus exploitation in alliance portfolio: performance implications of organizational, strategic, and environmental fit. Res Policy 40 (2):287–296

Yu N, Kats M, Pflügl C, Geiser M, Belkin MA, Capasso F et al (2009) Multi-beam multi-wavelength semiconductor lasers. Appl Phys Lett 95(16):1–3

Zaheer A, Bell GG (2005) Benefiting from network position: firm capabilities, structural holes, and performance. Strateg Manag J 26(9):809–825

Zaheer A, Soda G (2009) Network evolution: the origins of structural holes. Adm Sci Q 54 (1):1–31

Zahra SA, George G (2002) Absorptive capacity: a review, reconceptualization, and extension. Acad Manag Rev 27(2):185–203

Zollo M, Winter SG (2002) Deliberate learning and the evolution of dynamic capabilities. Organ Sci 13(3):339–351

Zollo M, Reuer JJ, Singh H (2002) Interorganizational routines and performance in strategic alliances. Organ Sci 13(6):701–713

Zucker LG, Darby MR, Armstrong JS (1998) Geographically localized knowledge: spillovers or markets? Econ Inq 36(1):65–86

Printed in the United States
By Bookmasters